Bridges

Mathematical Connections in Art, Music, and Science

2000

BRIDGES

Mathematical Connections
in Art, Music, and Science

Conference Proceedings
2000

Reza Sarhangi, Editor

Bridges: Mathematical Connections in Art, Music, and Science;
Conference Proceedings, 2000.

Reza Sarhangi, Editor
Department of Mathematics and Computer Science
Southwestern College
Winfield, Kansas 67156-2499

ISBN 0-9665201-2-2

ISSN 1099-6702

Printed by Central Plain Book Manufacturing, Winfield, Kansas, United States of America.

Front cover design: *Music of the Spheres,* Brent Collins

Back cover design: *The transition of a conceptual CAD model into a wood sculpture,* Carol Séquin

Cover layout: Karen Mages

Contents

Abstracts

Preface

"And yet you are almost three, and numbers
are only months away. Already you ask
for *free of em* when it's kisses or cookies.

Soon you'll be counting off one, two, three,
and learning to know time in a breath— "

from *Almost three* by Gar Bethel

There is a saying in Farsi that can be translated as "The game will not start until there are three." The idea of counting in some cultures is often expressed as *one, two, and many*. The Bridges Conference has achieved its goal by publishing its third proceedings. *Bridges* has survived the stage of vulnerability and finitude and the game has now begun; we may go forever!

There was a time that the Bridges Conference was only a tiny seed implanted in the minds of a small group of colleagues at Southwestern College. The seed was nurtured by attending the Arts and Mathematics Conference series at Albany, New York directed by Nat Friedman. Nat came to our campus a year before the first Bridges to convince administrators here that this small group of people located far from urban big brothers and multi-versities was able to bring the image to reality – the seed to fruition. "Nothing to it," as Nat would say and has said. He and *we* both know that there is a bit more than nothing involved. But the act of bringing together enough people with "child-like curiosities" (to quote Dan Daniel from "Bridges, June Bugs, and Creativity" and among the earliest to envision the conference we have now grown) is the most important thing *we* do. The conferees do the rest. And now we may celebrate the bright future of bridges on its third birthday. One, two, three, infinity!

The peer-reviewed Bridges proceedings for 2000 has attracted even more quality authors from around the world than ever before. The quantity of received papers forced us not only to be more sensitive to the selection of papers, but also to limit the number of pages for each article. Even with that we had to admit a significant growth in the number of papers accepted for publication. The increase in submissions, of course, caused more work for the referees. The 2000 Bridges proceedings reveals not only the quality of work of its authors but also the hard work of our referees. No words can express our appreciation for the work the referees have done to enable publication of the proceedings you have in your hands.

The cover of the book presents two images. One is a hand carving by well-known artist Brent Collins and the other is an image created with a computer by scholar Carlo Séquin. What Brent has conceived through his artistic intuition, Carlo has presented using computer languages and mathematics. This partnership offers one image of the collaborative adventure sought by the Bridges Conference. Michele Emmer, one of the first to call for a gathering of mathematicians and artists under one roof, clearly emphasizes the need for this merging of forces in his paper in this current volume.

To celebrate a new century and millennium, a group of faculty and students at Southwestern College decided to make a time capsule and fill it with materials for future undergraduates and faculty members. The capsule will be opened 100 years from now. Among other materials a volume of the 2000 Bridges Proceedings will be in that capsule. For a while, I wondered about the reaction of people in 100 years. Will they have understood the importance of uniting so many fields? Or will they have slipped back into honoring only narrow specializations? Or will they so take for granted the dialogue between and among fields they will not fully understand what we are doing. We can't determine what they will know, or feel; what they will wear or what they will have achieved in technology and science, but I am sure they will sense our excitement when they read this book -- and thrill at our early steps toward integration.

Finally, we do appreciate colleagues and friends of the conference from around the world who have carried our voices to such amazingly diverse parts of our planet.

Reza Sarhangi

Symmetry and Ornament

Slavik Jablan
Mathematical Institute
Knez Mihailova 35, P.O.Box 367
11001 Belgrade, Yugoslavia
E-mail: jablans@mi.sanu.ac.yu

Abstract

After the symmetry analysis of the Paleolithic and Neolithic ornamental art, it is given the evidence of different symmetry and antisymmetry groups that originated from it, and are preserved in the entire history of ornamental art as a kind of "ornamental archetypes".

Introduction

Throughout history there were always links between geometry and the art of painting. These links become especially evident when to the study of ornamental art we apply the theory of symmetry. Therefore, ornamental art is called by H. Weyl [39] "the oldest aspect of higher mathematics given in an implicit form" and by A. Speiser the "prehistory of group theory".

The idea to study ornaments of different cultures from the point of view of the theory of symmetry, given by G. Pólya [27] and A. Speiser [34], was supported by the intensive development of the theory of symmetry in the 20th century. This caused the appearance of a whole series of works dedicated mostly to the ornamental art of ancient civilizations, to the cultures which contributed the most to the development of ornamental art (Egyptian, Arab, Moorish, etc.) [2,14,16,17,25,37], and to the ethnical ornamental art [9,10,11]. Only in some recent works, research has turned to the very roots, the origins of ornamental art– to the ornamental art of the Paleolithic and Neolithic [22]. The Figures in this paper are adapted from the more extensive collection appearing in [22], where the reader will also find more detailed reference to their specific Paleolithic and Neolithic sources. The extensions of the classical theory of symmetry– antisymmetry and colored symmetry, made possible the more profound analysis of the "black-white" [19,20,39] and colored ornamental motifs in the ornamental art of the Neolithic and ancient civilizations.

This work gives the results of the symmetry analysis of Paleolithic and Neolithic ornamental art. It is dedicated to the search for "ornamental archetypes"– the universal basis of the complete ornamental art. The development of ornamental art started together with the beginnings of mankind. It represents one of the oldest records of human attempts to note, understand and express regularity– the underlying basis of any scientific knowledge.

The final conclusion is that most of the ornamental motifs, which have been discussed from the standpoint of the theory of symmetry, are of a much earlier date than we can expect. This places the beginning of ornamental art, the oldest aspect of geometric cognition, back to several thousands years before the ancient civilizations, i.e. in the Paleolithic and Neolithic.

Since ornamental art is mostly limited to the two-dimensional plane presentation of ornamental motifs, the subject of this art, regarded from the point of view of the theory of symmetry, are the plane symmetry groups: symmetry groups of rosettes, friezes and ornaments. The discrete symmetry groups of rosettes consist of two infinite classes: cyclic groups and dihedral groups. A cyclic group $\mathbf{C_n}$ is generated by the rotation of order n. A dihedral group $\mathbf{D_n}$ is generated by two reflections in lines crossing in the invariant point– the center of rotation of order n. Seven discrete frieze symmetry groups can be denoted by symbols: **11**, **1g**, **12**, **m1**, **1m**, **mg** and **mm**. In these concise symbols, the translation symbol **p** is omitted, **g** denotes glide reflection, **m** reflection, and **n** ($n=1,2$) a rotation of order n. All the symbols are treated in the coordinate sense. The elements of symmetry at the first position are perpendicular to the translation axis, and the elements of symmetry at the second position are parallel or perpendicular (exclusively for 2-rotations) to the direction of the translation.

Analogously, in the symbols of symmetry groups of ornaments, the symbol **p** denotes a two-dimensional translation subgroup, while the symbols **m**, **g**, **n** ($n=2,3,4,6$) have respectively the same meaning as in the case of symmetry groups of friezes. When we talk about the continuous groups of symmetry of friezes, the presence of a continuous translation is denoted by a subscript 0, while in the antisymmetry groups, antigenerators are denoted by '. Antisymmetry groups are presented also by group/subgroup symbols G/H [30].

By the term "prescientific period" we understand the Paleolithic and Neolithic epochs, covering the period from 25000-10000 B.C., till the end of the IV millenium B.C., when we have the signs of first alphabets.

In the absence of written sources, the study of geometry of the prehistoric period is based on the analyses of artifacts, which offer information on geometric knowledge in an implicit form. Among the artifacts mentioned we distinguish few kinds of them. The oldest ones are ornamental motifs realized in the form of bone engravings, carvings and drawings on stone from the Paleolithic and Neolithic. Later we have ornamental motifs in ceramics from the Neolithic phase, obtained by engraving, pressing, drawing or coloring, as well as architectural objects and constructions from the Neolithic period, so called megalithic monuments.

Rosettes

The simplest ornamental motifs are rosettes, symmetrical figures with an invariant point, that correspond to the symmetry groups $\mathbf{C_n}$ and $\mathbf{D_n}$. They are denoted in Shubnikov's notation by **n** and **nm**, respectively [30].

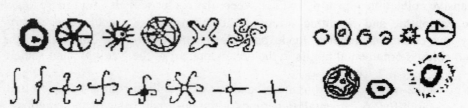

Figure 1. *Variations of the Sun symbol in the ornamental art of the Paleolithic and Neolithic.*

The continuous symmetry group of rosettes $\mathbf{D_\infty}$ ($\infty\mathbf{m}$) corresponds to the maximal symmetric rosette– a circle. Due to the maximal visual and constructional simplicity and maximal symmetry, a circle represents the primary geometric shape– geometric archetype. Within ornamental art it appears in the Paleolithic, as an independent rosette or in combination with some concentric rosette of a lower degree of symmetry, usually circumscribed or inscribed in a circle. Since the group $\mathbf{D_\infty}$ ($\infty\mathbf{m}$) contains all the other groups of symmetry of rosettes as subgroups, rosettes of a lower degree of symmetry are often derived by a desymmetrization of a circle. Owing to its visual-geometric properties: completeness, compactness, boundedness and uniformity of its structural segments, the circle may serve as a universal symbol of completeness and perfection. At the very beginning of ornamental art, the circle becomes the symbol of the Sun, remaining that throughout history (Figure 1).

The continuous symmetry group of rosettes C_∞ (∞) is the group of all rotations around a fixed point. A physical interpretation of it could be a circle uniformly rotating around the center, so in the static form this symmetry group is visually interpretable only by use of textures [31]: by using an asymmetric figure, statistically distributed in accordance with the desired symmetry C_∞ (∞).

The spiral is one of the oldest dynamic visual symbols. In the visual sense it suggests the rotational motion around the invariant point, and could be accepted as an adequate symbolic interpretation of the continuous symmetry group C_∞ (∞). In ornamental art, the spiral appeared already in the Paleolithic, as an independent ornamental motif, or in the form of a double spiral– a motif with symmetry group C_2 (**2**) generated by two-fold rotation.

Among the elementary geometric forms we have a line segment, usually placed in accordance with the basic natural directions– vertical and horizontal line. To a line segment corresponds the symmetry group D_2 (**2m**), generated by two reflections: one in the mirror line perpendicular, and the other in the reflection line collinear to the line segment. However, from the point of view of visual perception, due to the action of the visual and gravitational dominant, the vertical line, we visually experience the symmetry of a line segment as D_1 (**m**). In this case, the horizontal reflection is neglected. The combination of the vertical and horizontal line segment results in the cross form with symmetry group D_1 (**m**), D_2 (**2m**) or D_4 (**4n**). Rosettes with symmetry D_2 (**2m**) and D_4 (**4m**) possess another fundamental property: the existence of mutually perpendicular, vertical and horizontal reflection lines. The form of a cross with symmetry group D_4 (**4m**) is often subjectively, visually perceived as the symmetry D_2 (**2m**), neglecting the presence of four-fold rotation.

Static rosettes with symmetry group D_1 (**m**) or D_2 (**2m**) are linked to the plane symmetry of a man, its vertical attitude and perpendicularity to the base. Besides the rational mirror symmetry, which originated from motifs in nature, we have in the ornamental art different aspects of symbolic symmetry D_1 (**m**): the duplicated figures, two-headed animals, etc. These examples result mostly from the common use of vertical mirror symmetry as a visual dominant.

In Paleolithic ornamental art we have also rosettes with the symmetry group D_n (**nm**): D_3 (**3m**), D_4 (**4m**) and D_6 (**6n**), as well as the corresponding regular polygons: equilateral triangle, square and regular hexagon (Figure 2). For rosettes, the principle of crystallographic restriction (n=1,2,3,4,6) is not respected. Anyway, prevailing are rosettes with the symmetry group D_n (**nm**) for the mentioned values of n. In the later stage, in Neolithic we have also rosettes with the symmetry group D_5 (**5m**) with the use of regular pentagon and pentagram. The first appearance of pentagram is dated by H.S.M. Coxeter [7, pp. 8] in the VII century B.C. The visual characteristics of rosettes with the symmetry group D_n (**nm**) are stability, stationariness and absence of enantiomorphism. Enantiomorphism, the existence of a "right" and "left" modification of the same figure, appears with all figures possessing a symmetry group that does not contain indirect symmetry transformations.

In contradistinction to the static rosettes with the symmetry group D_n (**nm**), rosettes with the symmetry group C_n (**n**) (e.g. triquetra with the symmetry group C_3 (**3**), swastika with the symmetry group C_4 (**4**)) are visually dynamic rosettes. There exists the possibility for enantiomorphic modifications that suggest the impression of rotational motion (Figure 2).

In the next stage of the development of ornamental art, in Neolithic, after understanding the symmetry regularities on which the symmetry of rosettes is based and solving their elementary geometric constructions, the diversity of rosettes increases. This is followed by the application of plant and zoomorphic motifs and by varying the form of the fundamental region. Also, the superpositions of concentric rosettes resulting in a desymmetrization– a reduction to a lower degree of symmetry, are very common.

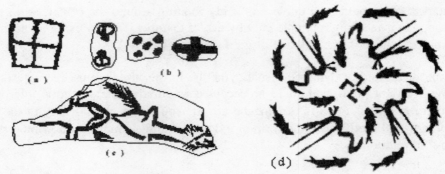

Figure 2. *Examples of rosettes with symmetry group C_n (n) and D_n (nm) in the ornamental art of the Paleolithic and Neolithic: (a) paleolithic of France, D_4 (4m); (b) Maz d' azil, D_2 (2m) and D_4 (4m); (c) Laugerie Basse, C_2 (2); (d) Neolithic ceramics of Middle Asia, C_4 (4),   around 6000 B.C.*

In the Neolithic, with two-colored ceramics, we have the antisymmetric "black-white" rosettes (Figure 3). In this case, antisymmetry can be treated either as the mode of desymmetrization for obtaining the subgroups of index 2 of a given symmetry group, or as an independent form of symmetry. In the table of antisymmetry groups, every group is denoted by the group/subgroup symbol G/H [30] and followed by a system of (anti)generators. The factor-group G/H is isomorphic to a cyclic group of order 2– the group of color change "black"-"white". Hence, we have the following antisymmetry groups of rosettes: $\mathbf{D_{2n}/D_n}$ **(2nm/nm) = (2n)'m; $\mathbf{D_n/C_n}$ (nm/n) = nm'; $\mathbf{C_{2n}/C_n}$ (2n/n) = (2n)'.**

In the case of antisymmetry groups, there is a possibility for interpreting the color change "black"-"white" as the alternating change of some physical or geometric bivalent property. In ornamental art color change mentioned introduces a space component, a suggestion of relations "in front"-"behind', "'up"-"down", "above"-"below". From the artistic point of view, it introduces the contrast between repeating congruent figures and specific equivalence of the "figure" and "background", thus expressing in a symbolical sense a dynamic conflict and duality.

Figure 3. *Neolithic antisymmetry rosette $\mathbf{D_8/C_8}$, Hajji Mohammed, around 5000 B.C.*

In ornamental art the use of color in the sense of regular coloring, i.e. antisymmetry and colored symmetry, opened and a large unexplored field. Hence, in the history of ornamental art, we can consider the Neolithic as its peak, a period in which after solving the basic technical and constructional problems, new possibilities for artistic research, imagination, variety of motifs and decorativeness were opened.

Friezes (one-dimensional patterns)

In the late Paleolithic (Magdalenian, about 25000-10000 B.C.) we find the oldest examples of the symmetry groups of friezes, plane symmetry groups without invariant points and with invariant line. We have the examples of all seven symmetry groups of friezes: **11**, **1g**, **12**, **ml**, **lm**, **mg**, **mm**, as well as two visually presentable continuous symmetry groups of friezes $_0$**ml** and $_0$**mm**.

Friezes are usually obtained by applying the rosettal method of construction, translational multiplication of an initial motif– a rosette, the symmetry of which directly conditions the symmetry of the frieze obtained. The other origin of friezes are models found in nature which, by themselves possess the symmetry of a frieze (Figure 4).

The way friezes are derived from models found in nature can be illustrated by examples: a herd of deer reduced to the frieze with the symmetry group **11**, the motif of cult-dance rendering the frieze with the symmetry group **m1**. Friezes with symmetry group **12** and **mg** can be considered as stylized waves. Models in nature with the symmetry groups **1g** and **1m** are found in the distribution of leaves of certain plants; they have served as the pretext for the construction of corresponding friezes in ornamental art. The importance of the plane symmetry in nature and the numerosity of rosettes with the symmetry group D_1 (**m**) and D_2 (**2m**) caused the appearance and frequent occurrence of friezes with symmetry group **mm**. These friezes can be derived by a translational multiplication of a rosette with the symmetry group D_2 (**2m**), where the translation axis is parallel with one reflection line of the rosette. The symmetry group of friezes **mm** is the maximal discrete group of symmetry of the friezes, generated by reflections. All the other symmetry groups of friezes are subgroups of the group **mm**. Hence the group **mm** can serve for derivation of all other symmetry groups of friezes by desymmetrization. Examples of all discrete frieze symmetry groups are found in Paleolithic ornamental art.

Besides friezes with a concrete meaning, which are based on material models found in nature, the appearance of certain friezes is caused also by the periodic change of many natural phenomena (the change of day and night, seasons, the tides, phases of the Moon, etc.). The corresponding friezes represent, at the same time, the oldest attempt to register the periodical change of natural phenomena, i.e. the first form of calendars. These friezes can also be understood as a way to register quantities, serving as tally boards, thus indicating the beginning of counting and recording the results of counting, i.e. the appearance of the set of natural numbers.

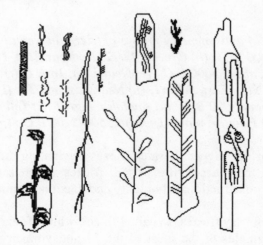

Figure 4. *Examples of frieze symmetry group 1g
in the ornamental art of the Paleolithic.*

Thanks to their symbolic meaning, certain "geometric" friezes became the means of visual communication. This is proved by the preserved names of friezes in the ethnical ornamental art [32]. This communication role of friezes, established in the Paleolithic, was partly preserved in the Neolithic. With the development of other communication forms, friezes lost their primary symbolic function, which was partly or completely replaced by their decorative function. The beginning of this process can be registered already in the Neolithic ornamental art.

The polarity, non-polarity and bipolarity of the translation axis of the friezes, the presence or absence of enantiomorphism implied by the presence or absence of indirect symmetries within the frieze symmetry group, etc. [22], represent some of the relevant geometric properties deserving more detailed geometrical consideration. At the same time, they define the visual characteristics of the friezes, thus conditioning also the spectrum of symbolic meanings which friezes with certain symmetry groups may possess.

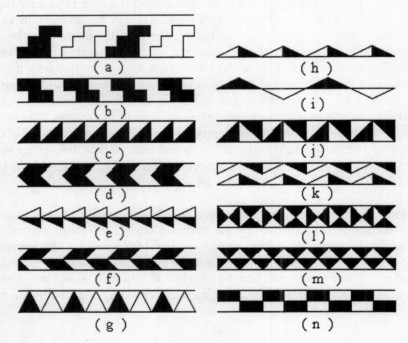

Figure 5. *Examples of 14 antisymmetry groups of friezes in Neolithic ornamental art:*
(a) Greece, 11/11, about 3000 B.C.; (b) Greece, 12/12; (c) Near East, 12/11, about 5000 B.C.;
(d) Near East, 1m/1m, about 5000 B.C.; (e) Near East, 1m/11; (f) Anatolia, 1m/1g,
around 5000 B.C.; (g) Near East, m1/m1; (h) Near East, m1/11; (i) Greece, mg/m1;
(j) Near East, mg/1g, about 5000 B.C.; (k) Anatolia, mg/12; (l) Tell el Hallaf, mm/mm,
about 4900-4500 B.C.; (m) Hacilar, mm/m1, about 5500-5200 B.C.; (n) Near East, mm/mg.

With regard to the frequency of occurrence, besides friezes originating directly from models found in nature, in the ornamental art of the prescientific period, friezes which satisfy the criterion of visual entropy [22]: maximal visual and constructional simplicity and maximal symmetry, are dominant.

The oldest examples of antisymmetry friezes, so called "black-white" friezes, date back to the Neolithic epoch, in which we have the examples of the most of the 17 antisymmetry groups of friezes. Further investigations should show whether or not from that period originate examples of all the 17 antisymmetry groups of friezes. With regard to the frequency of occurrence, the most numerous are "black-white" friezes derived from the most frequent classical-symmetry friezes by the use of antisymmetry desymmetrization method (Figure 5).

The frequency of occurrence of antisymmetry friezes depends also on the antisymmetry properties. Therefore, more frequent are antisymmetry friezes with oppositely colored adjacent fundamental regions. A domination of "geometric" antisymmetry friezes over antisymmetry friezes inspired by models found in nature is also evident, due to the absence of antisymmetry in nature among models from plant and animal life. In contradistinction to this, many natural alternating phenomena followed by bivalent changes (e.g. the change of day and night, etc.), already in Neolithic ornamental art are represented by antisymmetry friezes.

There is a possibility to treat the antiidentity transformation of order 2 (color change "black"-"white") as the way to represent in a plane the space symmetry structures– bands (three-dimensional symmetry groups with invariant plane and line contained in it, and without invariant points). The 31 antisymmetry groups of the friezes (7 generating + 7 senior + 17 junior antisymmetry groups) correspond to the 31 symmetry groups of bands. From the artistic point of view, that gives a possibility to suggest space in a flat plane of drawing. Besides this possibility, there are also many different geometric or non-geometric interpretations of the antiidentity transformation. In prehistoric ornamental art a primary symbolic function of "black-white" friezes, is evident.

Ornaments (two-dimensional patterns)

In the theory of symmetry and ornamental art, the most interesting field of study are the 17 groups of symmetry of ornaments, two-dimensional symmetry groups without invariant lines and points. The common characteristic of ornaments is the presence of discrete two-dimensional translation subgroup, generated by two independent translations. The difficulty of discovering and constructing examples of all the 17 symmetry groups of ornaments is shownby the fact that many cultures with a very rich ornamental art do not possess within their early ornamental art the examples of all these groups [16,17]. The same is proved by the fact that in the mathematical studies of symmetry, the complete derivation of the symmetry groups of ornaments can be found only in 1890, in the works of E.S. Fedorov, although this problem attracted also many other important mathematicians (for example, C. Jordan, L. Sohncke).

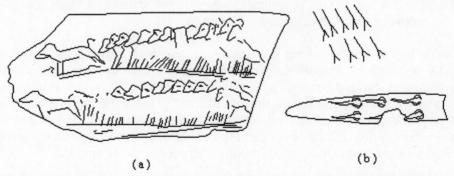

(a)   (b)

Figure 6. *Examples of ornaments with the symmetry group p1 in Paleolithic ornamental art: (a) Chaffaud; (b) bone engravings, Europe.*

This is the reason why it is rather surprising that already in the ornamental art of the Paleolithic we can find examples of the nine symmetry groups of ornaments: **p1, p2, pm, pmm, pmg, cm, cmm, p4m** and **p6m** [22]. In the Neolithic phase we have the appearance of five other symmetry groups of ornaments: **pg, pgg, p4, p4g** and **p6**. Examples of symmetry groups of ornaments **p3, p3m1** and **p31m** can be found in the early ornamental art of ancient civilizations, and probably also in the late Neolithic.

According to the stated presence of the corresponding symmetry groups **p4m** and **p6m** in the Paleolithic, all three regular tessellations: {4,4} with symmetry group **p4m**, {6,3} and {3,6} with symmetry group **p6m**, have been known. Besides the regular square, hexagonal and triangular lattices, in Paleolithic ornamental art we find the remaining two Bravais lattices: the lattice of parallelograms with symmetry group **p2** and the rhombic lattice with symmetry group **cmm**.

In the ornamental art of the Paleolithic and Neolithic, with regard to the construction methods used in obtaining the ornaments we distinguish four construction methods: multiplication of the friezes, multiplication of the rosettes, the method of Bravais lattices and the desymmetrization method. The first construction method is based on the translational repetition of a certain frieze by means of a discrete translation, non-parallel to a frieze axis. Because of the simplicity of this construction, and because of the existence of early examples of all seven discrete symmetry groups of friezes, this method was probably often used for the construction of ornaments. In the Paleolithic, it is probably used for the construction of ornaments with symmetry group **p1**, **p2**, **pm**, **(pg)**, **pmg** and **pmm**. The similar rosette method of construction is based on the multiplication of a rosette by two independent discrete translations. The symmetry of the ornament obtained is completely defined by the properties of these translations and by the symmetry group of the rosette. The appearance of the Bravais lattices in the Paleolithic and Neolithic ornamental art originates from the models in nature (e.g. honeycomb, different net structures). Another cause is a very high degree of visual and constructional simplicity of the Bravais lattices. The most frequent Bravais lattices, regular tessellations {4,4}, {6,3} and {3,6}, to which correspond the maximal symmetry groups of ornaments **p4m** and **p6m** generated by reflections, have often served as the basis for the application of the desymmetrization method. The importance of this construction method increases especially with the appearance of (two) colored ceramics in the Neolithic, i.e. with the beginning of antisymmetry and colored symmetry ornaments. All these construction methods probably were used in the ornamental art of the prescientific period.

Since they point to the very roots of ornamental art, ornaments from the Paleolithic, realized as bone engravings or stone carvings and drawings, deserve special attention.

Ornaments with the symmetry group **p1** are based on the multiplication of a frieze with the symmetry group **11** by a discrete translation, or on the multiplication of an asymmetric figure by two discrete translations. Because of a low degree of symmetry, they occur relatively seldom, and most often appear with stylized asymmetric models found in nature (Figure 6).

Ornaments with the symmetry group **p2** appear in the most elementary form: as a lattice of parallelograms. A highlight of Paleolithic ornamental art are ornaments with the application of the meander motif or double spiral, the rosette with symmetry group C_2 (**2**), which originates most probably from the territory of the Ukraine and Russia (Mezin, Mal'ta). That motifs will be, later on, often used in the ornamental art of almost all Neolithic cultures, mostly as a variation of the motif of waves. Because the forms with symmetry group **p2** are very rare in nature, ornaments with symmetry group **p2** are almost completely limited to geometric motifs or to symbolic stylized motifs (Figure 7).

Ornaments with the symmetry group **pm**, due to the presence of the reflections, belong to the class of static ornaments. Besides geometric motifs, there is a frequent use of models with mirror plane symmetry that are found in nature.

Although according to [22] no examples of ornaments with symmetry group **pg** have been found in Paleolithic, there are grounds to believe that they do appear in Paleolithic ornamental art, as there are examples of the frieze symmetry group **1g**.

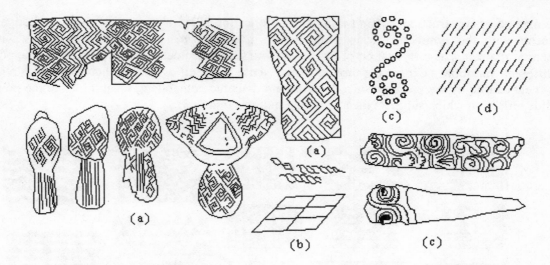

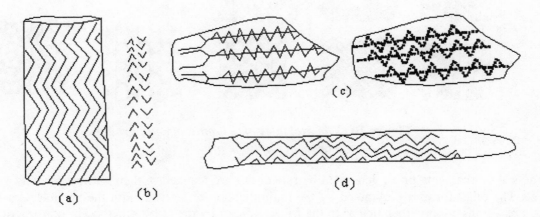

Figure 7. *Examples of ornaments with the symmetry group **p2** in Paleolithic ornamental art: Mezin, Ukraine, about 23000-18000 B.C.; (b) the Paleolithic of Western Europe; (c) the motif of double spiral, Mal'ta, Russia; (d) the application of the motif of double spiral, Arudy, Isturiz.*

Figure 8. *Examples of ornaments with the symmetry group **pmg** in Paleolithic ornamental art: (a) Mezin, Ukraine; (b) Western Europe; (c) Pernak, Estonia; (d) Shtetin.*

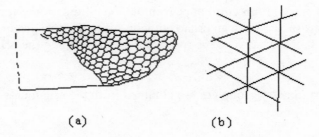

Figure 9. *Examples of the regular tessellations with the symmetry group **p6m**: (a) {6,3}, Yeliseevichi, Russia, 10000 B.C.; (b) regular tessellation {3,6}.*

Regarding the frequency of occurrence and their variety in Paleolithic and Neolithic ornamental art, ornaments with symmetry groups **pmg** and **pmm** prevail. Both of these ornaments can be obtained by the frieze method of construction, by translational multiplication of a frieze **mg** and **mm**, respectively. The ornaments with the symmetry group **pmg** appear in their primary form almost always within the geometric ornaments, as a stylization of the wave motif. The symmetry group **pmg** offers the possibility for different variations, expressing in the visual sense a specific balance between the static visual component caused by the presence of reflections and dynamic component, resulting from the presence of the glide reflection which suggests the alternating motion (Figure 8).

Figure 10. *The examples of 23 antisymmetry groups of ornaments in Neolithic ornamental art.*

The static ornaments **pmm** generated by reflections are realized in their earliest form as a rectangular lattice. The other forms are obtained by the multiplication of a frieze with the symmetry group **mm** by means of a translation perpendicular to the frieze axis, or by the rosette method of construction. There, a rosette with the symmetry group D_2 (**2m**) is multiplied by means of two translations perpendicular to the corresponding reflection lines of the rosette.

Ornaments with the symmetry group **cmm** appear in the Paleolithic in the form of the rhombic lattice. These ornaments can be constructed from an ornament with the symmetry group **pmm** by centering it, i.e. by the procedure in which the gaps between the rosettes D_2 (**2m**) forming the original ornament, are filled with the same rosettes.

The ornaments with the symmetry group **cm** are obtained from the ornaments with the symmetry group **pm** by the same procedure– by centering.

The symmetry groups of ornaments **p4m** and **p6m** correspond to the regular tessellations {4,4}, {6,3} and {3,6}. The regular tessellation consisting of regular hexagons, three of which are incident with each vertex of tessellation, most probably originates from its model in nature: the honeycomb (Figure 9). The regular tessellations {3,6} and {4,4} are from the same period, the Paleolithic.

The principle of visual entropy: maximal visual and constructional simplicity and maximal symmetry is a common, universal characteristic of all Paleolithic ornaments. Hence, among Paleolithic ornaments five of the nine existing symmetry groups of ornaments correspond to the Bravais lattices, seven of the nine groups contain reflections and belong to a class of static ornaments. In them, the almost complete absence of the dynamic elements of symmetry– polar translations, polar rotations and glide reflections, is evident.

In the Neolithic period we have the appearance of almost all the remaining symmetry groups of ornaments. The "black-white" ornaments, i.e. those having antisymmetry, have a special place in Neolithic ornamental art. Very many of the 46 antisymmetry groups of ornaments appear in the Neolithic ornamental art, in particular in the ornamental art of the Near and Middle East (Tel el Hallaf, Hacilar, Catal Hüjük). If we treat antisymmetry ornaments with the antisymmetry group **p6m/p3m1** as the classical-symmetry ornaments obtained by the method of antisymmetry desymmetrization, we can add to the list of symmetry groups of ornaments appearing in the Neolithic, also the symmetry group **p3m1**.

Neolithic ornamental art is one of the richest sources of different ornaments in all the history of ornamental art. The examples of the 14 symmetry groups of ornaments and 23 antisymmetry groups of ornaments (Figure 10) found in Neolithic ornamental art are the most complete testimony about the artistic creativity of Neolithic peoples.

Ornaments with the symmetry group **p3**, **p3m1** and **p31m** represent quite a problem with regard to their construction. In classical-symmetry sense, they first appear in the ornamental art of ancient civilizations, or maybe earlier, in late Neolithic ornamental art.

Very interesting and insufficiently explored fields related to the geometry of the prescientific period are still the following: dating of the appearance of all the plane symmetry structures and corresponding classical-symmetry, antisymmetry and color-symmetry groups, the registering of the most significant archaeological excavation sites from the point of view of the theory of symmetry and ornamental art, the links between the ornamental art of different cultures, the links between the friezes, natural numbers and calendars, etc. All these and many other similar questions relevant to the history of mathematics of the prescientific period should become a common field of research for mathematicians, archaeologists and specialists of different sciences.

References

[1] Akopyan I.D., *Simmetriya i asimmetriya v poznanii*, Akad. Nauk Armyanskoi SSR, Erevan, 1980.

[2] Belov N.V., Moorish Patterns of the Middle Ages and the Symmetry Groups, *Soviet Physics-Crystallography* **1** (1956), 482-483.

[3] Belov N.V., Neronova N.N., Mozaiki dlya 46 ploskih shubnikovskih grupp antisimmetrii i dlya 15 fedorovskih tsvetnih grupp, *Kristallografiya* **2**, I, (1957), 21-22.

[4] Brunes T., *The Secrets of Ancient Geometry*, I, II, Rhodos, Copenhagen, 1964.

[5] Coxeter H.S.M., *Introduction to Geometry*, 2nd ed., Wiley, New York, 1969.

[6] Coxeter H.S.M., *Regular Polytopes*, 3rd ed., Dover, New York, 1973.

[7] Coxeter H.S.M., *Regular Complex Polytopes*, Cambridge University Press, Cambridge, 1974.

[8] Coxeter H.S.M., Moser W.O.J., *Generators and Relations for Discrete Groups*, 4th ed., Springer Verlag, Berlin, Heidelberg, New York, 1980.

[9] Crowe D.W., The Geometry of African Art I. Bakuba Art, J. *Geometry* **1** (1971), 169-182.

[10] Crowe D.W., The Geometry of African Art, II. A Catalog of Benin Patterns, *Historia Math.* **2** (1975), 57-71.

[11] Crowe D.W., The Geometry of African Art III. The Smoking Pipes of Begho, In *The Geometric Vein*, ed. C.Davis, B.Grünbaum and F.A.Sherk, Springer Verlag, Berlin, Heidelberg, New York, 1981.

[12] Crowe D.W., Washburn D.K., Groups and Geometry in the Ceramic Art of San Ildefonso, *Algebras, Groups and Geometries* **3** (1985), 263-277.

[13] Eves H., *An Introduction to History of Mathematics*, Holt, Rinehart and Winston, New York, 1964.

[14] Garido J., Les groupes de symetrie des ornaments employes par les anciennes civilisations du Mexique, *C.R. Acad. Sci. Paris* **235** (1952), 1184-1186.

[15] Hilbert D., Cohn-Vossen S., *Geometry and the Imagination*, Chelsea, New York, 1952.

[16] Grünbaum B., The Emperor's New Clothes: Full Regalia, G string, or Nothing, *Math. Inteligencer* **6**, 4 (1984), 47-53.

[17] Grünbaum B., Grünbaum Z., Shephard G.C., Symmetry in Moorish and Other Ornaments, *Comput. Math. Appl.* **12B**, 3/4 (1986), 641-653.

[18] Grünbaum B., Shephard G.C., *Tilings and Patterns*, Freeman, San Francisco, 1987.

[19] Jablan S.V., Antisimetrijska ornamentika I, *Dijalektika* **1-4** (1985), 107-148.

[20] Jablan S.V., Antisimetrijska ornamentika II, *Dijalektika* **3-4** (1986), 13-56.

[21] Jablan S.V., *Teorija antisimetrije i visestruke antisimetrije u E^2 i $E^2\backslash\{0\}$*, Ph.D. Thesis, PMF, Beograd, 1984.

[22] Jablan S.V., *Theory of Symmetry and Ornament*, Mathematical Institute, Beograd, 1995.

[23] Lockwood E.H., Macmillan R.H., *Geometric Symmetry*, Cambridge University Press, London, New York, Melburne, 1978.

[24] Loeb A.L., *Color and Symmetry*, Wiley-Interscience, New York, 1971.

[25] Müller E., *Gruppentheoretische und Strukturanalytische Untersuchungen der Maurischen Ornamente aus der Alhambra in Granada*, Ph.D. Thesis, Univ. Zürich, Rüschlikon, 1944.

[26] Nicolle J., *La symmétrie dans la nature et les travaux des hommes*, Vieux Colombier, Paris, 1965.

[27] Pólya G., Über die Analogie der Kristallsymmetrie in der Ebene, *Z. Kristall.* **60** (1924), 278-282.

[28] Savelov A.V., *Ploskie krivie*, Gosizdat. fiz. mat. lit., Moskva, 1960.

[29] Shepard A.O., *The Symmetry of Abstract Design with Special Reference to Ceramic Decoration*, Carnegie Inst. of Washington, Publ. No. 575, Washington DC, 1948.

[30] Shubnikov A.V., Koptsik V.A., *Symmetry in Science and Art*, Plenum, New York, London, 1974.

[31] Shubnikov A.V., Belov N.V. *et al.*, *Colored Symmetry*, Pergamon, Oxford, London, New York, Paris, 1964.

[32] Smeets R., *Signs, Symbols, & Ornaments*, Van Nostrand, New York, Cincinnati, Toronto, London, Melbourne, 1975.

[33] Smith D.E., *History of mathematics*, I, II, Dover, New York, 1958.

[34] Speiser A., *Die Theorie der Gruppen von endlicher Ordnung*, 2nd ed., Berlin, 1927.

[35] Struik D.J., *A Concise History of Mathematics*, 2nd ed., Dover, New York, 1948.

[36] Washburn D.K., *A Symmetry Analysis of Upper Gila Area Ceramic Design*, Carnegie Inst. of Washington, Publ. No. 574, Washington DC, 1977.

[37] Washburn D.K., Symmetry Analysis of Ceramic Design: Two Tests of the Method on Neolithic Material from Greece and the Aegean, In *Structure and Cognition in Art*, Cambridge University Press, London, 1983.

[38] Washburn D.K., Crowe D.W., *Symmetries of Culture*, University of Washington Press, Washington, 1988.

[39] Weyl H., *Symmetry*, Princeton University Press, Princeton, 1952.

[40] Zamorzaev A.M., *Teoriya prostoi i kratnoi antisimmetrii*, Shtiintsa, Kishinev, 1976.

Hyperbolic Celtic Knot Patterns

Douglas Dunham
Department of Computer Science
University of Minnesota, Duluth
Duluth, MN 55812-2496, USA
E-mail: ddunham@d.umn.edu

Abstract

Centuries ago, Celtic knot patterns were used to decorate religious texts. Celtic knots are formed by weaving bands in an alternating over-and-under pattern. Originally, these were finite patterns on the Euclidean plane. Recently such patterns have also been drawn on spheres, thus utilizing a second of the three "classical geometries". We complete the process by exhibiting Celtic knot patterns in hyperbolic geometry, the third classical geometry. Our methods lead to a unified framework for discussing knot patterns in each of the classical geometries. Because of the precision and many calculations required to construct hyperbolic patterns, it is natural to generate such patterns by computer. Thus, the patterns we show are created by using computers, mathematics, and aesthetic considerations.

Introduction

In about the 6th century Irish monks started using what we now call Celtic knot patterns as ornamentation for religious texts. The monks also created spiral patterns, key patterns, zoomorphic patterns, and decorated lettering, but we will only consider knot patterns. Figure 1 shows a simple example of a knot pattern. The

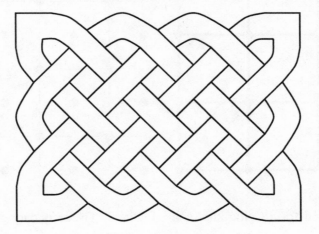

Figure 1: A simple Celtic knot pattern

use of this kind of decoration went out of style in about the 10th century, and the methods for creating such patterns were lost as well. Subsequently, people who wanted to make Celtic knot patterns had to copy existing patterns. That is, until the early 1950's when George Bain invented a method for creating such patterns [1].

In the late 1950's, the Dutch artist M. C. Escher became the first person to create hyperbolic art in his four *Circle Limit* patterns. The pattern of interlocking rings near the edge of his last woodcut *Snakes* (Catalog Number 448 of [6]) also exhibits hyperbolic symmetry. The goal of this paper is to take a first step toward combining Celtic knot art and hyperbolic geometry. Thus Celtic knot patterns will have been drawn on each of the three *classical geometries*: Euclidean, spherical (or elliptical), and hyperbolic geometry. Celtic knot patterns have also been drawn on convex polyhedra, which are very closely related to spherical patterns.

We will begin with a brief review of Celtic knots and hyperbolic geometry, followed by a discussion of regular tessellations, which form the basis for our hyperbolic Celtic knot patterns. Finally, we will develop a theory of such patterns, showing some samples, and indicate directions of future work.

Celtic Knot Patterns

Celtic knot patterns were used in the British Isles to decorate stonework and religious texts from the sixth through the tenth centuries. The methods used by monks to create such patterns have been lost. However, in 1951, George Bain published a method to create such patterns which he discovered after years of studying those ancient patterns. Later, his son, Iain Bain, published a simplified algorithm for making knot patterns in 1986 [2]. It is Iain Bain's method, as explained by Andrew Glassner [3], that we will discuss here.

The simplest knot patterns can be constructed from a rectangular grid of squares as shown in Figure 2. The set of vertices of this grid, thought of as a graph, form the starting point for Iain Bain's construction and is called the *primary grid* by Glassner. The center points of the squares form the vertices of another rectangular grid of squares as shown in Figure 3 (with its edges extended to the boundary of the primary grid). This is the *secondary grid*. The *tertiary grid*, shown in Figure 4, is formed by the union of the primary grid and the secondary grid. Thus, the tertiary grid is a grid of squares of half the edge width of the squares in the primary and secondary grids.

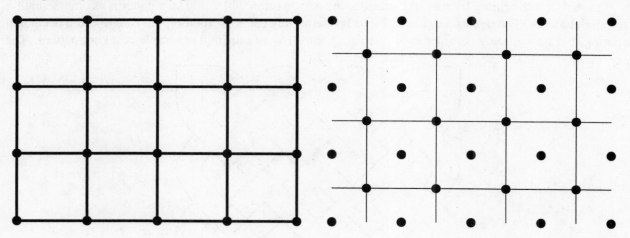

Figure 2: The primary grid for a Celtic knot construction.

Figure 3: The secondary grid for a Celtic knot construction.

Diagonal lines are drawn in each of the interior small squares of the tertiary grid using the lower-left to upper-right diagonal for the upper-left interior square, and then drawing the rest of the diagonals in an alternating pattern of lower-left to upper-right and upper-left to lower-right diagonals as in Figure 5. These diagonals will form what is called the *internal weaving*. The internal weaving in this example will be a plait, seen in the interior pattern of Figure 1.

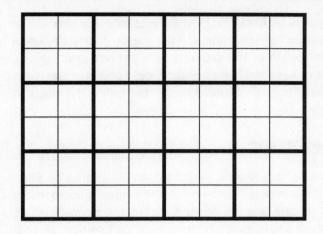

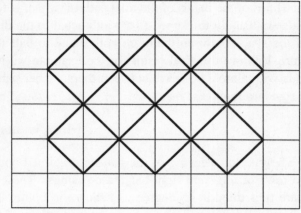

Figure 4: The tertiary grid — the union of the primary grid (heavy lines) and the secondary grid (light lines).

Figure 5: The diagonals in the interior of the tertiary grid that form the internal weaving.

Diagonals are also placed in the small edge squares of the tertiary grid in the same pattern but only going halfway to the outer edge as shown in Figure 6. These diagonals form the *external weaving,* which will connect the ends of the internal weaving. Next, at each of the tertiary grid points where four diagonals meet, form two paths by connecting the lower-left to the upper-right diagonal, and connecting the upper-left to the lower-right diagonal. Following one of the paths, let it go alternately above and below the paths it crosses. This can be done in a consistent way by using one kind of crossing on each row of crossing points and then using the other kind of crossing on the next row, as in Figure 7.

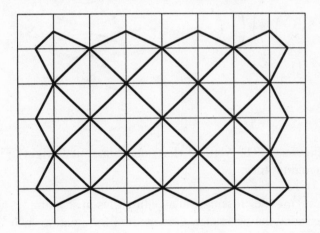

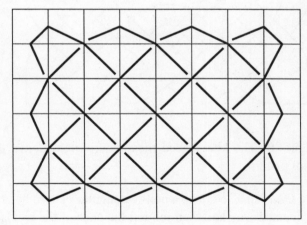

Figure 6: The outer diagonals that form the external weaving (in addition to the internal weaving).

Figure 7: The over-and-under specification of the path.

Using knot theory terminology, the over-and-under pattern formed by the diagonals is the *regular projection* (onto the plane) of a knot (a circle embedded in 3-space); it is regular because only two strands cross at a point. A "multi-knot" formed by more than one circle in 3-space is called a *link*. There will be only one path if the numbers of rows and columns of vertices in the primary grid are relatively prime. Most Celtic knot patterns are the regular projections of knots: there is only one path. The path or paths serve as the centerlines of the bands of the final pattern, which is formed by thickening the paths to form the bands. The bands are usually thickened to a width equal to the distance between them (so the standard band thickness and the space between them are both equal to half the length of the diagonal of a primary grid square). Figure 1 shows the final result for the example we have been studying. Some Celtic patterns use wider bands with almost no space between them. Other patterns use thin doubled bands that follow the edges of the standard thickness bands.

More General Patterns

The interior weaving of the pattern described above is very regular — it amounts to a tiling by alternate rows of left- and right-handed crossings. These crossings are enclosed in kite-shaped tiles, actually square tiles tilted at 45 degrees, as shown in Figure 8 (except that the center tile contains a non-crossing, as discussed below). The top and bottom vertices of these kite tiles are both primary grid vertices or both secondary grid vertices; the left and right vertices of the kites are vertices of the other grid.

To obtain more general patterns, one can replace some or all of the "crossing" tiles by either of the *avoiding* tiles shown in Figure 9. We call those tiles *vertical* or *horizontal* avoiding tiles because their paths avoid either the vertical or horizontal axis of their kite-shaped tile. Each such replacement may increase or decrease the number of loops in a link by one, or it may leave the number unchanged. If, after replacing all the crossing tiles by avoiding tiles, there is only one loop, it is called a *snake* by Glassner [4].

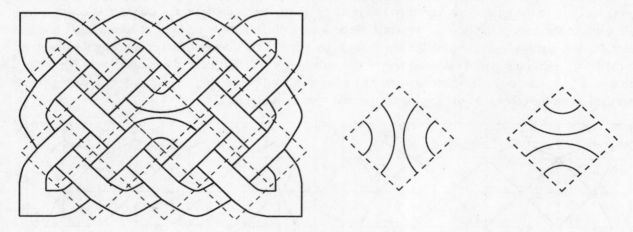

Figure 8: The kite-shaped tiles underlying a Celtic knot pattern.

Figure 9: The vertical (left) and horizontal (right) avoiding tiles.

One can also create a non-rectangular pattern by arranging the crossing and avoiding tiles in any simply-connected way and then joining the ends of the bands around the perimeter. One method for creating such patterns by hand involves lightly drawing the primary and secondary grids and then drawing more darkly some of the edges of either grid, with the rule that no dark edges may cross. These dark edges are *barrier* edges that the band cannot cross. In Figure 8 there is a horizontal barrier edge (not shown) connecting the left and right (secondary grid) vertices of the center kite. Glassner [3] and Christian Mercat [7] describe their versions of this method. Barrier edges are called *breaklines* by Glassner, and *longitudinal* and *transverse walls* by Mercat depending on whether they are edges of the primary or secondary grid.

With the goal of generalizing these techniques to the hyperbolic plane, we next discuss hyperbolic geometry, repeating patterns, and regular tessellations, which will form the basis for hyperbolic Celtic knot patterns.

Hyperbolic Geometry, Repeating Patterns, and Regular Tessellations

Among the classical geometries, the Euclidean plane, the sphere, and the hyperbolic plane, the latter is certainly the least familiar. This is probably due to the fact that there is no smooth distance-preserving embedding of the hyperbolic plane into ordinary 3-space, as there is for the sphere (and the Euclidean plane). However, there are *models* of hyperbolic geometry in the Euclidean plane, which must therefore distort distance.

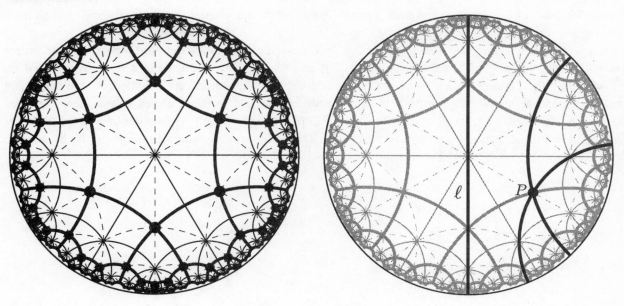

Figure 10: The regular tessellation $\{6, 4\}$ in heavy lines with dots at its vertices, its dual tessellation $\{4, 6\}$ in light lines, and common radii of the 6-gons and 4-gons in dashed lines.

Figure 11: An example of the hyperbolic parallel property: a line ℓ, a point P not on ℓ, and two lines through P not meeting ℓ.

One of these models is the *Poincaré circle model*, which has two useful properties: (1) it is conformal (i.e. the hyperbolic measure of an angle is equal to its Euclidean measure) — consequently a transformed object has roughly the same shape as the original, and (2) it lies entirely within a *bounding circle* in the Euclidean plane — allowing an entire hyperbolic pattern to be displayed. In this model, the hyperbolic points are the interior points of the bounding circle and the hyperbolic lines are interior circular arcs perpendicular to the bounding circle, including diameters. For example, all the arcs are hyperbolic lines in Figure 10.

By definition, (plane) hyperbolic geometry satisfies all the axioms of (plane) Euclidean geometry except the Euclidean parallel axiom, which is replaced by its negation. Figure 11 shows an example of this hyperbolic parallel property: there is a line, ℓ, in Figure 10 (the vertical diameter), a point, P, not on it, and more than one line through P that does not intersect ℓ.

Because distances must be distorted in any model, equal hyperbolic distances in the Poincaré model are represented by ever smaller Euclidean distances toward the edge of the bounding circle (which is an infinite hyperbolic distance from its center). All the curvilinear hexagons (actually regular hyperbolic hexagons) in Figure 10 are the same hyperbolic size, even thought they are represented by different Euclidean sizes.

A *repeating pattern* in any of the classical geometries is a pattern made up of congruent copies of a basic subpattern or *motif*. The motif for the pattern of Figure 10 is a curvilinear right triangle with a dashed hypotenuse and thick and thin lines for legs. Also, we assume that a repeating pattern fills up its respective plane. It is useful that hyperbolic patterns repeat in order to show their true hyperbolic nature.

An important kind of repeating pattern in any of the classical geometries is the *regular tessellation* by regular p-sided polygons, or *p-gons*, meeting q at a vertex; it is denoted by the Schläfli symbol $\{p, q\}$. We need $(p - 2)(q - 2) > 4$ to obtain a hyperbolic tessellation; if $(p - 2)(q - 2) = 4$ or $(p - 2)(q - 2) < 4$, one obtains tessellations of the Euclidean plane and the sphere, respectively. Figure 10 shows the hyperbolic tessellation $\{6, 4\}$ in heavy lines with a 6-gon centered in the bounding circle (the center of the bounding circle is not a special point in the Poincaré model, it just appears so to our Euclidean eyes). Figure 10 also shows the hyperbolic tessellation $\{4, 6\}$ in light lines with one of its vertices centered in the bounding circle. The dashed lines in Figure 10 do not form a regular tessellation, but when $p = q$ the analogous dashed lines form the regular tessellation $\{4, p\}$.

If we assume for simplicity that $p \geq 3$ and $q \geq 3$, there are five solutions to the "spherical" inequality $(p - 2)(q - 2) < 4$: $\{3, 3\}$, $\{3, 4\}$, $\{3, 5\}$, $\{4, 3\}$, and $\{5, 3\}$. These tessellations may be obtained by "blowing up" the Platonic solids: the regular tetrahedron, the octahedron, the icosahedron, the cube, and the dodecahedron, respectively, onto their circumscribing spheres. In the Euclidean case, there are three solutions to the equality $(p - 2)(q - 2) = 4$: $\{3, 6\}$, $\{4, 4\}$, and $\{6, 3\}$, the tessellations of the plane by equilateral triangles, squares, and regular hexagons. There are infinitely many solutions to the hyperbolic inequality $(p - 2)(q - 2) > 4$. This is summarized in Table 1 below.

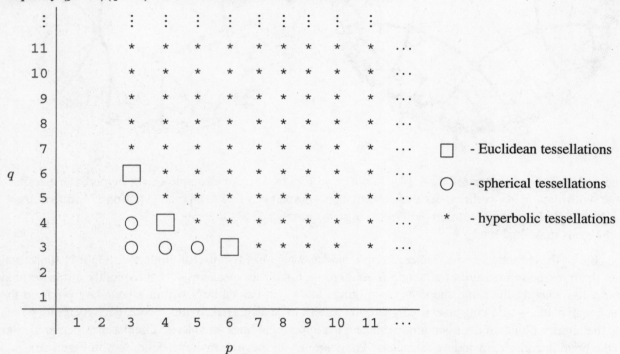

Table 1. The relationship between the values of p and q, and the geometry of the tessellation $\{p, q\}$.

For each tessellation $\{p, q\}$, its *dual tessellation* is $\{q, p\}$, whose vertices are at the centers of the p-gons of $\{p, q\}$ and whose edges are perpendicular bisectors of the edges of $\{p, q\}$. Figure 10 shows the tessellation $\{6, 4\}$ in heavy lines and its dual tessellation $\{4, 6\}$ in thin lines. Of course the dual of the dual of a regular tessellation is just the original tessellation. If $p = q$, the tessellation is self-dual: $\{3, 3\}$ is the spherical version of the regular tetrahedron, $\{4, 4\}$ is familiar Euclidean tiling by squares, and $\{5, 5\}$, $\{6, 6\}$, $\{7, 7\}, \ldots$ are hyperbolic.

This completes our discussion of hyperbolic geometry, repeating patterns, and regular tessellations. Next, we use these concepts to develop a theory of hyperbolic Celtic knot patterns, which is actually valid in all three of the classical geometries.

A Theory of Hyperbolic Celtic Knot Patterns

As we saw above, the method for creating knot patterns that was developed by Iain Bain and others is based on the regular tessellation of the Euclidean plane by squares. We extend that method to one based on any regular tessellation of one of the classical geometries. The tessellation $\{p, q\}$ itself serves as the primary grid, its dual, $\{q, p\}$, defines the secondary grid, and their union is the tertiary grid. In Figure 10, where $p = 6$ and $q = 4$, the primary grid is shown in heavy lines and the secondary grid in thin (solid) lines. The dashed lines in Figure 10 define a tessellation by kite-shaped tiles — rhombuses with vertex angles of $2\pi/p$, $2\pi/q$, $2\pi/p$, and $2\pi/q$ (with vertices alternately at the centers and vertices of p-gons of the tessellation $\{p, q\}$). If one starts with the Euclidean $\{4, 4\}$ tessellation, the rhombuses are actually squares tilted at a 45-degree angle, as shown in Figure 8.

Celtic knot patterns have two characteristics: (1) no more than two bands cross at a point, and (2) any one band goes alternately over and under other bands that it crosses. Such a pattern can be obtained if all the rhombuses are filled in only with left crossing tiles or only with right crossing tiles. Looking at a rhombus from a primary grid vertex, if the nearest band coming from the right is on top, it is a *right crossing tile*, otherwise it is a *left crossing tile*; both kinds are shown in Figure 12 (the rhombuses shown are the ones to the right of the center of the bounding circle in Figure 10). Figure 13 shows a complete pattern composed of right crossing tiles based on the $\{4, 5\}$ tessellation. Such a Celtic pattern is called a *regular weaving* or *plait*. The central pattern in Figure 1 is another example — of the standard Euclidean weaving.

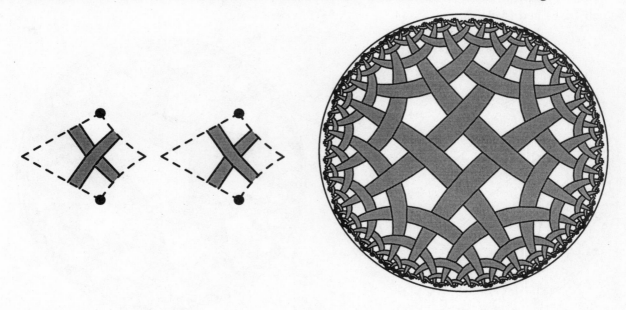

Figure 12: A left crossing tile (left) and a right crossing tile (right), with dots at the primary grid vertices.

Figure 13: A regular weaving or plait based on the $\{4, 5\}$ tessellation.

The bands of a regular weaving based on the tessellation $\{p, q\}$ follow the edges of the *uniform tessellation* $(p.q.p.q)$ (also called *Archimedean* or *semiregular* tilings by some authors). The edges of $(p.q.p.q)$ are formed by connecting the midpoints of adjacent edges of the p-gons of $\{p, q\}$. Those midpoints serve as the vertices of $(p.q.p.q)$, each of which is surrounded by a p-gon, a q-gon, a p-gon, and a q-gon (which explains the notation). Figure 14 shows the uniform tessellation $(4, 5, 4, 5)$ underlying the regular weaving knot pattern of Figure 14. Since p-gon edge midpoints are also q-gon edge midpoints in the dual tessellation $\{q, p\}$, a regular tessellation and its dual produce the same regular weaving — which is not surprising since p and q play symmetrical roles in $(p.q.p.q)$.

There are three regular spherical weavings, which are based on the self-dual tessellation $\{3,3\}$, and on the two pairs of duals, $\{3,4\}$ and $\{4,3\}$, and $\{3,5\}$ and $\{5,3\}$. The weaving based on $\{3,3\}$ traces the edges of the "uniform" tessellation $(3.3.3.3)$, which is actually the regular tessellation $\{3,4\}$, the blown-up version of the octahedron. There is a band in each of three mutually perpendicular planes through the center of the sphere containing the $\{3,3\}$. These three bands are linked, forming Borromean rings. Glassner shows such a weaving based on a cube rather than an octahedron (Figure 10a of [5]). The octahedron is the intersection of the tetrahedron $\{3,3\}$ and its dual, which together form the stella octangula. In fact, the regular weaving based on any self-dual tessellation $\{p,p\}$ traces the edges of the regular tessellation $\{p,4\}$. The weaving based on the pair $\{3,4\}$ and $\{4,3\}$ traces the edges of uniform tessellation $(3.4.3.4)$, which is the spherical version of the cuboctahedron. Last, the weaving based on the pair $\{3,5\}$ and $\{5,3\}$ traces the edges of uniform tessellation $(3.5.3.5)$, which is the spherical version of the icosadodecahedron. Glassner shows a version of this weaving in Figure 18 of [5].

There are only two regular Euclidean weavings, which are based on the self-dual tessellation $\{4,4\}$, and on the dual pair $\{3,6\}$ and $\{6,3\}$. The weaving based on $\{4,4\}$ is just the standard Euclidean weaving seen in the center of Figure 1, which is the basis for most Celtic knot patterns. The weaving based on $\{3,6\}$ and $\{6,3\}$, with its triangular and hexagonal holes, is sometimes seen in the caning for the seats of chairs.

There are infinitely many regular hyperbolic weavings. They are either based on the self-dual tessellations $\{p,p\}$ for $p \geq 5$, or on the dual pairs $\{p,q\}$ and $\{q,p\}$, where $p \neq q$ and $(p-2)(q-2) > 4$. Figure 15 shows the weaving based on the self-dual $\{5,5\}$ tessellation, and Figure 13 shows the weaving based on the pair $\{4,5\}$ and $\{5,4\}$.

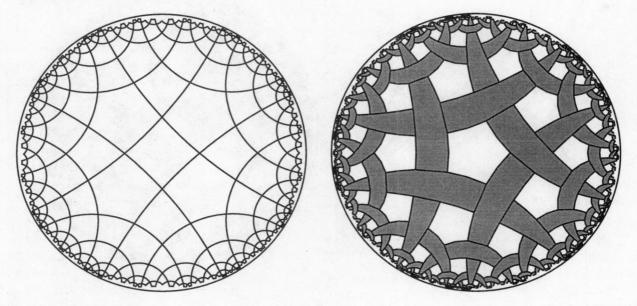

Figure 14: The uniform tessellation $(4.5.4.5)$ underlying the regular weaving of Figure 13.

Figure 15: The regular weaving based on the tessellation $\{5,5\}$.

More general Celtic knot patterns may be obtained by replacing some of the crossing tiles of a regular weaving with *avoiding tiles*. Figure 16 shows the two kinds of avoiding tiles, which are distinguished by the diagonal of the tile rhombus that their paths avoid (as in Figure 12, the rhombuses shown are the ones to the right of the center of the bounding circle in Figure 10); Figure 9 shows the avoiding tiles for the standard Eulidean weaving (based on $\{4,4\}$). One of the diagonals of each rhombus is an edge from the underlying tessellation $\{p,q\}$, and the other diagonal is an edge from the dual tessellation $\{q,p\}$. Figure 17

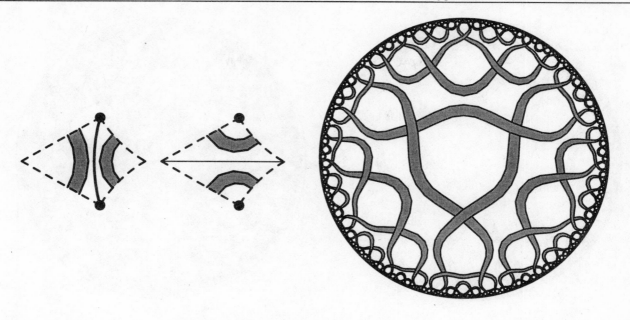

Figure 16: A p-gon edge avoiding tile (left) and a q-gon edge avoiding tile (right).

Figure 17: A Celtic knot pattern of right crossing tiles and p-gon edge avoiding tiles.

shows a pattern of alternating right crossing tiles and p-gon edge avoiding tiles. Figure 18 shows a pattern of alternating right crossing tiles and q-gon edge avoiding tiles.

One of the rules of Celtic knot patterns is that paths cannot avoid both diagonals of a rhombus (there would be no way to connect the ends the paths coming into that tile). Thus, if we want to construct Celtic knot patterns from rhombus tiles, our collection of basic tiles is complete, consisting of the two kinds of crossing tiles and the two kinds of avoiding tiles.

It is possible to further generalize the methods above to apply to non-rhombic quadrilateral tiles. For any quadrilateral, there are only four ways to connect ends of bands coming into it across each of its four sides: the two kinds of crossing configurations and the two kinds of avoiding configurations. Glassner has used non-rhombic quadrilaterals to construct several of his patterns in [5]. As an example, if we have a pattern of triangles upon which we would like to draw a knot pattern, we could first subdivide the triangles into three quadrilaterals by connecting the triangle's center to the midpoints of its sides.

We will apply this method to the construction of what we call *Celtic ring patterns* — rings interlocked in the over-and-under pattern characteristic of Celtic knots. We start by subdividing the p-gons of the tessellation $\{p, q\}$ into p isosceles triangles with angles $2\pi/p$, π/q, and π/q, as shown in the central p-gon in Figure 19 (where $p = 6$ and $q = 4$). Then we subdivide each triangle into three quadrilaterals (shown for one of the isosceles triangles in Figure 19). Finally, we place a crossing tile in each of the quadrilaterals, producing the final ring pattern of Figure 19. Note that the crossing is pushed as far as possible toward one vertex of the quadrilateral. Figure 19 shows the pattern of interlocking rings that Escher used near the edge of his last woodcut, *Snakes* (Catalog Number 448 of [6]).

This finishes our discussion of the theory of hyperbolic Celtic knot patterns and the methods for creating them. Of course, the theory and methods also apply to each of the three classical geometries as well. In the final section, we indicate directions of future work.

Future Work

We have presented a theory of Celtic knot patterns and methods for creating such patterns in each of

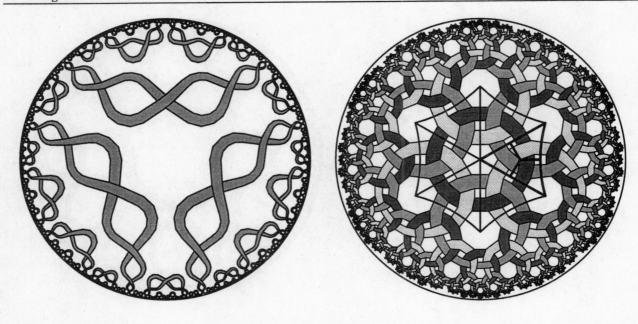

Figure 18: A Celtic knot pattern of right crossing tiles and q-gon edge avoiding tiles.

Figure 19: An interlocking "Celtic ring" pattern showing part of the underlying $\{6, 4\}$ and some of the triangles used in the construction, with one of them subdivided into three quadrilaterals.

the three classical geometries. Some natural directions of future work include extensions to hyperbolic knot patterns not based on regular tessellations, and the creation of hyperbolic versions of other kinds of Celtic patterns, such as key patterns, spiral patterns and zoomorphics.

Acknowledgment

I would like to thank my student, Alark Joshi, for preparing many of the figures in this paper, and the reviewer for a number of helpful suggestions.

References

[1] G. Bain, *Celtic Art : The Methods of Construction,* Dover Publications, June 1973.

[2] I. Bain, *Celtic Knotwork,* Sterling Publishing Co., Inc., New York, 1986.

[3] A. Glassner, Celtic Knotwork, Part 1 (Andrew Glassner's Notebook), *IEEE Computer Graphics and Applications* Vol. 19, No. 5, pp. 78–84, September/October 1999.

[4] A. Glassner, Celtic Knotwork, Part 2 (Andrew Glassner's Notebook), *IEEE Computer Graphics and Applications* Vol. 19, No. 6, pp. 82–86, November/December 1999.

[5] A. Glassner, Celtic Knotwork, Part 3 (Andrew Glassner's Notebook), *IEEE Computer Graphics and Applications* Vol. 20, No. 1, pp. 70–75, January/February 2000.

[6] J. L. Locher, editor, *M. C. Escher, His Life and Complete Graphic Work*, Harry N. Abrams, Inc., New York, 1982.

[7] C. Mercat, *Cours d'enluminure à base d'entrelacs,* http://www.bok.net/kri/Celte/. English version: S. Abbott, *How to draw Celtic knots,* http://www.abbott.demon.co.uk/mercatmethod.html

" – To Build a Twisted Bridge – "

Carlo H. Séquin

Computer Science Division, EECS Department
University of California, Berkeley, CA 94720
E-mail: sequin@cs.berkeley.edu

Abstract

This is a follow-up on a discussion at the 1999 Bridges conference; it is a topological / geometrical study of how the structure of a Moebius band might be applied to bridges and/or buildings, with the possibility of creating an intriguing construction on the Campus of Southwestern College, to commemorate the series of annual Bridges conferences started in 1998.

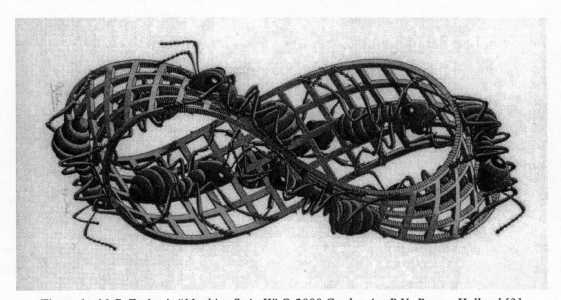

Figure 1: *M.C. Escher's "Moebius Strip II" © 2000 Cordon Art B.V., Baarn, Holland [3].*

1. Introduction

At the Bridges'99 conference, Jason Barnett gave an inspiring presentation "A Bridge for the Bridges" [1]. He showed scale models of spaces reminiscent of Escher structures that combine many different perspectives in surprising ways. However, Barnett's buildings were realized in 3-space and were designed to be readily navigable by human observers. In that session, the idea was put forward that it might be desirable to construct a monument on the Campus of Southwestern College commemorating the series of annual Bridges conferences. Given the interests of the typical conference participant, such a structure might well be Escher-like or reminiscent of the shapes of Klein bottles or Moebius bands. Perhaps – and most fittingly – such an intriguing shape might be applied to some form of pedestrian bridge crossing a little creek or connecting two tall buildings. Inspired by those discussions, this paper investigates several ways in which the structure of a Moebius band could be imposed onto a functional bridge (Section 2) or onto a habitable and usable building (Section 3). We conclude with some Moebius structures that have a purely aesthetic purpose (Section 4).

The title of this treatise pays homage to a delightful science fiction short story by Robert A. Heinlein, "–And He Built a Crooked House–" [4], in which an architect builds a 4-dimensional house in the shape of a hypercube. After an earthquake, the house traps its occupants who are then doomed to run around in the 3-dimensional surface of this structure. Contrary to this unfortunate outcome, our focus is on finding structures that are not only conceptually intriguing, but also sound and of practical use.

2. Moebius Bridges

The first idea that comes to my mind when I hear the term "Moebius" is that of a belt twisted through 180 degrees. However, if this idea is applied in a straightforward manner to the construction of a bridge, we end up with a surface that is difficult to walk on; after crossing less than half the length of the span, travellers would slide off the bridge (Fig.2). Thus the first task is to look for modifications of the basic concept to create a more or less level walking surface along the whole length. After that I will explore the possibilities of transforming the topology of a closed Moebius band into a shape that can serve as a bridge.

Figure 2: *A slab with a longitudinal twist -- a key element for a Moebius bridge -- is difficult to walk on.*

2.1 Twisted Surfaces You Can Walk On

The first and fundamental challenge is to find geometries that are clearly twisted, but on which one can still walk, and which will even admit wheelchairs, i.e., surfaces that have a continuous strip of a nearly horizontal surface along the whole span of the bridge. A first inspiration may come from ruled surfaces, in particular from a single-shell hyperboloid, which exhibits a rather twisted look, but can be composed entirely of a bundle of straight lines. The wish to keep a more or less "level path" across the chasm to be bridged implies that we also maintain horizontal tangents in the lateral direction. We may model these constraints with a Bézier patch that gives us sufficient degrees of freedom (Fig.3).

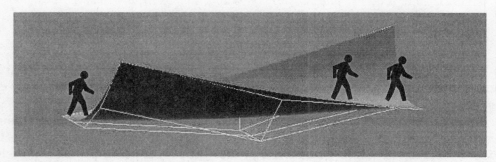

Figure 3: *A longitudinal twist can be simulated with a suitably distorted Bézier patch.*

Using five control points in the lateral direction, allows us to create a "flat" zone of adjustable width. Along the length of the bridge span this flat zone shifts from one edge of the patch to the other, i.e., at one end of the bridge the right-hand wall is pulled high, and at the other end the left-hand wall. This will create the impression that the slab has been flipped through almost 180 degrees. In the longitudinal direction, three or four rows of control points are sufficient to define the transition of the cross section from one end to the other, and perhaps to impose a slight arching of the bridge for stability and for aesthetic purposes.

Contemplation of the shape in Figure 3 makes one realize that the main reason that travelers do not slide off this bridge is the fact that its cross section is a "hollow" C-shape which offers a spot with a horizontal tangent for all of its orientations. We can exploit this idea more directly by sweeping a semicircular cylindrical element along the arch of the bridge. We can twist this cross section through 180 degrees and still maintain along its whole length a continuous path with a horizontal lateral tangent (Fig. 4). It should be noted that on such a surface one could walk either on the inside or on the outside of the C-shape; in the latter case we would twist the cross section in such a way that, in the middle of the bridge, the opening of the "C" points downwards. However, for a large portion of the span, travellers may not even realize that they are walking on an open C-profile; it may just feel like walking along the top of a giant cylinder. I believe the sense of "twistiness" of the bridge would be mostly absent in this option. In Figure 4, I show my preferred solution, where the profile is oriented like a "U" at the mid-point, and in which travellers across the bridge are naturally sheltered most of the way. Of course, towards the ends of the span, some railings must be added for safety.

Figure 4: *Swept C-shaped cross section with a longitudinal twist of 180° also offers a level walking surface.*

2.2 Closing the Moebius Loop

To shape such a twisted bridge surface into a closed Moebius band, we have to somehow form an end-to-end loop and close it off with the appropriate orientation. This closure could occur in three fundamentally different ways: below the walkway, and thus symbolically acting as the foundation or support for the walkway (Fig.5a); above the walkway, as a large arch from which the walkway could be "suspended" in some way (Fig.5b); and, third, besides the walkway -- perhaps forming a separate alternative walkway.

(a)   (b)

Figure 5: *Closing the loop of the Moebius bridge: (a) as a supporting structure, (b) as a suspension arch.*

The first alternative leads to a rather straightforward implementation (Fig.5a). This is just a simple, singly-twisted Moebius band. All the twist is in the "return path" and support structure. Aesthetic goals and engineering functionality might be nicely combined if the "bulge" created by the twisting slab where it passes through its vertical orientation is made to touch the middle of the horizontal portion of the slab that serves as the actual bridge; it can then act as center support and thus allows the slab to be thinner. Access to the bridge surface is also easy in this configuration, since the slab turns downwards at both ends of the span. Light-weight, transparent-looking "on-ramps" would connect the main structure to the slopes of the trench to be bridged.

The second alternative develops from an upside-down version of the first. The "roll" of the slab through its vertical orientation can now serve as a robust suspension arch from which the walkway could be suspended with many thin steel cables (Fig.5b). One difficulty with this arrangement is that the travellers will now walk on the "inside" of the structure, and access to this surface has to be provided in some way. One solution is to cut openings through the band itself at both ends of the horizontal bridge section where the band turns upward. This will produce two doorways, each framed by two columns formed by the outer flanges that support the slab.

Another possibility is to bend the twisted surface laterally away from the main walkway, thereby giving straight access onto the walkway. If the walkway happens to be in a concavity of the main Moebius band (Fig.4), then such a bending away to the sides is reminiscent of the swept C-section ribbon sculptures of Brent Collins [2] in which he aims to orient the C-shaped profile in such a way that its opening always points away from the bending direction of the space curve, i.e., in the negative normal direction of its Frenet frame. This results in a ribbon surface of consistently negative curvature. However, a C-shaped profile is clearly two-sided; it has an inside and an outside, and it can therefore not readily be closed into a single-sided Moebius configuration. To achieve this goal, we may gradually straighten out the C-profile into a flat slab, which can then readily be connected back-to-front. This may again lead to a configuration where the return path forms a suspension arch (Fig.6).

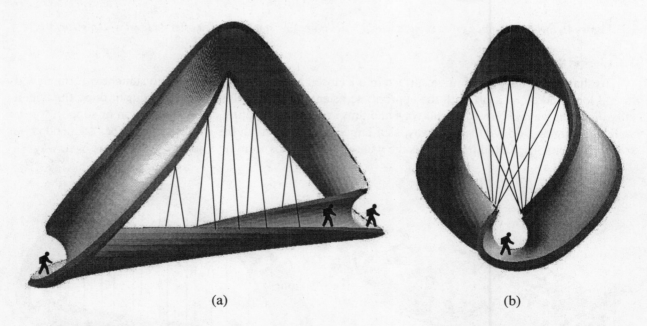

(a)   (b)

Figure 6: *Closing the Moebius loop through a non-planar space curve and varying its cross section, (a) side view, and (b) view from one end looking down the walkway.*

2.3 Function Follows Form

So far, all my constructions started with a functional walkway, then I tried to close those twisted surface pieces into Moebius loops. Now I will start from various geometrical Moebius configurations and then see how such a shape might serve as a bridge. For this purpose, the work of M.C. Escher serves as a great source of inspiration. In "Band van Möbius II" he draws a simply twisted band in grid form that serves as a climbing structure for nine ants (Fig.7a). Let's study this shape and see in which way it may be turned into a bridge. The portion facing backwards in Figure 7a offers a reasonably flat surface that runs most of the length of the whole object. It may serve as a walking surface if the whole structure is oriented suitably in space and the twist is redistributed somewhat differently around the loop (Fig.7b).

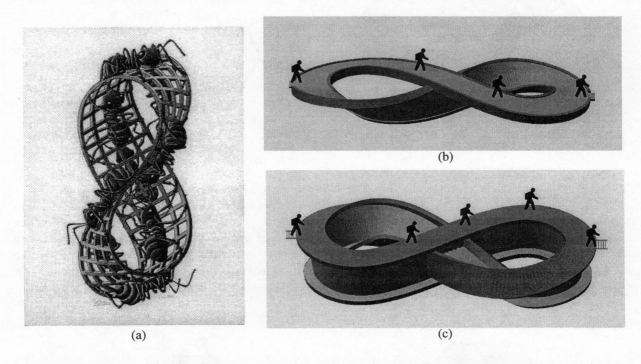

(a)   (b)   (c)

Figure 7: *Use of the geometry of Fig.1 [3] (a) for a bridge design, (b) by walking on part of the surface, and (c) by walking on the widened edge of the Moebius strip.*

Alternatively, we could orient this surface downward and thereby turn the front-facing portion of the rim of this object into the proper orientation so that it could be walked on. The edge would have to be widened to make a safe and comfortable walkway. This could be accomplished by making the whole band much thicker, or by using it only as a support structure and by adding a perpendicular flange onto it to serve as the actual walking surface. Since this flange would of course run around the whole Moebius band, it would convert the cross section of the band into that of an I-beam. Upon closer inspection of Escher's object, it appears that the backward-facing edge of this same part of the Moebius band has a somewhat larger longitudinal extent than the front-facing rim, and might thus be more suitable to serve as a bridge over a trench or ravine of a given width. But as is evident from (Fig.7b), some part of that edge is obstructed by the crossing slab that we originally considered as the walking surface. This can easily be remedied if we a modify the structure so that it intersects itself at the central point (Fig.7c). This self-intersection may also increase the overall strength of the whole structure.

M.C. Escher also drew another Moebius band that exhibits three-fold symmetry and which has a built-in twist of 540 degrees (Fig.8a). This basic arrangement also appears in the familiar recycling symbol (Fig.8b) created by Container Corporation of America during Earth Day 1970. A contest for graphic art students to design a symbol representing paper recycling attracted over a thousand entries which were judged at the Aspen Institute for Humanistic Studies. The winning entry submitted by Gary Anderson, an art student at U.C. Berkeley, was modified by William Lloyd into the well-known "chasing arrows" design.

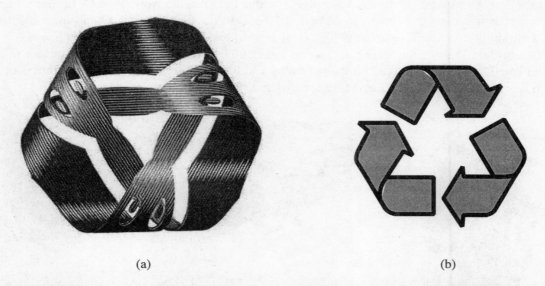

(a)   (b)

Figure 8: *(a) M.C. Escher's "Moebius Strip I" © 2000 Cordon Art B.V., Baarn, Holland [3]; (b) recycling symbol with same three-fold symmetry.*

Sometimes one encounters a modified form of the recycling symbol that has an overall twist of only 180° (Fig.9a). We will explore the usefulness of this structure to create a pedestrian bridge. We follow the possibility shown in Figure 7a of walking on the edge of the Moebius band, rather than on its flat surface. Figure 9b shows how one inner section of the edge might be used as a walkway.

(a)   (b)

Figure 9: *(a) Asymmetric recycling symbol; (b) conceptual application to a bridge design.*

However, there is an intriguing possibility to also use the "suspension arch" as a secondary walkway. Reminiscent of a strongly arched bridge in a Japanese tea gardens, steps may lead up on the outer edges of the suspension arch. On the very top, where the mostly vertically oriented band needs to flip over to make the desired Moebius connection, a flat spot might be created suitable for a small observation platform.

3. Moebius Buildings

In 1992 Peter Eisenman designed a Moebius building (Fig.10a). One gets the impression, that this particular building was purely a mental exercise in the strictest tradition of an architectural design paradigm known as "function follows form." It is not immediately clear how this particular form can be structured internally to make a set of convenient and usable floors and suites. Another "Moebius Haus" was conceived by Van Berkel & Bos who try to capture the endless figure-8-type movement possible in this topology in some internal unstructured space (Fig. 10b).

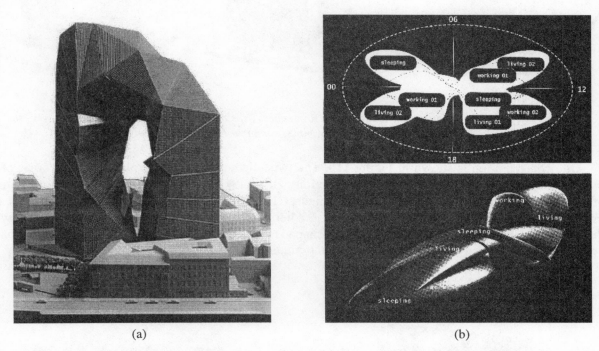

(a) (b)

Figure 10: *Moebius buildings: (a) a proposal by Peter Eisenman, (b) a concept by* Van Berkel & Bos.

3.1 Form Follows Function

For this study we assume that the client wants a reasonably traditional office building, say, for a Mathematics Institute, but one that clearly reflects the shape of a Moebius band. We are not interested in just a thin mathematical surface, but in a prismatic structure of substantial volume that exhibits the desired "twist." We start our analysis with a cubic module of 32 feet on a side. This is large enough to accommodate two 12 foot deep offices and an eight foot corridor "across" the building, and two stories in the vertical dimension. We first will explore ways of stacking these useful modules so that the essence of a Moebius strip is conserved, i.e., if we follow along one face of the prismatic toroid, we expect that it will take more than one loop around the ring before we come back to the starting point and that we will visit "other" faces of the prismatic structure on this path.

Three such cube modules in sequence can form either a straight row or a small symmetrical L-shape. Four cubes in sequence can form a planar L-shape with one leg twice as long as the other or a planar "Z"/"S" jog structure. But they can also form a twisted, 3D Z-structure, which may appear as a right-handed or as a left-handed version. It is the appearance of this twist element that carries a path on the surface of the prismatic structure from "one face" to "another". The simplest "twisted" loop can be built from just ten such cube-modules (Fig. 11a). A path on one of the prism surfaces incurs a twist of 90 degrees on every loop around the structure. It thus takes four loops to come back to the starting point, and during this traversal we will have visited all the exposed cube faces on the whole structure.

But let's assume that the client wants a "normal" Moebius band, in which we return to the starting point after only two revolutions. There are many basic configurations in which two twist elements can be arranged so that their effect adds up, rather than cancels out, and so that the total prismatic twist around the loop becomes 180 degrees.

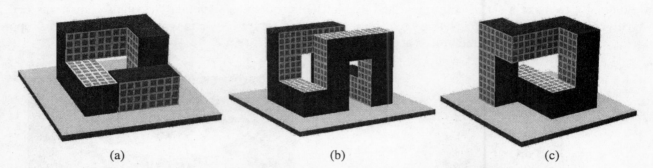

(a)   (b)   (c)

Figure 11: *Twisted configurations created from unit cubes: (a) simplest configuration with a 90º twist, (b) and (c) two configurations with a complete 180º Moebius twist.*

One configuration is shown in Figure 11b. The twist elements have been placed at the corner of the loop and have been stretched in the vertical direction so that they form entrances into the courtyard / atrium of the building. Another possibility, shown in Figure 11c, introduces a 3D crossing in the form of a sky-bridge in order to close the loop with a 180º twist. Alternatively, this bridge can be straightened out, while the connecting path at ground level becomes S-shaped (Fig.12a). Figure 12b gives an impression what an actual building, derived from this basic configuration, might look like.

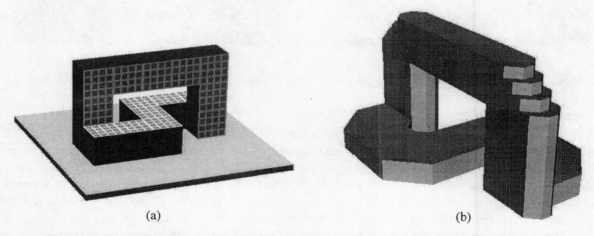

(a)   (b)

Figure 12: *Transformation of Figure 11c into (a) a related structure, and (b) into a building sketch.*

3.2 Function Follows Form.

Now that we have seen that it is indeed possible to create Moebius shapes from rather conventional building elements, we might want to make the Moebius shape more apparent. First, and foremost, we would like to use a more band-like structure, i.e., the aspect ratio of the rectangle swept along the loop should be at least 2:1. In addition, we want to make the small sides and the long sides of the cross section visibly distinct by using different materials. The large faces could be made from glass and steel, which would be natural for the window fronts, and the narrower faces could be made from concrete, or could be covered with dark, opaque glass.

However, just sweeping a 32' by 64' cross section along a circle with continuous twist (Fig.13a) will not lead to a practical building geometry. Figure 13b shows a modification of the basic Moebius shape in which one end has been squared to form the base of a building and the rest of the loop is connected to it with two right-angle turns. This shape can be further deformed until it resembles the bridge structure of Figure 12b with an S-shaped return path underneath, but now the return section stands on its small edge (Fig. 13c). As a further refinement, the impractical wide span of the bridge can be shortened, by narrowing the turns in the S-shaped building section underneath.

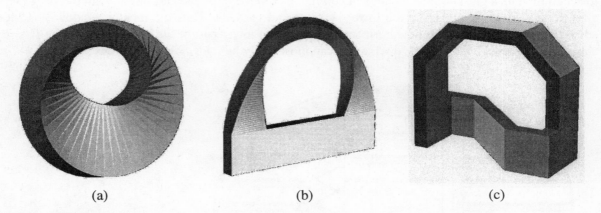

(a)   (b)   (c)

Figure 13: *Deforming a basic (impractical) Moebius loop into a usable building geometry, (a) through (c).*

Now it is time to look more closely at the internal organization and the resulting requirements for the surfaces of these structures. Clearly the more lightly colored vertical walls of the S-shape should be in glass, since they are natural window surfaces. The narrow faces of the prism could then be kept fairly opaque. To enhance the visual difference between the two types of surfaces even further, the narrow surfaces could also be reshaped into wedges, which would show off the Moebius property most dramatically. On top of the S-structure, where the wedge runs horizontally as a "roof," it could hide air-handling units, and along the vertical edges of the two towers it could accommodate fire stairs or elevator shafts (Fig.15d). The problem with the structure in Figure 13c is that this opaque face or wedge runs on the sides of the bridge on the facades where one would prefer to have a set of good windows; this prime real-estate should not be obscured for the purpose of maintaining continuation of the Moebius edge. In Figure 14, I explore a different way to work around this problem by reversing the roles of the wide and narrow sides of the swept cross section.

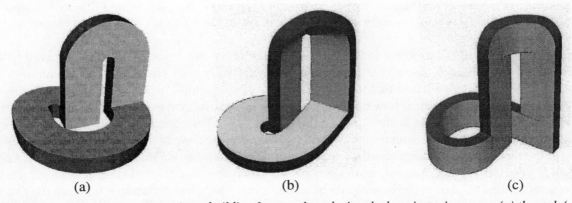

(a)   (b)   (c)

Figure 14: *Starting form a prominent building feature, then closing the loop in various ways (a) through (c).*

In Figure 14a I started with a dramatic vertical loop cut from a 32' thick slab that has plenty of good window spaces on both sides. The narrow Moebius edge runs vertically at the ends of the floors in the two towers and also sweeps over the top of the building. Figure 15 illustrates how the space in this prominent arch can be put to good use and provide much office space with generous windows. To make the desired Moebius closure for the overall structure, we need to connect the front side of one tower with the backside of the other. This can be achieved with a spiral loop formed by a low horizontal building branch (Fig.14a). In our envisioned Math Institute, this part could accommodate public functions such as the class-rooms, the library, the cafeteria, and the reception area. The latter three functions can make good use of skylights to accommodate a glass roof as the logical continuation of the glass facades of the main building loop. The class rooms could be placed in the lower story or underground. Thus the mostly window-less Moebius edge, that needs to run around the horizontal faces of this lower part of the building, could lie just slightly above ground level, where vertical windows can most easily be traded off for copious skylights.

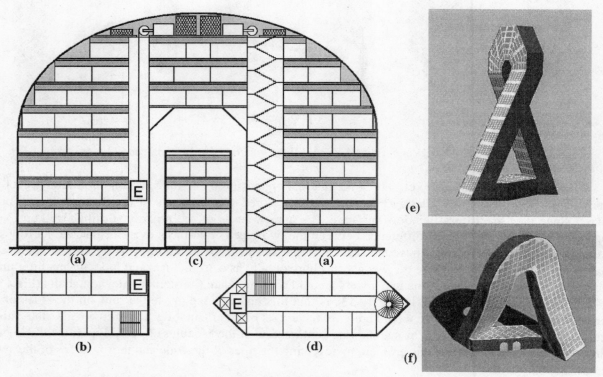

Figure 15: *(a) Elevations, and (b) floor plans of the arch in Fig. 14a; (c) cross section of an "on-edge" branch of the basic rectangular profile, e.g., as used in Fig. 14c; (d) a variant of the profile with wedges at the ends. (e) and (f) two different views of another use of such an arch in a tall building with slanted towers.*

The bottom portion of the structure in Figure 14a is rather larger. We can reduce its area by angling the two towers at 90 towards each other. This could change the originally semi-circular slab at the top into a conical shape (Fig.14b) – a structure that might be difficult to build and to outfit with elevators all the way to the top floor. Also it is not clear how usable the assignable floor spaces would be in this conical part of the building. Thus the concept is further developed in Figure 14c. The conical top section has been replaced with just a short, bent bridge with perfectly level floors and vertical walls. In contrast to the bridges in Figure 13, the elongated cross section is now swept in a vertical orientation through this part. This keeps the Moebius edge on top and bottom, and maintains good window fronts on the vertical faces. This same approach can also be used at the bottom end of the building. The spiral connection between the two towers can be made with a vertically oriented sweep, providing much good window front (Fig.14c).

One possible drawback with this geometry is that it may lead to rather tall buildings. Even if the swept rectangular cross section is reduced to only 30' by 60', an arch tall enough to loop over its own cross section put on edge would be at about 140' tall. Still I believe this arched configuration of the sweep has a lot of potential and can be shaped into attractive and usable buildings. Figure 15e and f show another way of closing this dominant arch into a Moebius loop. The legs are bent forwards and backwards like the two leaning towers in the "Gate of Europe" in Madrid. This spread must provide enough separation at the bottom to accommodate a fairly direct connection between the back-side of the front leg and the front-side of the back leg. Again this low, horizontal portion would be provided with continuous skylights that offer dramatic views of the arch 15 to 20 stories above.

4. Moebius Sculptures

A functional bridge or building may be beyond the means that Southwestern College is willing to expend on a commemorative construction. Perhaps a pure art object is a viable alternative. In this section we explore the realm of artistic Moebius bands and spaces.

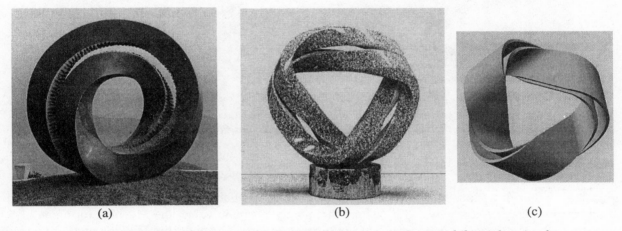

(a) (b) (c)

Figure 16: *Split Moebius sculptures by Keizo Ushio: (a) singly twisted (b) triply twisted.*

Keizo Ushio [5] is a sculptor who has celebrated twisted band structures in many different forms. While I am not sure whether he has ever carved an elementary, singly-twisted Moebius band, he has created doubly and triply twisted bands, and dozens of split Moebius bands (Fig.16a,b). Often his Moebius strips are produced by a rectangular cross section swept along a simple space curve with the proper twist so that the "front" side of the slab connects to the "back" side after one travel around the loop. Most often Keizo Ushio takes such a Moebius band only as the starting point, and then adds extra drama to it by splitting it along a center line.

In Keizo Ushio's sculptures typically the large dimension of a wide band is split down the middle, resulting in two nearly square prismatic shapes travelling side by side. This is a natural approach for sculptures carved from stone. Even Escher's triply twisted Moebius band is split in the same way (Fig. 8a). However, for different materials, say, steel, it may be just as natural to split the band from its narrow sides, leading to two even thinner cross sections travelling "on top of one another" (Fig.16c). This has the intriguing effect of creating a much more pronounced enclosed space between the two parts. That space itself now has the topological properties of a Moebius band, even though its bounding material band is orientable, i.e., double-sided (one side faces the Moebius space and the other side is pointing outward). Since the major portion of this paper is concerned with the creation of twisted Moebius spaces accessible to humans, I will now focus on a sculpture that emphasizes the enclosed space rather than its bounding material walls.

My proposed "Moebius Space" sculpture is derived from a "split torus" in which the cutting gap makes a 180 degree twist while sweeping around the torus. Since this space is the intended focus of our attention, I have widened it and also bulged it out to make it look more like a cave. To draw even stronger attention to it, I have given its wall a shiny silvery color, while keeping the outside of the torus from which it is carved in matte black (Fig.17).

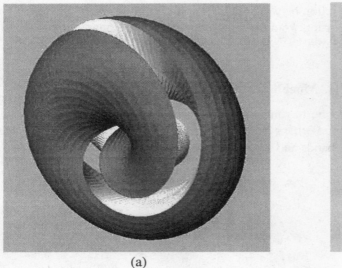

(a)   (b)

Figure 17: *The proposed "Moebius Space" sculpture, (a) computer-generated view, and (b) a physical maquette made with a Fused Deposition Modeling machine.*

5. Conclusions and Acknowledgments

The single-sided Moebius loop is a fascinating geometrical shape and leads readily to a plethora of aesthetically pleasing sculptures. But beyond this purely aesthetic value, the Moebius form can also be transformed into practical functions such as buildings or bridges. Given the tight linking of arts and mathematics exhibited in the Moebius shape, it would be a very fitting symbol for a commemorative construction for the Bridges Conferences at Southwestern College in Winfield, Kansas.

This work was partially supported by the National Science Foundation under a research grant to study the designer/fabricator interface for rapid prototyping. The synthetic figures and sculpture models have been constructed with SLIDE, a modeling and rendering system built by Jordan Smith [6] and Jane Yen.

References

[1]   J. F. Barnett: "A Bridge for the Bridges," Bridges 1999, Special Session VI, Aug.1, 1999.

[2]   Brent Collins: "Merging Paradigms," Bridges 1999, Conference Proceedings, pp 205-210.

[3]   M. C. Escher; with permission from Cordon Art B.V., Baarn, Holland. All rights reserved.

[4]   R. A. Heinlein: "——And He Built a Crooked House——," Berkeley Publishing Group, NJ.

[5]   K. Ushio, sculptor, homepage: http://www2.memenet.or.jp/~keizo/

[6]   J. Smith: "SLIDE: Scene Language for Interactive Dynamic Environments," http://www.cs.berkeley.edu/~ug/slide/

Sections Beyond Golden

Peter Steinbach
Albuquerque Technical-Vocational Institute
Albuquerque NM 87106
psteinbach@tvi.cc.nm.us

Over the centuries, the Fine Arts have celebrated several special numerical ratios and proportions for their visual dynamics or balance and sometimes for their musical potential. From the golden ratio to the sacred cut, numerical relationships are at the heart of some of the greatest works of Nature and of Western design. This article introduces to interdisciplinary designers three newly-discovered numerical relations and summarizes current knowledge about them. Together with the golden ratio they form a golden family, and though not well understood yet, they are very compelling, both as mathematical diversions and as design elements. They also point toward an interesting geometry problem that is not yet solved. (Some of these ideas also appear in [5] in more technical language.)

I. Sections and proportions

First I want to clarify what is meant by golden ratio, golden mean, golden section and golden proportion. When a line is cut (sected) by a point, three segments are induced: the two parts and the whole. Where would you cut a segment so that the three lengths fit a proportion? Since a proportion has four entries, one of the three lengths must be repeated, and a little experimentation will show that you must repeat the larger part created by the cut. Figure 1 shows the golden *section* and golden *proportion*. The larger part, *a*, is the *mean* of the extremes, *b* and *a* + *b*.

As the Greeks put it: *The whole is to the larger part as the larger is to the smaller part.* To the Greeks this was a harmonious balance, an ideal asymmetry. Later, when geometry was being translated and relearned in Europe, medieval philosophers would divide a perception into perceiver and perceived and divide an ideal society into smaller ruling and larger serving classes. Still later, Renaissance artists would be drawn into many questions of means between extremes.

Ignoring a reflection of Figure 1 (*b* could be on the left and *a* on the right) this cut is unique, which can be proved if we find the golden *ratio* — the only possible value of *a:b*. The starred pentagons in Figure 2 illustrate the same proportion using similar triangles, naming the golden ratio ϕ (phi, pronounced *fee*). To derive a value for ϕ, follow the algebra suggested by Figure 2:

$$\frac{a}{b} = \frac{a+b}{a} = \frac{\phi}{1} \quad \Rightarrow \quad \frac{a}{b} = \frac{a}{a} + \frac{b}{a} \quad \Rightarrow \quad \phi = 1 + \frac{1}{\phi} \quad \Rightarrow \quad \phi^2 - \phi - 1 = 0.$$

The positive solution of this quadratic equation is $\phi = (1 + \sqrt{5})/2 \approx 1.618$, the golden ratio. Thus the section determines a single ratio.

The Greek fascination with proportion inspired a design tradition in which a harmonious arrangement of elements is defined as one that realizes some special ratio and repeats it in proportion. Often this repetition is potentially endless, as seen in the progressions of rectangles and triangles in Figure 3. There is a long tradition of religious art associated with the square root of 3 — a diagonal of the unit-sided hexagon — and Robert Lawlor's *Sacred Geometry* [3] describes some uses of this number. Renaissance architects identified properties of a diagonal of the unit-sided octagon, $\theta = 1 + \sqrt{2}$, later called the Sacred Cut and recently described by Kim Williams [6] (see Figure 4). These numbers and others are involved not only in human artifice but in the growth of natural forms.

What's new in ratio and proportion? The new discoveries are partial answers to the question: What proportions can be made by multiple cuts of a segment? If we cut twice, we create six segments: the three parts, the sums of the middle and the left or right part, and the whole. Is there some extended proportion that can be satisfied by six such lengths? The answer is in Figure 5. Quite by accident I found that the unit heptagon's diagonals form a 3-by-3 proportion that describes a *tri*section of a segment. It remained, then, to prove that this is best possible, that this is the unique optimally proportional trisection analogous to the golden bisection. By *optimal* I mean: just as there is no cut other than the golden section that fits a non-trivial proportion, so there is no pair of cuts other than the

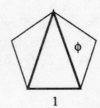

Figure 1.
Golden section & proportion

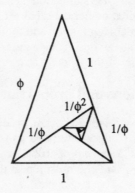

Figure 2.
Derivation of φ

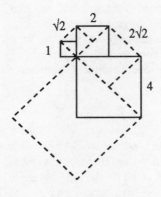

Figure 3.
Repeating proportions

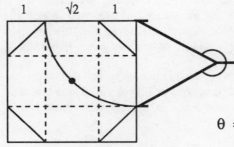

Figure 4.
Sacred Cut construction
of octagon

$$\theta = \frac{1 + \sqrt{2}}{1} = \frac{1 + \sqrt{2} + 1}{\sqrt{2}}$$

heptagonal type that yields an equal or greater harvest of proportions.

To explain this it will help, first, to explain what a 3-by-3 proportion is. A triple ratio — say 6:9:12 — might be a ratio of height-to-width-to-length of a shoebox. And one can see that a proportion of three triple ratios —

$$
\begin{aligned}
&\ \ 2: \ 3: \ 4 \\
= &\ \ 6: \ 9: 12 \\
= &\ 10: 15: 20
\end{aligned}
$$

— suggests three geometrically similar boxes. To cross multiply such a thing (say, if you wanted to solve for an unknown), set a copy of the proportion on its right.

$$
\begin{array}{ccc ccc}
2 & 3 & 4 & \mathbf{2} & \mathbf{3} & \mathbf{4} \\
6 & 9 & 12 & \mathbf{6} & \mathbf{9} & \mathbf{12} \\
10 & 15 & 20 & \mathbf{10} & \mathbf{15} & \mathbf{20}
\end{array}
$$

Find the six diagonals and multiply the three numbers along each diagonal.

$$2 \cdot 9 \cdot 20 = 3 \cdot 12 \cdot 10 = 4 \cdot 6 \cdot 15 = 4 \cdot 9 \cdot 10 = 2 \cdot 12 \cdot 15 = 3 \cdot 6 \cdot 20 = 360.$$

Figure 5. Optimal sections and proportions

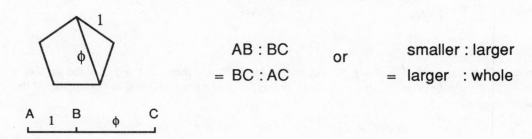

AB : BC   or   smaller : larger

= BC : AC   = larger   : whole

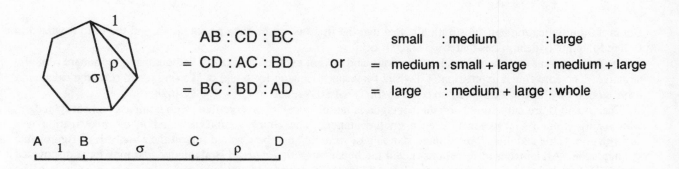

AB : CD : BC   small   : medium   : large

= CD : AC : BD   or   = medium : small + large   : medium + large

= BC : BD : AD   = large   : medium + large : whole

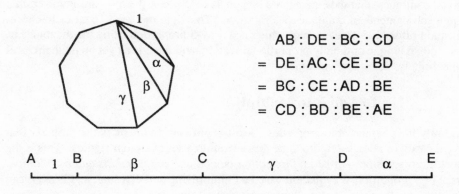

AB : DE : BC : CD

= DE : AC : CE : BD

= BC : CE : AD : BE

= CD : BD : BE : AE

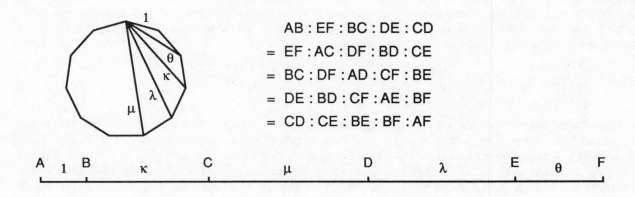

AB : EF : BC : DE : CD

= EF : AC : DF : BD : CE

= BC : DF : AD : CF : BE

= DE : BD : CF : AE : BF

= CD : CE : BE : BF : AF

All six products are equal. (As we used to say in industrial design, there's nothing like a consistent product.)

To deal with the problem of uniqueness, it will help to simplify the notation a little. Let the parts of the golden section be 1 and x, with $1 < x$. Then

$$1 : x$$
$$= x : x + 1.$$

Cross multiplication reveals that $x^2 = x + 1$, or $x^2 - x - 1 = 0$, and we have already seen that $x = \phi$. Let the parts of a trisection be $1 < x < y$. These can be arranged in any of three orders, depending on which is in the middle of the segment, so there are three cases:

order 1, x, y

$1 : x \quad : y$
$= x : 1 + x : x + y$
$= y : x + y : 1 + x + y$

order x, 1, y

$1 : x \quad : y$
$= x : 1 + x : 1 + y$
$= y : 1 + y : 1 + x + y$

order 1, y, x

$1 : x \quad : y$
$= x : 1 + y : x + y$
$= y : x + y : 1 + x + y$

But cross multiplication and substitution reveal that the first two cases have no solutions, and the third (realized in Figure 5) uniquely determines that $x = \rho$ and $y = \sigma$.

Rho and sigma are solutions of cubic equations, and are not expressible exactly without using the square root of a negative. The convenient expression $(1 + \sqrt{5})/2$ for ϕ has no analog for ρ and σ. To find ρ and σ on the calculator, use trigonometry: as $\phi = 2 \cos(\pi/5)$, so $\rho = 2 \cos(\pi/7) \approx 1.80194$, and $\sigma = \rho^2 - 1 \approx 2.24698$.

Since ρ and σ are cubic numbers, the heptagon is not classically constructible (with compass and straightedge), which may explain the ancients' silence on the matter. The Greek geometers' method of investigation was construction. This and their limited understanding of irrational numbers would inhibit their analysis of figures like the heptagon. Archimedes at least constructed the heptagon with a *marked* straightedge and may have discovered more. (Dijksterhuis [1] cites evidence of a lost Archimedean manuscript entitled *On the Heptagon in a Circle*.) The derivation of ϕ and its properties by similar triangles (Figure 2) has been known since ancient times, and one would think that the Greeks would have applied the same reasoning to other figures despite their inconstructibility.

Figure 5 also shows that this trend continues, that the enneagon's diagonals — $1 < \alpha < \beta < \gamma$ — are involved in a 4-by-4 proportion describing the ten subsegments of a unique quadrisection. This time there are 16 cases to check, and one is realizable. The pentasection offers 125 cases, one of which is a 5-by-5 proportion using the diagonals of the unit 11-gon. This is the largest known unique optimally proportional section, and there is as yet no mathematical proof that uniqueness continues indefinitely. Ideas are welcome.

II. To Add is to Multiply

The numbers in this optimal family have remarkable properties. Most significant to artists is the property that allows similar figures to be arranged easily or allows a figure to be dissected into a set of similar figures. This is the origin of the Greek idea of geometric progression — that *multiplication can be accomplished instead by addition*. Similar figures are similar by virtue of a proportion, an equation based on multiplying and dividing, and if these can be accomplished by adding and subtracting, then the figures allow repeated similarity. For instance, it is well known that the square of the golden ratio is one more than itself (written above as $\phi^2 = \phi + 1$) and that one less than ϕ is its reciprocal ($1/\phi = \phi - 1$). The first relation equates multiplication with addition, while the second accomplishes division through subtraction. These relations generate one of the drawings in Figure 3, as well as the famous golden spiral or nautilus construction (shown in Figure 6 as a double spiral, the fiddlehead). The heptagonal ratios also behave in surprising ways — adding to multiply, subtracting to divide.

(A)
$$\rho^2 = 1 + \sigma \qquad \sigma/\rho = \sigma - 1 \qquad 1/\sigma = \sigma - \rho$$
$$\rho\sigma = \rho + \sigma \qquad \rho/\sigma = \rho - 1 \qquad 1/\rho + 1/\sigma = 1$$
$$\sigma^2 = 1 + \rho + \sigma \qquad 1/\rho = 1 + \rho - \sigma$$

Another application of "to add is to multiply" is shown in Figure 7, which depicts a visual analog of the optimal 2-by-2 and 3-by-3 proportions, where the areas of the small rectangles are the proportions' entries. P. H. Scholfield, in his *Theory of Proportion in Architecture* [4], used the square with golden sected sides (Figure 7, top) to illustrate a problem in design economy. If a rectangle is dissected by one vertical and one horizontal line arbitrarily, then in general nine differently shaped rectangles (no two similar) are formed. There are many ways to reduce this number

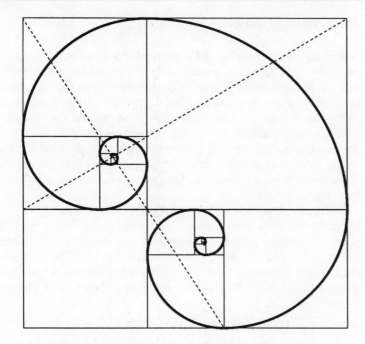

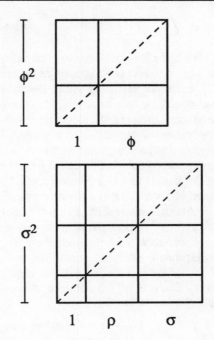

Figure 6. Golden double spiral (fiddlehead)   **Figure 7.** ϕ and σ dissections

by introducing symmetries in the figure. But if we are allowed only one symmetry, a diagonal reflection, then there is only one solution that yields the minimum of three different rectangles (three similarity classes). The ϕ^2-square in Figure 7 is the solution, and its rectangle types are 1-by-ϕ, 1-by-ϕ^2, and square.

Now if two vertical and two horizontal lines cut a rectangle, then in general 36 rectangles are formed. Scholfield points out that the sacred-cut square (Figure 4) reduces this to five non-similar rectangles, and the square with side length $\phi + 1 + \phi$ has only four. Both of these figures have four axes of reflective symmetry, and Scholfield does not ask for less symmetric or less repetitious solutions. Figure 7 also shows the square with sides trisected non-optimally as $1 + \rho + \sigma$, which, remarkably, yields only nine similarity classes. It is unknown whether this is best possible for so little symmetry and no repetition.

Another manifestation of "to add is to multiply" involves area. The areas of the sected squares in Figure 8 (with side lengths ϕ and σ) are dissectible. Since $\phi^2 = \phi + 1$, we find areas of 1 and ϕ inside a ϕ-by-ϕ square — in two different ways. And since $\sigma^2 = \sigma + \rho + 1$, the trisected σ-by-σ square is dissectible into contiguous areas of 1, ρ, and σ — this time in six essentially different ways. An analogous square with side length γ (recall the enneagon, Figure 5) and optimally quadrisected sides has area $\gamma^2 = \gamma + \beta + \alpha + 1$, and can be dissected into these four contiguous areas in 17 different ways.

Figure 8. Area dissections

III. Panels and Quasiperiodicity

The early 20th century saw many efforts to formalize the use of ratios like ϕ for the repetition of similar figures in design. (A short bibliography at the end lists some of them.) These efforts culminated in Le Corbusier's Modulor [**2**], a scale of segments whose lengths form a geometric progression of powers of ϕ. Another supplementary scale had double or half these lengths. These two scales (or two of the same scale), placed perpendicular to each other, produced an array of rectangles with rampant proportionality. Le Corbusier wanted to realize the ultimate application of "to add is to multiply" for design economy, to produce any number of similar but different-sized objects that pack together with no loss of space. Toward this end he used the additive properties of these rectangles (or "panels") for what he called *panel exercises*: choose three or more panels from the array, and use multiple copies of the panels to tile a given rectangle or square in as many ways as possible. The panel exercise was a brilliant teaching tool, and it is a shame that so few wanted to take the lesson.

Another interesting array of panels (Figure 9) makes use of a new idea in mathematics — *quasiperiodic sequences*. A periodic sequence is generated by repetition of a subsequence. For example, in the sequence *abcbabcbabcb...* the subsequence *abcb* is repeated. A quasiperiodic (QP) sequence, generated by an iterated replacement rule, has repetition, but when or how the repetition occurs is in many ways still a mystery.

Using the fact that $\phi^2 = 1 + \phi$, generate a sequence on the characters 1 and ϕ by multiplying them by ϕ over and over. The rule is: 1 becomes ϕ, ϕ becomes 1ϕ. When iterated, this rule generates an infinite sequence. From the initial word "1" we have:

$$1 \rightarrow \phi \rightarrow 1\phi \rightarrow \phi1\phi \rightarrow 1\phi\phi1\phi \rightarrow \phi1\phi1\phi\phi1\phi \rightarrow 1\phi\phi1\phi\phi1\phi1\phi\phi1\phi$$
$$\rightarrow \phi1\phi1\phi\phi1\phi1\phi\phi1\phi\phi1\phi1\phi\phi1\phi \rightarrow 1\phi\phi1\phi\phi1\phi1\phi\phi1\phi\phi1\phi1\phi\phi1\phi1\phi\phi1\phi\phi1\phi1\phi\phi1\phi \rightarrow \bullet\bullet\bullet$$

The infinite QP sequence formed in this way is not periodic, and yet *any* subsequence will occur infinitely often!

Using the relations (**A**) we can generate eight different sequences on the characters 1, ρ, σ. Here is one. Begin by applying the following rule, amounting to multiplication by σ: 1 becomes σ, ρ becomes $\rho\sigma$, σ becomes $1\sigma\rho$. Then

$$1 \rightarrow \sigma \rightarrow 1\sigma\rho \rightarrow \sigma1\sigma\rho\rho\sigma \rightarrow 1\sigma\rho\sigma1\sigma\rho\rho\sigma\rho\sigma1\sigma\rho \rightarrow \bullet\bullet\bullet$$

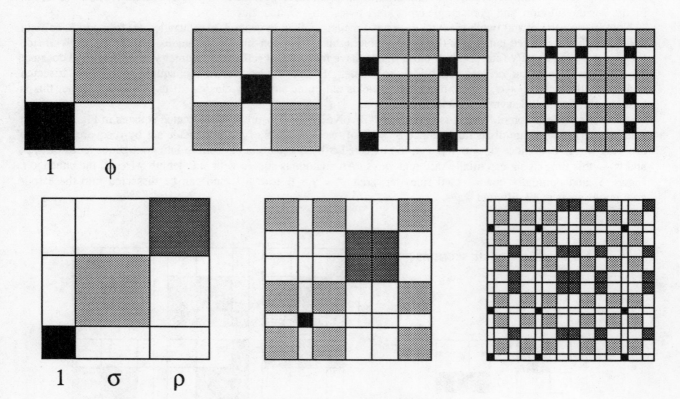

Figure 9. ϕ- and σ-aperiodic tilings

Figure 9 shows the aperiodic tilings created by applying either the φ-rule or the σ-rule to the sides of a square. The top sequence employs Modulor panels and could have been drawn by Le Corbusier or Mondrian. These tilings have a profoundly balanced asymmetry and a wealth of proportionality, as revealed in Figure 10. Diagonal regulating lines reveal a few of the similar rectangles and their corresponding proportions. (Figure 10 modifies the heptiling of Figure 9 by using the rule: σ becomes 1ρσ. This is best possible.) To get an idea of how densely the similarities are packed in Figure 10, consider that a square sected as σρσ1ρσ contains 52 rectangles of the similarity class ρ:σ alone! This density of similarity is accomplished not only by the ratios' natural tendencies, but also by the QP replacement rules, which spread the three values as evenly as possible, allowing them to interact with each other everywhere. There are analogous constructions for quadrisectional and pentasectional systems.

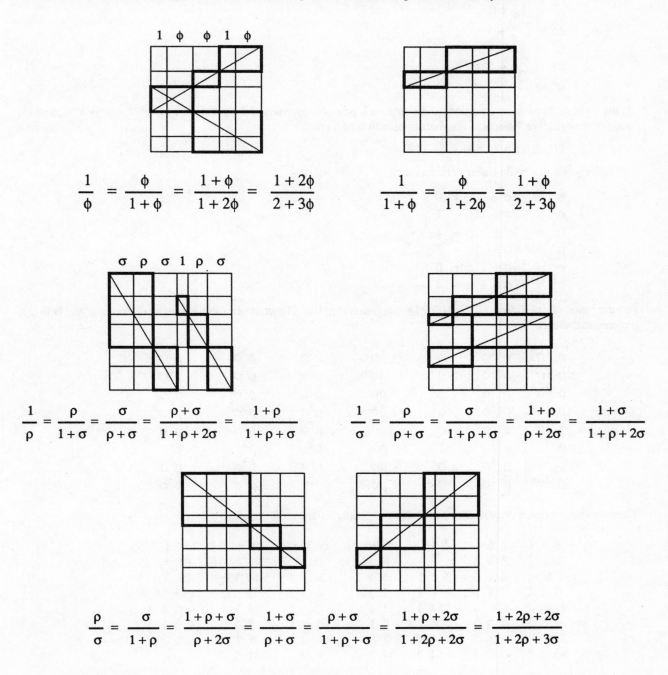

Figure 10. Similarities and proportions in the golden and heptagonal systems.

Three-dimensional structures and packings are virtually unexplored. The 1-by-ρ-by-σ box (the pack o' Luckies) is a very pleasant and useful shape, and the 1-by-ρ-by-σ ellipsoid looks like every cobblestone on every beach. Have we overlooked ρ and σ in Nature?

IV. Rational Approximants

The Fibonacci sequence $(0,1,1,2,3,5,8,13,21,\ldots)$ is connected to the golden ratio in three senses: (1) Powers of ϕ are expressible as linear combinations with Fibonacci numbers —

$$\phi = \phi + 0$$
$$\phi^2 = \phi + 1$$
$$\phi^3 = 2\phi + 1$$
$$\phi^4 = 3\phi + 2$$
$$\phi^5 = 5\phi + 3 —$$

(2) the right sides of these equations show that each power is the sum of the previous two: $\phi^{n-1} + \phi^n = \phi^{n+1}$; and (3) ratios of consecutive Fibonacci numbers approach ϕ as a limit:

$$1/1, \ 2/1, \ 3/2, \ 5/3, \ 8/5, \ 13/8, \ldots \to \phi$$

Similarly, we can write linear combinations for powers of σ —

$$\sigma = 1\sigma + \ 0 + 0$$
$$\sigma^2 = 1\sigma + 1\rho + 1$$
$$\sigma^3 = 3\sigma + 2\rho + 1$$
$$\sigma^4 = 6\sigma + 5\rho + 3$$
$$\sigma^5 = 14\sigma + 11\rho + 6$$
$$\sigma^6 = 31\sigma + 25\rho + 14 —$$

and the ratios of coefficients (e.g. 31:25:14) approach σ:ρ:1 as a limit. Better yet, arrange powers of ρ and σ in a 2-dimensional array.

$\rho^{-2}\sigma^5$	$\rho^{-1}\sigma^5$	σ^5	$\rho\sigma^5$	$\rho^2\sigma^5$	$\rho^3\sigma^5$	$\rho^4\sigma^5$
$\rho^{-2}\sigma^4$	$\rho^{-1}\sigma^4$	σ^4	$\rho\sigma^4$	$\rho^2\sigma^4$	$\rho^3\sigma^4$	$\rho^4\sigma^4$
$\rho^{-2}\sigma^3$	$\rho^{-1}\sigma^3$	σ^3	$\rho\sigma^3$	$\rho^2\sigma^3$	$\rho^3\sigma^3$	$\rho^4\sigma^3$
$\rho^{-2}\sigma^2$	$\rho^{-1}\sigma^2$	σ^2	$\rho\sigma^2$	$\rho^2\sigma^2$	$\rho^3\sigma^2$	$\rho^4\sigma^2$
$\rho^{-2}\sigma$	$\rho^{-1}\sigma$	σ	$\rho\sigma$	$\rho^2\sigma$	$\rho^3\sigma$	$\rho^4\sigma$
ρ^{-2}	ρ^{-1}	1	ρ	ρ^2	ρ^3	ρ^4
$\rho^{-2}\sigma^{-1}$	$\rho^{-1}\sigma^{-1}$	σ^{-1}	$\rho\sigma^{-1}$	$\rho^2\sigma^{-1}$	$\rho^3\sigma^{-1}$	$\rho^4\sigma^{-1}$
$\rho^{-2}\sigma^{-2}$	$\rho^{-1}\sigma^{-2}$	σ^{-2}	$\rho\sigma^{-2}$	$\rho^2\sigma^{-2}$	$\rho^3\sigma^{-2}$	$\rho^4\sigma^{-2}$

Then write their linear combinations $a\sigma + b\rho + c$ simply as $a \ b \ c$ in the array below:

5 3 1	8 6 3	**14 11 6**	25 20 11	45 36 20	81 65 36	146 117 65
1 2 2	3 3 2	**6 5 3**	11 9 5	20 16 9	36 29 16	65 52 29
2 0 -1	2 1 0	**3 2 1**	5 4 2	9 7 4	16 13 7	29 23 13
-1 1 2	0 1 1	**1 1 1**	2 2 1	4 3 2	7 6 3	13 10 6
2 -1 -2	1 0 -1	**1 0 0**	1 1 0	2 1 1	3 3 1	6 4 3
-2 1 3	**-1 1 1**	**0 0 1**	**0 1 0**	**1 0 1**	**1 2 0**	**3 1 2**
3 -2 -3	1 0 -2	**1 -1 0**	0 1 -1	1 -1 1	0 2 -1	2 -1 2
-3 1 5	-2 2 1	**0 -1 2**	-1 2 -1	1 -2 2	-1 3 -2	2 -3 3

For example, one reads on the two arrays that $\rho^2\sigma^5 = 45\sigma + 36\rho + 20$. Now notice two remarkable phenomena.

First, 45 36 20 = 20 16 9 (below) + 25 20 11 (left). In fact, each entry in this array of integer triples is the coordinate sum of the entry below and the entry to the left — that is,

$$\rho^m \sigma^n = \rho^{m-1} \sigma^n + \rho^m \sigma^{n-1} .$$

Second, 45:36:20 is an excellent approximation to σ:ρ:1, and the better approximants are farther right and/or up in the array. But where? Some directions yield better improvements than others. For instance, the triple 45:36:20 is a better approximant than any other triple up to its radius from the origin (0 0 1) of the array. Considering these triples as the lattice points of a grid, the best approximants *seem* to lie near a line through the origin with slope $1 + \rho$. This has not been explained or proved.

A more concise arrangement of all these integer sequences is given by the *golden matrices*. The relation

$$\begin{bmatrix} 0 & 1 \\ 1 & 1 \end{bmatrix} \begin{bmatrix} a \\ b \end{bmatrix} = \begin{bmatrix} b \\ a+b \end{bmatrix}$$

defines a transformation $(a, b) \to (b, a+b)$ that approaches the ratio 1:ϕ for non-negative values of a and b. Powers of the key matrix —

$$\begin{bmatrix} 0 & 1 \\ 1 & 1 \end{bmatrix}, \begin{bmatrix} 1 & 1 \\ 1 & 2 \end{bmatrix}, \begin{bmatrix} 1 & 2 \\ 2 & 3 \end{bmatrix}, \begin{bmatrix} 2 & 3 \\ 3 & 5 \end{bmatrix}, \begin{bmatrix} 3 & 5 \\ 5 & 8 \end{bmatrix}, \begin{bmatrix} 5 & 8 \\ 8 & 13 \end{bmatrix}, \begin{bmatrix} 8 & 13 \\ 13 & 21 \end{bmatrix}, \dots$$

— are Fibonacci approximants to the golden proportion $\begin{bmatrix} 1 : & \phi \\ = \phi : & 1+\phi \end{bmatrix}$. The relation

$$\begin{bmatrix} 0 & 0 & 1 \\ 0 & 1 & 1 \\ 1 & 1 & 1 \end{bmatrix} \begin{bmatrix} a \\ b \\ c \end{bmatrix} = \begin{bmatrix} c \\ b+c \\ a+b+c \end{bmatrix}$$

defines a transformation $(a, b, c) \to (c, b+c, a+b+c)$ that approaches the ratio 1:ρ:σ for non-negative values of a, b, and c. Powers of the key matrix —

$$\begin{bmatrix} 0 & 0 & 1 \\ 0 & 1 & 1 \\ 1 & 1 & 1 \end{bmatrix}, \begin{bmatrix} 1 & 1 & 1 \\ 1 & 2 & 2 \\ 1 & 2 & 3 \end{bmatrix}, \begin{bmatrix} 1 & 2 & 3 \\ 2 & 4 & 5 \\ 3 & 5 & 6 \end{bmatrix}, \begin{bmatrix} 3 & 5 & 6 \\ 5 & 9 & 11 \\ 6 & 11 & 14 \end{bmatrix}, \begin{bmatrix} 6 & 11 & 14 \\ 11 & 20 & 25 \\ 14 & 25 & 31 \end{bmatrix}, \begin{bmatrix} 14 & 25 & 31 \\ 25 & 45 & 56 \\ 31 & 56 & 70 \end{bmatrix}, \dots$$

— are (3rd-order) Fibonacci approximants to the proportion $\begin{bmatrix} 1 : & \rho : & \sigma \\ = \rho : & 1+\sigma : & \rho+\sigma \\ = \sigma : & \rho+\sigma : & 1+\rho+\sigma \end{bmatrix}$.

Numbers in the array at the bottom of the previous page occur either in these last matrices or as sums of matrix entries. This second sequence of matrices contain what are now called the "3rd-order Fibonacci numbers." The 4th- and 5th-order Fibonacci numbers — approximating the diagonals of the 9-gon and 11-gon — are obtained by extending the 0/1 key matrices above to 4×4 and 5×5. *Mathematica* will give you a page full of matrices for the following line. Adjust for larger matrices.

Table[MatrixForm[MatrixPower[{{0,0,1},{0,1,1},{1,1,1}}, n]], {n,1,15}]

After Scholfield surveyed the history of proportion from Vitruvius to Le Corbusier, the art world lost interest in the subject. Today it is often said that no one *designs* anymore — they just *express* themselves. Twenty years after Scholfield's survey, quasi-crystals were discovered (see Grünbaum & Shephard), and mathematicians, who had thought that everything meaningful had been said about proportion centuries ago, suddenly reopened the subject. In their efforts to explain the new phenomena they took a new and closer look at the golden ratio and found themselves asking questions that designers had asked and forgotten. Le Corbusier and others of his generation wanted to know whether the golden "key to the door of the miracle of numbers" was a unique phenomenon, and perhaps someone is still waiting for an answer. The mathematician's answer is emphatically *no*; there is an infinitude of irrational numbers whose geometric sequences of powers have additive properties useful for the repetition of similar figures. Crystallographers have recently named the silver, copper, and bronze ratios, among others. And there are three new and unapplied patterns belonging to the golden family — the optimal trisection, quadrisection, and pentasection — that concentrate maximal repetition of ratio into the least space. The possibilities await our exploration.

References

[1] E. J. Dijksterhuis, *Archimedes*. Copenhagen: Ejnar Munksgaard, 1956.

[2] Le Corbusier, *The Modulor*. Cambridge: Harvard University Press, 1954.

[3] Robert Lawlor, *Sacred Geometry*. London: Thames & Hudson, 1992.

[4] P. H. Scholfield, *The Theory of Proportion in Architecture*. Cambridge: The University Press, 1958.

[5] Peter Steinbach, *Golden Fields: A Case for the Heptagon*. MAA Mathematics Magazine, February 1997.

[6] Kim Williams, *The Sacred Cut Revisited: The Pavement of the Baptistery of San Giovanni, Florence*. The Mathematical Intelligencer 16:2 (1994) 18.

Related Reading

György Doczi, *The Power of Limits*. Boulder: Shambala, 1981.

Matila Ghyka, *The Geometry of Art and Life*. New York: Dover, 1978.

Grünbaum & Shephard, *Tilings and Patterns*. New York: Freeman and Co., 1987.

Jonathan Hale, *The Old Way of Seeing*. New York: Houghton Mifflin, 1994.

Jay Hambidge, *The Elements of Dynamic Symmetry*. New Haven, 1948.

Gyorgy Kepes, *Module, Proportion, Symmetry, Rhythm*. New York: George Braziller, 1966.

Nelson, Joseph, & Williams, *Multicultural Mathematics*. Oxford University Press, 1993.

Dan Pedoe, *Geometry and the Visual Arts*. New York: Dover, 1976.

M. C. Escher's Association with Scientists

J. Taylor Hollist
Department of Mathematical Sciences
State University of New York
Oneonta, NY 13820-4015
E-mail: HOLLISJT@ONEONTA.EDU

Abstract

Mathematicians, crystallographers, engineers, chemists, and physicists were among the first admirers of Escher's graphic art. Escher felt closer to people in the physical sciences than he did to his fellow artists because of the praise he received from them. Some of Escher's artwork was done more like an engineering project using ruler and compass than in a free spirit mode. Mathematicians continue to promote his work, and they continue to use his periodic patterns of animal figures as clever illustrations of symmetry.

Introduction

For more than 40 years, scientists have been impressed with the graphic art of M. C. Escher, recognizing with fascination the laws of physics contained within his work. Psychologists use his optical illusions and distorted views of life as enchanting examples in the study of vision. Mathematicians continue to use his periodic patterns of animal figures as clever illustrations of translation, rotation and reflection symmetry.

Escher's visual images relate directly to many scientific and mathematical principles. Some of his drawings give visual examples of the infinite process. Many scientists see in his work visual metaphors of their scientific theories. My interest in Escher stems from my many years of teaching geometry and the fact that some of Escher's work gives excellent examples of translation symmetry, rotational symmetry, glide reflection symmetry, and line reflection symmetry. Also, some of his work relates to models in non-Euclidean geometry. For example, the first time I saw a print of Circle Limit III, I said to myself, "that is the most beautiful example that I have ever seen of the Poincare circle model for Hyberbolic Geometry." Little did I realize at the time, that Escher got his original idea for this woodcut from a geometric figure in an article written by the mathematician H. S. M. Coxeter.

A question that might be asked, "Why is this paper being written by a mathematician rather than an artist?" Well for one thing, M. C. Escher is not considered one of the great artists of the world. In 1998, a New York Times art critic stated, "Escher's work tends to be viewed, at best, as art for beginners, an esthetic first love" [1]. Escher's name is usually mentioned in an art history course mainly because of his popularity. Also, people in the physical sciences (mathematicians, chemists, physicists, and crystallographers) were the first admirers of Escher's work, and they continue to enjoy his work [2].

Mathematicians continue to be promoters and admirers of Escher's work. For example, in 1998, I gave a paper at an International Escher Conference in Italy. Members of the Department of Mathematics at the University of Rome 'La Sapienza' organized the conference. Doris Schattschneider, a mathematics professor at Moravian College in Pennsylvania, was also a major organizer of the conference. I consider her to be the world authority on M. C. Escher.

At the above mentioned conference held in Italy, an Art Exhibition was held where contemporary artists displayed their work which they claimed was influenced by Escher. Displayed were examples of impossible objects, variations on tessellations, sculptures of many kinds, reflections in cylinders, reflections in spheres, approaches to infinity, depth perception, periodic space division, multi-dimensional space graphics, fractal tiles, etc. [3]

During the later part of Escher's life, he developed a close association with people in the physical sciences. He gave a lecture at Massachusetts Institute of Technology. The Union of Crystallographers invited him to speak at one of their conferences. Escher often visited his grandchildren in Canada, and during one of these visits, H. S. M. Coxeter of the Mathematics Department at the University of Toronto arranged for Escher to give some lectures on connections between mathematics and art.

The following two quotes demonstrate how M. C. Escher felt about his relationship to mathematicians. Escher said, "…the sad and frustrating fact remains that these days I'm starting to speak a language which is understood by very few people. It makes me feel increasing lonely. After all I no longer belong anywhere. The mathematicians may give me a fatherly pat on the back, but in the end I am only a bungler to them" [4].

Another quote: "I never got a pass mark in math. The funny thing is I seem to latch on to mathematical theories without realizing what is happening. No indeed, I was a pretty poor pupil at school. And just imagine—mathematicians now use my prints to illustrate their books. Fancy me consorting with all these learned folk, unaware of the fact that I'm ignorant about the whole thing" [5].

The Roosevelt Collection

Long before the artwork of M. C. Escher became popular, Cornelius V. S. Roosevelt (1915 – 1991) began collecting Escher's work. He was a grandson of American president Theodore Roosevelt, and a graduate of Massachusetts Institute of Technology in Mining Engineering. For most of his professional career, he worked for the United States Central Intelligence Agency.

Beginning in 1952, Roosevelt began collecting Escher art. The National Gallery of Art in the United States has a collection of over 200 original Escher prints that Cornelius Roosevelt donated to the gallery. To celebrate the centennial of the birth of Escher in 1998, the National Gallery held an Escher art exhibit that drew large crowds, over twice the average for a small National Gallery show. Approximately half of the prints displayed were from the Roosevelt Collection. Roosevelt purchased most of his prints directly from Escher.

Beginning with a written correspondence in 1957, Roosevelt's friendship with Escher developed into a warm and cordial relationship. Roosevelt acted as an advisor for Escher in the United States and took care of many copyright problems for him. People who wanted an Escher print placed in a publication, often went to Roosevelt's residence in Washington, D. C. to look over his collection.

I never met Roosevelt in person, but did talk to him on the telephone several times and corresponded with him. I asked Roosevelt for his picture standing next to some Escher prints, and he sent me a copy of a 1974 edition of *Holland Herald*, a news magazine of the Netherlands, which contained Roosevelt's picture in an article titled "The Roosevelt Eschers."

All the written materials that Roosevelt accumulated were carefully catalogued and placed into binders and these volumes were donated to the National Gallery Library. From a historical point of view, this written material is the most valuable part of the Cornelius Roosevelt Collection. Many of the examples cited in this paper are from this collection.

In a letter dated 24 December 1966, M. C. Escher requested that Roosevelt handle on his behalf inquiries which he received from publishers in the United States who wished to use Escher prints in books and magazines. He repeated his request in a letter dated 16 December 1968. The correspondence in the Roosevelt Collection contains many letters of requests to have an Escher print placed in a publication. I will cite only a few such requests here. The Roosevelt Collection at the National Gallery of Art Library contains many more examples. The Unitarian Universalist Association requested permission to use the print *Relativity* in a teacher's workbook. They also made some requests to use some prints for a curriculum course for preadolescent children.

In the Roosevelt correspondence there are many letters from well-known publishing houses asking for permission to use an Escher print in a publication. There are also requests from Nobel Laureates and from college professors. Requests were made by Harvard University Biological Labs, the Department of Psychology of Cornell University, a Harvard University psychology professor, American Hotel Association, Colorado State University, etc.

Since so many have given so many different interpretations to Escher's work, perhaps a note might be appropriate as to how he felt about all this. A letter from C. V. S. Roosevelt to John Graham in November 1976 sheds some light on this: "It is always a pleasure to hear from another admirer of the late M. C. Escher, and I will do my best to answer some of your questions. Bear in mind that on several occasions Escher said that he never disagreed with someone who explained what he saw in one of Escher's prints, and many of them were used by Nobel Laureates to explain some point (perhaps in physics, chemistry or mathematics) that Escher never had in mind when he made the print. What he <u>did</u> object to was somebody trying to explain what Escher had in mind when he made the print. In a few cases, he agreed, but in general he thought that those who claimed to recognize Zen Bhuddism or an anticlerical bias in his works were making themselves ridiculous."

Coxeter and the Circle Limit Prints

It is well documented in written correspondence that the mathematician H. S. M. Coxeter influenced Escher's work. The initial ideas for Escher's *Circle Limit* prints came from a drawing in a volume that Coxeter sent him, and Escher recognized this when he wrote on 5 December 1958: "Did I ever thank you for sending me (more than half a year ago) *A Symposium on Symmetry?* Though the text of your article … is much too learned for a simple plane patterned-man like me, some of your text-illustrations…gave me quite a shock. Since a long time I am interested in patterns with 'motives' getting smaller and smaller till they reach the limit of infinite smallness. The question is relatively simple if the limit is a point in the center of the pattern. Also a line-limit is not new to me, but I was never able to make a pattern in which each 'blot' is getting gradually smaller from a center towards the outside… Are there other systems besides this one to reach a circle-limit? Nevertheless I used your model for a large woodcut (of which I executed only a sector of 120 degrees in wood, which I printed 3x)." Here Escher is talking about *Circle Limit I* for which he carved one-third of the interior of circle from a block of wood (as though someone were having a huge piece of pie) and printed it three times in a circle by rotating the same wood block. This gives the figure rotational symmetry.

Coxeter responded to Escher on 29 December, 1958: "In answer to your question 'Are there other systems besides this one to reach a circle limit? I say yes, infinitely many! This particular pattern is denoted by [4, 6] because there are 4 white and 4 shaded triangles coming together at some points, 6 and 6 at others. But such patterns [p, q] exist for all greater values of p and q, and also for p = 3 and q = 7, 8, 9, … . A different but related pattern, called {p, q}, is obtained by drawing new circles through the 'right-angle' points, where just 2 white and 2 shaded triangles come together. I enclose a spare copy of {3, 7}… After the Colloquium of the Edinburgh Mathematical Society in St. Andrews, my wife and I are hoping to visit Holland at the beginning of August, and we would like very much to come and see you. With all good wishes for 1959, Sincerely yours, H. S. M. Coxeter"

Whenever Coxeter gave Escher a mathematical explanation of anything, Escher always responded that he did not understand. For example, Escher wrote to Martin Gardner on January 30, 1961: "I enclose herewith a copy of my letter to Prof. Coxeter, together with his answer of May 1960. His theoretical explanations are, no doubt, more comprehensible to you than to me. I am and shall ever be a perfect layman in the mathematical field. It is true that I never could have made this picture if I hadn't seen a schematic figure in one of Coxeter's publications, but as soon as he starts to argue abstractly, with formulas, I'm completely lost. I think he won't believe it, but it's a fact."

Impossible Figures and the Penroses

It was Escher who brought drawings of impossible objects to the attention of the general public with his lithographs *Belvedere* (1958), *Ascending and Descending* (1960) and *Waterfall* (1961). The ideas for the prints *Ascending and Descending*, and *Waterfall* were inspired by an article published by L. S. Penrose and R. Penrose (father and son). Escher recognized the Penroses contribution when he wrote to Roosevelt on November 30, 1961: "Waterfall...which is brand new, is also based on an idea of the same two Penroses. It's another of their exciting 'impossible' objects, which I copy here underneath for you."

Roger Penrose, British mathematician and inventor of the now famous Penrose tiles (kites and darts), visited Baarn, and Escher wrote Roosevelt the following about the visit on March 8, 1962: "Young Dr. Roger Penrose, son of the London prof. paid me a very nice visit with his wife. We had so many things to discuss and so much to tell each other that they lost their plane back to England. I am often struck by the simplicity and childish playfulness of most of these leaned scientists and that is why I like them and feel more at my ease with them than with my own colleagues."

Scientific American

Escher's popularity gradually became widespread in the United States because of some national coverage beginning with *Time* magazine in April 1951, *Life* magazine in May 1951, *Saturday Evening Post* in 1961 [6], *Scientific American* magazine in 1961 and in 1966, and *This Week* magazine in 1966. Martin Gardner who wrote a monthly column called "Mathematical Games" in *Scientific American* featured Escher in his April 1966 column. These articles resulted in the artist receiving a large volume of mail concerning his prints. An indication of this is stated in a letter on May 6, 1966 from Escher to Roosevelt: "After Mr. Gardner's article, my customers, especially in America, give me no peace."

Influence of Polya

George Polya, who was a research mathematician, published an article in a crystallographic journal in 1924 that influenced Escher's work. This article pictured examples of the 17 symmetries of the plane sometimes called the wallpaper groups because their resemblance of the periodic patterns in wallpaper. Each figure of Polya's table has translation symmetry but also contains different combinations of line reflection, rotation and glide reflection symmetry to complete the classification. Escher studied the figures, which helped him produce animal shapes that fit into tile-like patterns that cover the plane in a regular interlocking manner [7]. Polya verified his influence on Escher when he wrote the following on August 3, 1977 in a letter to

Doris Schattschneider: "Some time after the publication of my article, I received some drawings from a then unknown Dutch artist accompanied by a nice letter in which he said that my article was very useful to him—the artist was Escher. Unfortunately, in my wartime moving from Switzerland to the U. S. the letter and drawings were lost—your discovery is some consolation for this loss."

Crystallographers

A crystallographer is someone who studies the science of crystal structure and phenomena. Members of the IUC (International Union of Crystallography) were among the early admirers and promoters of Escher's work. P. Terpstra was the first crystallographer to give published recognition to Escher. In 1955 he used an Escher drawing as a front-piece on one of his textbooks. In 1960, Escher gave a lecture to an overflow crowd at the Congress of the IUC in Cambridge, England. Prior to this in 1959, Escher had some visits from crystallographers to his home in Baarn, Holland. Escher wrote the following about these visits: "the nicest thing to have happened to me recently was a visit yesterday afternoon—which is why I couldn't write to you—from a lady, Professor MacGillavry, who lectures on crystallography at the University of Amsterdam. She came with her sister-in-law, who was also somehow interested in divisions of the plane, and the two of them had their eyes glued to my prints from half past two till after half past five. What a pair of smart ladies! It's such a relief to have visitors at last who don't stare at my creatures uncomprehendingly, but who can chuckle with amusement at anything worth chuckling at. How they gazed at some of those prints. A few months ago I received a Belgian-American colleague of hers, a Professor Donnay, who lectures somewhere in the U S, and apparently she put Mrs MacG. on to me. She will try to have me give a lecture in Cambridge in August 1960, where there is a conference of about seven hundred crystallographers. In fact, that Professor Donnay had also mentioned it and it now looks as though it will really happen. There will be an exhibition of my prints as well, and my travelling and hotel expenses will be reimbursed" [8]. Because of the great interest in Escher's work at the Congress, the IUC commissioned Carolina H. MacGillavry to write a book to help crystallography students learn about symmetry using Escher's periodic drawings. Escher was delighted with the idea. Professor MacGillavry worked with Escher and selected a variety of regular periodic plane patterns. She found that some the patterns that crystallographers use in their classification of crystals were missing, and requested Escher to create some new drawings. Escher also improved some of his drawings that were selected by MacGillavry [9]. The book was published in 1965 and reprints of it today are titled *Fantasy & Symmetry: The Periodic Drawings of M. C. Escher.*

Other Scientists

Scientists from many different fields have related their abstract scientific theories to Escher's work. The fact that so many people have used his prints in their books, articles, and lectures attest to Escher's genius in being able to express graphically so many scientific ideas. Escher did not know he was expressing these theories, he just created what was in his head. First, a few examples given by Doris Schattschneider from her book *Visions of Symmetry*.

In 1957, Escher's drawing of the horsemen was placed on the cover of a book by Nobel-prize-winning physicst C. N. Yang.

Physist J. H. de Boer related his theories about the structure of matter to Escher's *Sky and Water*. He said, "The question which of the two, atoms or vacancies, is the most important in crystals, is well-demonstrated by the Dutch artist M. C. Escher."

The biologist E. P. Whitehead related close packing in molecular biology to Escher's metamorphosis drawings.

Nobel-prize-winning chemist Melvin Calvin used Escher's print *Verbum* to illustrate his thinking on chemical evolution. He said that it "struck me as representing, in artistic form, the essence of what I had been thinking about the nature of evolution and living processes...."

J. W. Wagenaar used Escher's prints *Sky and Water*, and *Day and Night* in a 1952 article "The Importance of the Relationship 'Figure and Ground' in Fast Traffic."

In 1968, Jane Abercrombie used Escher's prints *Sky and Water*, *Metamorphosis II*, and *Liberation* for a lecture on group analytic psychotherapy. She said this about the first print, "These designs picture the closely interlocking relationships which are established in a group, and the way the participants draw apart, reestablish their own contours, and become individuals again as they leave it" [10].

My bookshelf contains a popular book "Mathematics A Human Endeavor" by Harold R. Jacobs, which was published by W. H. Freeman and Company in 1970. The front and back covers of this book show Escher's woodcut *Horseman*.

I have always thought that Escher's lithograph *Waterfall* is an excellent example of a machine of perpetual motion. A retired colleague of mine, Albert Read of the Physics Department, thinks that *Waterfall* is an excellent example of "conservation of energy." Professor Read received permission directly from Escher to place a copy of *Waterfall* in one of his publications. A letter dated March 11, 1970 from M. C. Escher to A. J. Read says: "I am giving you permission to print 'Waterfall', however yust this once, in the physics textbook you are writing on condition that you will send me a copy of the book." The spelling of "yust" is left as in the original letter.

A few years ago I was wearing an Escher tie with a butterfly pattern (Figure 1), and a friend of mine, Greg Weed, a PhD chemist who works for DuPont, after studying the tie said, "I see polymers in your tie, does that make sense?" I responded immediately with "yes." Even though I know very little about chemical polymers, and I do not know much about the background history of the pattern in the tie, I still feel confident that my answer was correct, "yes it does make sense for you to see polymer patterns in that tie." The only thing I know about the construction of the butterfly pattern is that it was initially drawn on isometric paper [11]. Isometric paper is graph paper made up of small equilateral triangles.

Mathematicians have used Escher's repeating symmetrical patterns that tessellate (fill the plane without gaps) as examples in formal group theory.   They let G be the group of symmetries when color is ignored, and H the normal subgroup of symmetries that preserves all colors. Then it turns out that G/H is a group of order N, where N is the number of different colors or a multiple of the number of colors. The original colored butterfly pattern shown in Figure 1 has 3 different colors, and has quotient group P6/P3 which has order 6. The order or degree of a finite group is the number of elements in the group. The groups P6 and P3 are individually infinite groups whose elements are rotations and translations of butterflies in the figure [12].

Computer Scientists have written programs that will create Escher-like repeating patterns that will tile or tessellate the plane. Douglas Dunham of the University of Minnesota at Duluth has used a computer plotter to produce beautiful hyperbolic patterns like Escher's "Circle Limit" Series [13].

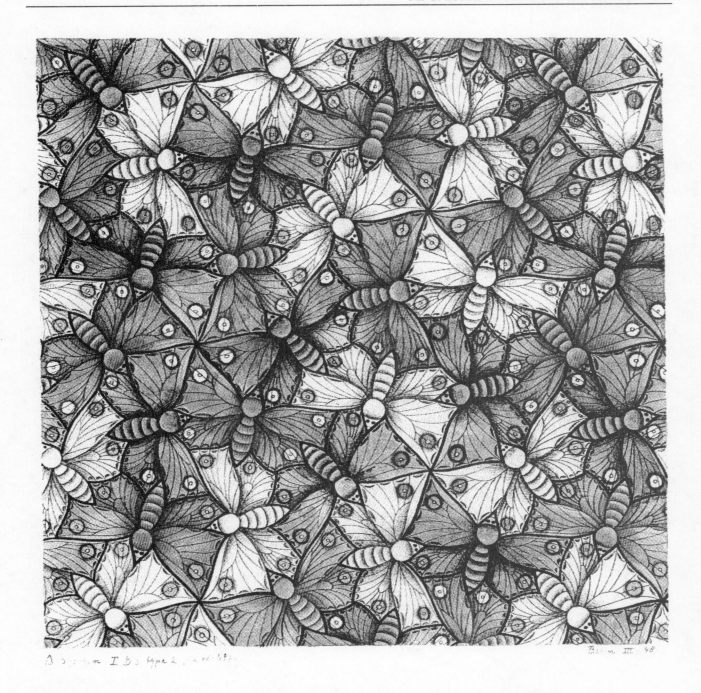

Figure 1: *Escher's Butterfly Pattern. The original is colored in red, blue, and green so that no two adjacent butterflies have the same color. Two colors alternate around the butterfly wing tips to form a six-fold center of rotation (red & blue around one point, red & green around another point, and blue & green around another point). Copyright-fee paid to Cordon Art B. V. of Baarn, Holland to include this print in this article. Cordon Art is the exclusive worldwide representative of the M. C. Escher copyrights.*

References

[1] Roberta Smith, "Just a Nonartist in the Art World, But Endlessly Seen and Cited", *New York Times*, page E1. January 21, 1998.

[2] J. Taylor Hollist, "Escher Correspondence in the Roosevelt Collection", *Leonardo,* Vol. 24, No. 3, pages 329-331. 1991.

[3] Michele Emmer & Doris Schattschneider, *Homage To Escher*, (Department of Mathematics, University of Rome 'La Sapienza') June 1998.

[4] J. L. Locher, *M. C. Escher: His Life and Complete Graphic Works* (New York: Harry N. Abrams), page 93. 1981.

[5] Bruno Ernst, *The Magic Mirror of M. C. Escher* (New York: Barnes & Noble Books), page 24. 1984.

[6] See [4] pages 67, 120.

[7] Doris Schattschneider, *Visions of Symmetry: Notebooks, Periodic Drawings, and Related Work of M. C. Escher* (New York: W. H. Freeman), pages 23-26. 1990.

[8] See [4] page 94.

[9] See [7] pages 42, 72, 103, 115, 275.

[10] See [7] pages 277-281.

[11] See [7] page 114.

[12] H. S. M. Coxeter, "Coloured Symmetry", *M. C. Escher: Art and Science* (New York: Elsevier Science Publishing Company) pages 15 - 33. 1986.

[13] Douglas J. Dunham, "Creating Hyperbolic Escher Patterns", *M. C. Escher: Art and Science* (New York: Elseveir Science Publishing Company) pages 241- 248. 1986.

The Art and Science of Symmetric Design

Michael Field
Department of Mathematics
University of Houston
Houston, TX 77204-3476
E-mail: mf@uh.edu
URL: nothung.math.uh.edu/~mike/

Abstract

We describe some of the issues – both philosophical and technical – involved in the evolution of computer software used for the design and realization of aesthetically appealing symmetric patterns. We illustrate how the choice of algorithm can dramatically affect the final design.

1. Introduction

The images shown in this paper were all designed and constructed using a software package called *prism* (an acronym for PRograms for the Interactive Study of Maps) that I started to develop about eleven years ago. In recent years I have used *prism* as the basis for a course on 'Patterns, Designs and Symmetry' that I teach to Junior/Senior level art students at the University of Houston, and also in a seminar for teachers held under the auspices of the *Houston Teachers Institute*. These teaching experiences, and the possibilities for using this computer software in the classroom, are described in more detail in [3] and, more especially, [4]. Although working with *prism* in a teaching situation has often provided a strong stimulus to make the interface more user friendly, the main impetus for developing *prism* has always been my personal interest in finding new and effective ways to design, color and realize attractive symmetric patterns.

In this article, I want to share some speculations and experiences generated by recent attempts to produce 'art' from silicon and chaos. All of the images that are shown result from an ongoing project to develop a library of algorithms for the design and artistic realization of the one- and two-color wallpaper patterns. Necessarily, the pictures shown here are grey scale versions of the color originals[1]. For this reason, the emphasis will be on algorithms for *one*-color designs rather than on coloring algorithms for 2-color designs. Throughout, we shall follow the standard notational conventions for wallpaper patterns as described, for example, in Washburn and Crowe [6]. We refer to [1,5] for general background on symmetry and the use of methods from chaos and dynamical systems in design. For 2-color designs, see [1,2]. Many colored images can be found at the *URL:* nothung.math.uh.edu/~mike/. Colored versions of some of the images shown in this article may be found at the *URL:* nothung.math.uh.edu/~mike/bridges2000.

In Figure 1, we show a characteristic wallpaper pattern designed using *prism*. I feel that the development of the program that led to the creation of this image as an *integral* part of the

---

[1]The images prepared for the article were, however, *colored* in grey scale rather than being transformed into grey scale from the original color image.

Figure 1: *A wallpaper pattern of type* **p31m**

creative process. That is, the mathematical algorithms, the computer program, and the design and reproduction (realization) of the images are just different facets of the same underlying process. In this regard, I am not so interested in creating pictures using a generic graphics software package. For me, the process of creation and design involves developing the tools and techniques in parallel with their application. Indeed, the development of these tools is largely governed by the character of the image I want to design and the way in which I want to realize the final image (for example, on photographic paper). Of course, sculptors and artists have always developed their own technology – whether it be pigments or the engineering and machining of contemporary sculpture. In spite of the apparent demise of the Renaissance man and the increasing compartmentalization of knowledge, I believe a creative artist – or scientist – cannot be detached from the tools with which he works. Sometimes, he or she needs to be an engineer or computer programmer.

In the following sections, we show some of effects that we can achieve with different types of algorithm. Partly for reasons of space, we avoid any detailed description of the explicit mathematical form of the algorithm (but see [3,5]).

2. Algorithms for symmetric design: I

Let us start with a hypothetical situation. Suppose that a designer wants to create a wallpaper pattern based on lions. What type of algorithm should be used? Is it possible to design a 'lion algorithm' that will reliably produce 'lion-like' images and which can be adjusted to vary the shape and textures of the lion? What type of symmetry should the designer use to convey, for example, a sense of motion or speed?

In one sense it is relatively straightforward to implement a lion algorithm. Take a scanned image of a lion and symmetrize over the appropriate symmetry group. This is the kaleidoscopic method employed so effectively in Kevin Lee's *KaleidoMania!* program. However, symmetric images produced by symmetrization techniques typically lack global coherence. More precisely, the process of symmetrization typically introduces discontinuities in the image – unless one avoid overlaps (see [3]). This, of course, is not a criticism of the kaleidoscopic method. Rather, it an observation that to create artistically satisfying images one needs more than a kaleidescope.

Although our approach has not yet led to a simple recipe for producing a 'lion algorithm', one class of algorithms we have experimented with can lead to images with a pronounced naturalistic flavor. In Figure 2 we show one such pattern that would resonate with any visitor to the Australian outback.

Figure 2: *Fly Quilt*

The quilt shown in Figure 2 is a wallpaper pattern of type **cm** and was constructed using ideas based on iterated function systems. In particular, the algorithm used was non-deterministic. The choice of symmetry **cm** is apposite as this pattern has just one direction of symmetry. We might also have used the symmetry **pm** – then the flies would have lined up wing-tip to wing-tip. The generating function used was a polynomial map in two variables. By making small changes in the coefficients of the generating function, it is easy to make incremental changes in the shape of the flies. We show one possible evolutionary sequence in Figure 3.

3. Algorithms for symmetric design: II

In another direction, we have developed algorithms that can generate abstract angular patterns.

Figure 3: *Evolving flies*

In Figure 4, we show two examples of these patterns. The pattern on the left is of type **p4**, that on the right of type **pmm**. Both patterns were produced using a non-deterministic algorithm and used the same continuous piecewise linear generating function. However, when we attempt to color this type of image, we find that there are drawbacks to using a non-deterministic algorithm. The reason for this is somewhat technical to explain but is related to the fact that the associated invariant measures (see [3]) are often rather uniform. This can make effective coloring quite difficult. It turns out that this problem is less severe when we use a *deterministic* algorithm. Initial investigations of piecewise linear deterministic algorithms with symmetry **p4m** have resulted in a number of quite promising angular designs – some of which I hope to show at *Bridges 2000*.

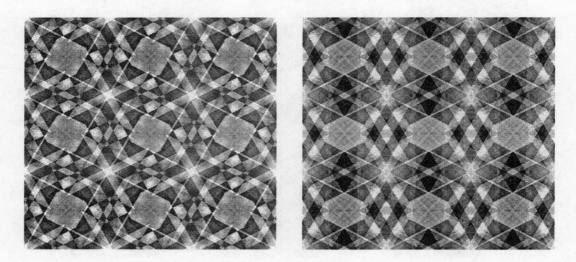

Figure 4: *Abstract designs of types* **p4** *and* **pmm**

4. Algorithms for symmetric design: III

Although it is relatively easy to construct quilt patterns using methods based on iterated function systems, there is no doubt that the most aesthetically pleasing patterns are generated using deterministic algorithms associated to (trigonometric) polynomial mappings. In part this is because deterministic algorithms of this type generally lead to images with sharply defined edges. In addition, there is usually an abundance of fine structure in the image - often highly localized. Experimentation with the algorithm, symmetry and coloring, leads to a symmetrically related set of curves associated to the singular set of the mapping. In Figure 5 we shown an image of type **pm** that was constructed using a deterministic algorithm. The symmetry type **pm** is well adapted to giving a sense of direction and motion to the image - in this case from left to right.

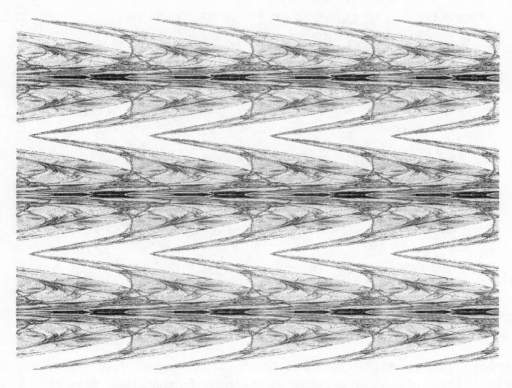

Figure 5: *A wallpaper pattern of type* **pm**

Figure 6: *A wallpaper pattern of type* **cmm**

In Figure 6, we show a deterministic quilt pattern of type **cmm**. This example has a particularly rich geometric structure that is strikingly shown using a grey scale coloring.

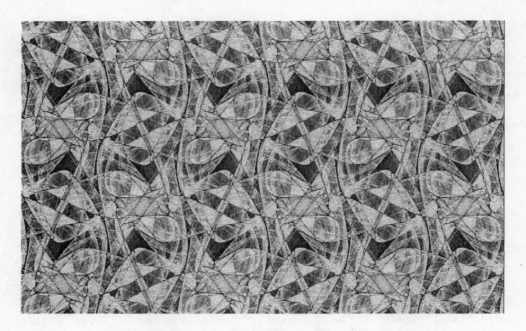

Figure 7: *A wallpaper pattern of type* **pgg**

The design of wallpaper patterns offers additional challenges. It is well known that some symmetry types seem much more attractive than others. Further, there appears to be a cultural bias in our response to symmetry. (We refer the reader to [6, Chapter 1] for a more extensive discussion of cultural and psychological responses to different types of symmetry.) In Western culture, designs with symmetry **p4m** (square tiling) or **p6m** (hexagonal tiling) are ubiquitous. On the other hand, patterns of type **pgg** are rather rare. One can speculate that it is the absence of reflection symmetry in **pgg** patterns that is responsible for their lack of appeal. However, patterns of type **pg** – also without reflection symmetry – are quite common. Generally, we have been rather unsuccessful in designing attractive patterns of type **pgg** using non-deterministic algorithms. We have had more success using deterministic algorithms – indeed, the search for algorithms that would yield appealing versions of patterns of type **pgg** was one of the reasons for our implementation of deterministic algorithms for all the 1- and 2-color quilt patterns. In Figure 7, we show an example of a quilt pattern of type **pgg** designed using a deterministic algorithm (a colored version of this quilt was shown at the 1999 Bridges meeting). Although the symmetry and fine structure of the pattern is quite complex, the overall effect is quite pleasing.

5. Conclusions

Perhaps the main criticism levelled against those who create art using computer graphics or other modern technology is that somehow what is done is not art. My own view is that creative artists are often innovative engineers who develop their chosen medium (I do not claim the converse). Thinking about computer graphics and art in this way leads to many intriguing and challenging problems. In art, technology and science. *Plus ça change, c'est le même chose.*

References

[1] M J Field. 'Harmony and Chromatics of Chaos', In: Proc. of the 1999 Bridges Conference, Southwestern College, Kansas.

[2] M J Field, 'Color symmetries in chaotic quilt patterns', In: Proc. ISAMA 99 (eds: N Friedman

and J Barrallo), Universidad del Pais Vasco, 1999, 181-187.

[3] M J Field. 'Designer Chaos', to appear in special issue of CAD.

[4] M J Field. 'Mathematics through Art - Art through Mathematics', preprint, 2000.

[5] M J Field and M Golubitsky, *Symmetry in Chaos*, Oxford University Press, New York and London, 1992.

[6] D Washburn and D Crowe. *Symmetries of Culture*, University of Washington Press, Seattle, 1988.

Mathematical Building Blocks for Evolving Expressions

Gary R. Greenfield
Department of Mathematics & Computer Science
University of Richmond
Richmond, VA 23173, U.S.A.
E-mail: ggreenfi@richmond.edu

Abstract

We consider the problem of analyzing visual imagery obtained using Sims' method of evolving expressions from the perspective of the mathematical building blocks, or function primitives, that are used to form such expressions. We survey previously used sets of building blocks and then give the design principles and motivation for our latest set of such building blocks. This includes a new construction of building blocks starting from arbitrary continuous functions. We conclude by incorporating our building blocks into a coevolutionary system and show examples of images that emerged from the "primordial ooze" defined by our building blocks.

1. Introduction

Richard Dawkins formulated the principle of interactive evolution for use with populations of primitive drawing routines in his *Biomorphs* program [2]. Dawkins' key concept was "fitness by aesthetics," which meant that the survivability of an image was determined by the user. Karl Sims expanded upon the idea by turning interactive evolution into an art *medium* [12]. The new feature Sims introduced was to free image generation from procedural drawing routines by encoding images as expression trees. Methods for organizing and maintaining expression trees as evolving populations had been developed previously by Hillis [6] and Koza [7]. Therefore, perhaps Sims' most significant contribution was his recognition of the importance of the mathematical building blocks that needed to be incorporated into such a system in order to create imagery using the principles of artificial evolution and fitness by aesthetics. Those following in Sims' footsteps were forced to design and implement their own systems based on evolving expressions entirely from scratch due to the computational limitations underlying Sims' original methods. Because there are no standards and only fragmentary descriptions of how the essential mathematical building blocks are defined — they often constitute the proprietary component of the system — comparison of the art that has been produced, as well as evaluation of the capabilities of the image generating systems, has been difficult. In this paper we will review what is known about previous efforts and we will discuss the mathematical building blocks we designed for a new system to generate images using coevolution *i.e.* where aesthetic fitness decisions are algorithmic thanks to *two* populations, a population of hosts which produce images and a population of parasites which consume images.

This paper is organized as follows. In section two we recall the technical details for producing images from expressions. In section three we discuss the building blocks used by previous designers of systems of evolving expressions. In section four we describe how we designed our latest set of building blocks. In section five we discuss some coevolution experiments using our set-up.

2. Images from Expressions

A (symbolic) expression is a rooted tree. The nodes of the tree contain the function primitives that are used to build the expression. We use the terms function primitive, basis function, building block, or operator interchangeably. Function primitives are distinguished by their *arity*, the number of arguments the function requires. Typically, one limits the arity to at most three. The leaves of the tree contain the inputs to the expression so they must be chosen from the set of basis functions of arity zero. Any function $F(u, v)$ from, say, the unit square to the unit interval can be written as an expression tree in one of three ways: using prefix notation $(F\ u\ v)$, using infix notation $(u\ F\ v)$, or using postfix notation $(u\ v\ F)$. Adopting the familiar prefix notation, any expression E can therefore be thought of as a function from the unit square to the unit interval that is formed by functional composition as, for example, $G(H(v), K(v, u))$. The external representation of the expression E that we choose to adopt does not affect its value $E(u, v)$. By resolving the unit square into a grid of points (u_i, v_j), calculating the expression $E(u_i, v_j)$ over all points on the grid, and then mapping those values to colors, we obtain an image from the expression. In artificial life terminology the genotype (expression) yields the phenotype (image). The above example clearly demonstrates why the set of primitives wholly dictates the type of imagery that it will be possible to discover when exploring the resulting image space of expressions. The richer the set of building blocks, the more interesting the images. There is, however, one further complication. Images from expressions may be based on either intrinsic color or extrinsic color. For intrinsic color the expression's value is either a vector of colors in a standard color space such as HSV space or RGB space, while for extrinsic color an expression's value is used as a look-up index into a color table for either one channel in the color space or the full palette of colors.

3. Previous Examples

Sims [12] actually describes *three* sytems for producing images: one for 2d imagery, one for 3d imagery, and one for animation. We will consider only his 2d system based on extrinsic color. We denote by A_i the set of primitives of arity i. It appears that the functional primitives Sims used are:

$$A_0\ =\ \{X, Y, \text{together with various constants}\},$$
$$A_1\ =\ \{round,\ expt,\ log,\ atan,\ sin,\ cos,\ invert(?),\ if,\ ifs\},$$
$$A_2\ =\ \{+, -, *, /, mod,\ min,\ max,\ and,\ or,\ xor,\ bw\text{-}noise,\ color\text{-}noise\},$$
$$A_3\ =\ \{hsv\text{-}to\text{-}rgb,\ vector,\ transform\text{-}vector,\ warped\text{-}color\text{-}noise,\ blur,\ band\text{-}pass,\ grad\text{-}mag$$
$$grad\text{-}dir,\ color\text{-}grad(?),\ bump,\ warped\text{-}ifs,\ warp\text{-}abs,\ warp\text{-}rel,\ warp\text{-}by\text{-}grad\}.$$

The reason we are uncertain about some of the placements in this classification is that because Sims' system is extrinsic it is not always clear which functions are defined componentwise and which are not. To cite one example, since *ifs* stands for iterated function system, one presumes *if* stands for iterated function, and in either case these could be true vector functions or vector functions induced by being defined componentwise. Also, some primitives only appear in Sims' examples. (E.g. *invert* and *color-grad* only appear in conjunction with Figure 9 on page 325 of [12].) Because of the sophisticated use of blurring, warping and other image processing functions, the implementation details of which are not given, it is not surprising that Sims' successors have seldom matched his image making prowess.

The next system about which we have some information is a bare-bones implementation by Baluja et al [1] which was used as an experimental testbed for attempting to train an artificial neural net to guide the interactive evolution of the images. Their function set is given as

$$A_0 = \{x, y\},$$
$$A_1 = \{reciprocal,\ natural\ log,\ common\ log,\ exponent,\ square,\ square\ root,\ sine,$$
$$hyperbolic\ sine,\ cosine,\ hyperbolic\ cosine\},$$
$$A_2 = \{average,\ minimum,\ maximum,\ addition,\ subtraction,\ multiplication,\ division,$$
$$modulo,\ random\}.$$

They remark that *random* is a function for randomly selecting either the first or second argument of the function and is not a function for generating random values. Here is a convenient spot to point out that designers using the *divide* function use the aptly named "protected divide" to prevent division by zero.

In the user's manual for an X-windows system available on the web, Unemi lists his functional primitives [13] as

$$A_0 = \{XY0, YX0, \text{together with various constants}\},$$
$$A_1 = \{-, abs,\ sign,\ sin,\ cos,\ log,\ exp,\ sqrt,\ image\},$$
$$A_2 = \{+, -, *, /, pow,\ hypot,\ max,\ min,\ and,\ mdist,\ mix\}.$$

The *image* function signifies that one must use a source image provided at run time. Presumably *mix* is a blending operator. One interesting feature of the system is that one can set on/off switches to delimit the set of primitives that will be available for use in the genotypes.

In [4] we gave a complete listing of the function primitives for a variant of a Sims' style system. We reproduce the complete listing here although without the primitive's definitions.

$$A_0 = \{u, v, c, e\}$$
$$A_1 = \{sin,\ cos,\ exp,\ log,\ abs,\ sqrt,\ square,\ cube,\ not\}$$
$$A_2 = \{and,\ add,\ multiply,\ mod,\ min,\ max,\ pwr,\ vee,\ cir\}$$

The leaf node symbol c signifies that a constant is stored at the node, while e, instead of accessing a source image, is a re-direction operator to a source expression (see [4] for details). Most of the binary primitive functions will be discussed further in the next section. [Note: For future reference *cir* will become *cone* and *vee* is a variant of *zabs*.]

For 2d systems these are the only complete listings of building blocks that we are aware of. Though we keep an archive of sample genomes from systems that have been implemented, they do not yield the complete function primitive sets, so our subsequent discussion has many gaps. One system, which for reasons easily explained, produces imagery that most would describe as "fractal" is the system found on the web http://www.cs.cmu.edu/~jmount/g3.html) by Mount

Figure 1: (a) A linear black-to-white color ramp. "Smoothing" operators for the ramp are defined using (b) the squaring function and (c) the square root function.

and Witbrock [14]. It uses quaternion operators including rotations, inversions, and conjugation together with the usual arithmetic operators. Similar remarks apply to the genotypes built from function primitives defined on the complex plane in a system that appears on the web by Yoshiaki [15]. Steven Rooke has for many years been producing images from his evolving expressions system. A sample description of one of his genomes reveals heavy use of both complex and quaternionic primitives, iterated functions, and of course the usual complement of arithmetic, boolean, and trigonometric functions [11]. Rooke also makes significant use of random constants. On the other hand, Musgrave, in constructing a prototype Sims' style system, followed Sims' original image processing theme more closely by including many noise, turbulence, and wave image processing functions but also added significant image generating capability by including fractal Brownian motion primitives [10]. Finally, Ibrahim [8] adopts a design where all internal nodes have arity four, but the primitive functions are Renderman shaders, so comparison of the primitives used is not justified.

4. On Designing a Set of Building Blocks

We will restrict ourselves to the problem of designing primitives for 2d image generating systems based on extrinsic color. We remind the reader that for us this implies an expression E maps the unit square to the unit interval. The set of terminals of arity zero must include variables for the coordinate axes. To avoid confusion with systems we have discussed earlier we shall use generic coordinates (V_0, V_1) instead of the usual choices of (x, y) or (u, v). The next decision to be made is whether or not to allow constants. We will include distinct generic constants $C_1, \ldots, C_n$ lying in the unit interval. The problem of introducing background imagery, source imagery, or secondary imagery is a thorny one with many factors to consider. It is intimately connected with image resolution, language dependencies, and choice of direct or indirect access to the source imagery. To avoid unnecessary complexity, we will not consider it here. Thus we will fix our arity zero set as:

$$A_0 = \{V_0, V_1, C_1, \ldots, C_n\}.$$

We now turn to unary operators of arity one. These operators are perhaps the most misunderstood ones. What is their purpose? If these basis functions are continuous, then we think of them as either smoothing operators or redistribution operators. Why? Let us suppose we are using a simple color *ramp* from black (zero) to white (one) having 255 increments such as $F(V_0) = V_0$ (Fig.

Figure 2: Redistributing the linear color ramp using (a) an inversion operator, (b) a parabolic map, and (c) a sine function.

1a). Then, $F(V_0) = V_0 * V_0$ (Fig. 1b) and $F(V_0) = \sqrt{V_0}$ (Fig. 1c) smooth the ramp towards black and white respectively. On the other hand, the inversion operator $F(V_0) = 1 - V_0$ (Fig. 2a), the parabolic operator $F(V_0) = 4*(V_0-0.5)^2$ (Fig. 2b), and the sine operator $F(V_0) = 0.5+0.5\sin(2\pi V_0)$ (Fig. 2c) redistribute the ramp. Some authors report considerable success with unary magnification operators [14], however, in our experience a preponderance of unary operators does not help image generation; it clogs image space. The principal danger in using too many unary operators is that through iteration a chain of unary operators can bleed to a constant or, worse yet, if an operator is its own inverse a chain can create an identity operator. In either case, the difficulty is that chains of unary operators fill the expression with dead weight. Thus we favor a very limited set of unary operators

$$A_1 = \{sqr, \; sqrt, \; not, \; parab, \; sin\},$$

and we avoid clogging expressions with unary operators by limiting the number that can appear in an expression.

We now turn to the set of binary primitive functions of arity two. Arithmetic operators are efficient and easy to implement, but because we require values to lie in the unit interval, normalization is often necessary. Thus, under the heading of addition, we include operators

$$avg\text{-}sum(V_0, V_1) = (V_0 + V_1)/2 \; \text{(Fig. 3a)}$$

and

$$mod\text{-}sum(V_0, V_1) = (V_0 + V_1) \bmod 1 \; \text{(Fig. 3b)}.$$

For subtraction, our choice is

$$sub(V_0, V_1) = |V_0 - V_1| \; \text{(Fig. 3c)}.$$

The *multiply* and the *power* functions borrowed from the standard mathematical library implementations require no discussion. The *min* (Fig. 4a) and *max* operators are straightforward, and although the method is somewhat implementation dependent, the bitwise logical operators can be used to produce fractals (Fig. 4b). The cone over the unit square (Fig. 4c) is included for its radial diffusion characteristics.

Another inspiration for binary primitives comes from using drawing routines as primitives. We adapted three from Maeda's book [9], two of which were based on the vector drawing algorithms

Figure 3: (a) An addition primitive based on averaging; (b) an addition primitive obtained by using the fractional part of the sum; and (c) a subtraction primitive.

Figure 4: (a) The *min* function; (b) the "fractal" bitwise *and* operator; and (c) the cone over the unit square.

written in his design language. They depend heavily on his use of unusual metrics. We give the results of converting Maeda's primitives [9, page 129 and page 242] to coordinate form.

$$maeda1(V_0, V_1) = \begin{cases} 1 - V_1 & \text{if } V_0 < V_1 \\ V_0 & \text{if } V_0 > V_1 \end{cases} \quad \text{(Fig 5a)}$$

$$maeda2(V_0, V_1) = \begin{cases} 0.5 * (1 + (V_1 - 0.5)^2/(V_0 - 0.5)^2) & \text{if } V_1 < V_0 \\ 0.5 * (1 + (V_0 - 0.5)^2/(V_1 - 0.5)^2) & \text{if } V_1 > V_0 \end{cases} \quad \text{(Fig 5b)}$$

$$maeda3(V_0, V_1) = \begin{cases} 0.5 * (1 + V_1^2/(1 - V_0)^2) & \text{if } V_1 < 1 - V_0 \\ 0.5 * (1 + (1 + (V_0 - 1)/V_1)^2) & \text{if } V_1 > 1 - V_0 \end{cases} \quad \text{(Fig 5c)}$$

Finally, we give our construction for converting *any* continuous function $F(X)$ defined on the interval $[-0.5, 0.5]$ into a binary primitive function. Our goal is to define a function

$$B(V_0, V_1) : [0, 1] \times [0, 1] \longrightarrow [0, 1].$$

By setting $V_i' = V_i - 0.5$, and letting $B(V_0, V_1) = B'(V_0', V_1')$, thanks to simple translation of axes, it suffices to construct $B'(V_0', V_1') : [-0.5, 0.5] \times [-0.5, 0.5] \longrightarrow [0, 1]$. We start with the continuous function $V_1' = F(V_0')$ defined on $[-0.5, 0.5]$. Parameterizing by c, the family of vertical translations $\{V_1' = F(V_0') + c\}$, foliate the strip $[-0.5, 0.5] \times \mathbb{R}$. Choose c_{max} and c_{min} such that over the interval $[-0.5, 0.5]$ the minimum of $V_1' = F(V_0') + c_{max}$ is 0.5 while the maximum of $V_1' = F(V_0') + c_{max}$ is -0.5. The reason for doing so is that now

$$c_{min} \leq V_1' - F(V_0') \leq c_{max}$$

on the translated unit square, $[-0.5, 0.5] \times [-0.5, 0.5]$. Now choose a "height map"

$$Z : [c_{min}, c_{max}] \longrightarrow [0, 1]$$

and set

$$B'(V_0', V_1') = Z(V_1' - F(V_0')).$$

The height map can be used to intensify the primitive's contrast.

EXAMPLE 4.1 Let $F(X) = |X|$. Then $V_1' = |V_0'|$ and $c_{max} = 1/2$ while $c_{min} = -1$. Since $Z(c) = \frac{2}{3}(c + 1)$ maps $[-1, 1/2]$ onto $[0, 1]$ we obtain the primitive

$$zabs(V_0, V_1) = 0.666667 * ((V_1 - 0.5 - |V_0 - 0.5|)) + 1) \text{ (Fig. 6a)}.$$

EXAMPLE 4.2 Let $F(X) = 0.5 - X^2$. Then $V_1' = 0.5 - (V_0')^2$. Again $c_{max} = 1/2$ and $c_{min} = -1$, but this time we use $Z(c) = \frac{2}{3}(1 - c)$ to obtain

$$zparab(V_0, V_1) = 0.666667 * (1.5 - V_1 + (V_0 - 0.5)^2) \text{ (Fig. 6b)}.$$

EXAMPLE 4.3 Let $F(X) = 0.5 + \sin(2\pi(x - 0.5))$. Then $V_1' = 0.5 * \sin(2\pi(V_0' - 0.5))$ yields $c_{max} = 1$ and $c_{min} = -1$. We intensify the sine curve by setting $Z(c) = 1 - c^2$. This gives

$$zsin(V_0, V_1) = 1 - (V_1 - 0.5 - 0.5 * \sin(2\pi(V_0 - 1)))^2 \text{ (Fig. 6c)}.$$

This completes the set of binary primitives that will be used in our next section. They are

A_2 = {*multiply, subtract, avg-add, mod-add, power, min, max, and, cone, maeda1, maeda2, maeda3, zabs, zprb, zsin*}.

5. Coevolutionary Images

The primitives that were described in the previous section have been incorporated into a coevolutionary system. Details will appear elsewhere [5], so we give only a cursory description in order to provide some context for our examples. The rationale for making the system coevolutionary is to test ideas for automating the user-guided fitness by aesthetics technique for evolving images that are defined by expressions. The coevolutionary model we will consider is that of a predator-prey model using hosts and parasites. The purpose of parasites is to digitally filter the image in order to detect structures that look significantly different after filtering, and thus whose spatial organization might suggest that the image is visually interesting.

Figure 5: Vector graphic images designed by Maeda which we converted to coordinate form for use as binary primitives: (a) *maeda1*, (b) *maeda2*, (c) *maeda3*.

Figure 6: Conversions of (a) an absolute value function, (b) a parabola, and (c) a sine curve to binary primitives using the new construction.

The expressions, or rather the phenotypes of the expressions, serve as hosts to parasites. Parasites are 3×3 digital filters defined by scaling a 3×3 array of integers, with each entry in the interval $[-15, 15]$, by the reciprocal of the absolute value of the array sum. For convenience, we assume the resolution of the hosts is 100×100. At a fixed number of sites on each host, parasites are attached to the host. The host-parasite interaction is quantified by convolving the digital filter with a small 10×10 patch of the host. The patch is therefore simply a neighborhood of the location where the parasite is attached. In general, the parasite is successful at preying upon the host if the convolved image matches the underlying patch to within a specified tolerance, while the host repels the parasite if the convolved patch is significantly different from the underlying patch (thereby exposing the parasite). To quantify this, the comparison is made on a pixel by pixel basis within the patch so that the host is assigned a fitness between zero and one hundred. Of course, when a host has multiple parasites, each attached at a different site, the host fitness is the fitness averaged over all parasites.

The host population evolves in the usual way: Randomly selected pairs of hosts chosen from the pool of fittest hosts produce progeny by swapping subexpressions. Mutation of the host progeny occurs by mutating the building blocks within the progeny genomes on a primitive by primitive basis with arity conserved. We devised our own artificial genetics for parasites based on cloning and mutation. More precisely, we clone (*i.e.* copy) parasites selected from the pool of fittest parasites to replace the weakest parasites, preserve only the very best parasites (elitism), and then subject all the remaining parasites to a suite of mutation operators which, for example, could cause rows or columns in their arrays to be swapped, replace randomly selected array entries, etc. The motivation

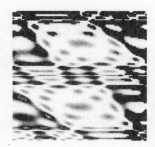

Figure 7: Three images from one run of the coevolutionary simulation using the building block primitives developed in the text. Their diversity is striking. Shown left to right are the most fit images after 5500, 6500, and 7000 time steps respectively. A maximum of fifty primitives is allowed per image.

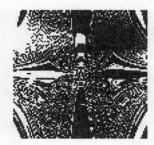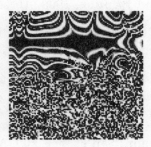

Figure 8: Images from three separate coevolutionary runs. The number of time steps is limited to 1500, however the upper bound on the number of primitives allowed per image has been increased to one hundred.

for specifying computational aesthetics using our set-up is that when the hosts repel the parasites they are alerting our digital filters — our eyes — that there might be something visually interesting taking place at the parasite's locations. Since parasites attack the hosts *locally*, while hosts, whose only counter measure is new genotypes whose primitives are defined on the entire unit square, can only respond *globally*, evolutionary pressure exerted by the parasites "chases" the host population phenotypes into new regions of image space. The sample coevolved hosts of Figure 7 were coevolved starting from small random populations of hosts and parasites using very small host genomes. Thus they represent images that evolved from the primordial ooze. More typically an evolutionary run uses thirty hosts whose genome expressions each consist of approximately 75 primitives. Each host has three parasites attached, and coevolution lasts for 1500 generations with the most fit host being culled every 200 generations. Examples obtained under these conditions are shown in Figure 8. We are in the early stages of investigating what kind of imagery will result from our coevolutionary setup. Further experimentation needs to be done before drawing any meaningful conclusions about the limitations of coevolved images based on computational aesthetics.

References

[1] Shumeet Baluja, Dean Pommerleau & Todd Jochem, Towards automated artificial evolution for computer-generated images, *Connection Science*, Volume 6, Numbers 2 & 3, 1994, 325–354.

[2] Richard Dawkins, The evolution of evolvability, *Artificial Life*, Christopher Langton (ed.), Addison Wesley, Reading, MA, 1989, 201–220.

[3] Gary Greenfield, New directions for evolving expressions, *Bridges: Mathematical Connections in Art, Music, and Science; Conference Proceedings 1998* (ed. R. Sarhangi), Gilliland Printing, 1998, 29–36.

[4] Gary Greenfield, On understanding the search problem for image spaces, *Bridges: Mathematical Connections in Art, Music, and Science; Conference Proceedings 1999* (ed. R. Sarhangi), Gilliland Printing, 1999, 41–54.

[5] Gary Greenfield, Art and artificial life — a coevolutionary approach, *Artificial Life VII*, to appear.

[6] Danny Hillis, Co-evolving parasites improves simulated evolution as an optimization procedure, *Artificial Life II*, C. Langton et al (eds.), Addison-Wesley, Reading, MA, 1991, 313–324.

[7] John Koza, *Genetic Programming III : Darwinian Invention and Problem Solving*, Morgan Kaufmann, San Francisco, CA, 1999.

[8] Aladin Ibrahim, GenShade, *Ph.D. Dissertation*, Texas A&M University, 1998.

[9] John Maeda, *Design by Numbers*, MIT Press, Cambridge, MA, 1999.

[10] F. Kenton Musgrave, *personal communication*.

[11] Steven Rooke, *personal communication*.

[12] Karl Sims, Artificial evolution for computer graphics, *Computer Graphics*, **25** (1991) 319–328.

[13] Tatsuo Unemi, sbart, *User's manual*.

[14] Miichael Witbrock & Scott Neil-Reilly, Evolving genetic art, in *Evolutionary Design by Computers*, P. Bentley (ed.), Morgan Kaufmann, San Francisco, CA, 1999, 251–259.

[15] Ishihama Yoshiaki, http://www.bekkoame.ne.jp/~ishmn/gallery/.

Symbolic Logic with a Light Touch

Charles C. Pinter

Department of Mathematics

Bucknell University

Lewisburg, PA 17837

cpinter@bucknell.edu

Abstract

In this paper, we discuss a unit on Symbolic Logic which has been designed in the context of a course entitled "Mathematics from a Humanist Perspective." The challenge of such a unit is that it must keep a light tone, avoid the use of heavy deductive machinery, and have relevance in the eyes of students. The objective of the unit is to bring about an understanding of the process of formal reasoning by using deductive rules which are elementary and well motivated. This paper contains two innovations: First, we have devised a deductive system which is very easy to use. A second innovative feature is the introduction of natural-language logic puzzles whose translation into symbols is quite straightforward, and whose solution by symbolic processing is easier to carry out than a solution by verbal reasoning. This last fact is especially useful in demonstrating the value of formal reasoning to students.

For about ten years I have been teaching a course entitled "Mathematics from a Humanist Perspective", which is open to non-science students at Bucknell University. Over the years I have experimented with different mixes of topics and a variety of course formats. Despite my familiarity with the demands of the course and the profile of my students, it is absolutely a surprise, every time, to discover what topics ignite the interest of students and what other topics leave them cold. The successful topic may be taught again the following year to see if the effect persists. And if it does persist --- if five or six successive student populations learn the topic and love it --- one yields to the conclusion that for reasons which transcend understanding, this topic "works": It is an authentic connection between the mathematical imagination and the open curiosity of active learners. It is a kind of conducting rod between mathematics and the arts.

One of these surprises was the fact that students from every walk of the academic spectrum could be made to enjoy learning symbolic logic. Admittedly, it was a somewhat unconventional take on symbolic logic, but it was neither watered down nor was its rigor diminished in any way. The Unit that I designed was confined to the propositional calculus; but the approach lends itself easily to various extensions, in particular to the addition of individual variables and quantifiers.

Few things are more important today, for students who must understand the key ideas --- the truly generative ideas --- of the ambient scientific culture, than to grasp the difference between empirical knowledge and knowledge attained by deduction. Generally, empirical knowledge is clear. But the process of formal reasoning --- the fact that it begins with unproved assumptions, and uses mechanical rules to derive conclusions from the assumptions --- that it does not produce any "new" knowledge except what is implicit in the premises, somehow these facts are not widely understood. The purpose of a Unit on Logic is to impart this understanding and make it fully explicit. It is our belief that this understanding is fundamental for relating mathematics to the humanities. Moreover, it is a big step toward clarifying the sense in which computers are able to generate music and art --- by starting with explicit 'aesthetic premises' and procedural rules and using these to build formal constructions blindly and mechanically.

The Unit is built around a set of logical puzzles which were originally designed to be solved by insight and deft reasoning, but which may also be solved symbolically by rigorous use of logical rules. Such problem sets have been around for a long time, and in most mathematics libraries you may find dusty volumes of logic puzzles, some published before the turn of the century; surprisingly you will notice a considerable overlap in the contents. But without a doubt the most charming and inventive re-creation of logic puzzles may be found in the books of Raymond Smullyan, which shine with new relevance. A selection of these puzzles, quoted from Raymond Smullyan, will be used in this article. I will begin with an example:

"THE ISLAND OF KNIGHTS AND KNAVES. *There is a wide variety of puzzles about an island in which certain inhabitants called 'knights' always tell the truth, and others called 'knaves' always lie. It is assumed that every inhabitant of the island is either a knight or a knave.* **Problem**: In this problem, there are only two people, **a** and **b**, each of whom is either a knight or a knave; **a** makes the following statement: 'At least one of us is a knave'. What are **a** and **b** ?" (Smullyan [1]).

In order to solve the problem symbolically, each statement in the problem must be represented by a distinct letter. For example, "*a is a knight*" is symbolized by the letter A. Therefore, "**a** is *not* a knight", in other words "*a is a knave*", is symbolized by ¬A (not-A). Similarly, "*b is a knight*" is symbolized by B, hence "*b is a knave*" is ¬B. So for example,

"**a** is a knight and **b** is a knight"   is symbolized by $A \wedge B$.

"if **a** is a knight, then **b** is a knight"   is symbolized by $A \Rightarrow B$.

"**a** is a knight, or **b** is a knave"   is symbolized by $A \vee \neg B$.

In our problem, the statement which   **a** makes is symbolized as follows : ¬A $\vee$ ¬B. (This is the symbolic form of "not-A or not-B").   Of course, the statement is not necessarily true. It is true if **a** happens to be a knight. But if **a** happens to be a knave, the statement is false, that is, its negation is true. The negation of ¬A $\vee$ ¬B is A $\wedge$ B. So, in order to express this problem in symbols, we must write the fact that: *if a is a knight,* then ¬A $\vee$ ¬B is true. And *if a is a knave,* then the opposite is true, that is, A $\wedge$ B is true. These two facts are the premisses of this problem: They are the facts given at the start, with which we must work to deduce the solution.

Premisses:

(P1) $A \Rightarrow (\neg A \vee \neg B)$

(P2) $\neg A \Rightarrow (A \wedge B)$

Solution:

	Reason
1. Assume $\neg A$.	
2. $\therefore A \wedge B$.	From premiss (P2) and rule MP (modus ponens).
3. $\therefore A$	Step 2 and Rule CON (rule of conjunction).
4. *Contradiction with Step 1.*	
$\therefore \neg A$ cannot be true.	
$\therefore A$	
5. $\therefore \neg A \vee \neg B$	Step (4), premiss (P1) and rule MP.
6. $\therefore \neg B$	Steps (4) and (5), and Rule MP' (see next page).

Our conclusions are: A, and $\neg B$. In other words, **a** is a knight, and **b** is a knave.

Note that the suggested solution is organized like a proof of elementary geometry. But it is more abstract, because the lines of the solution are entirely symbolic, and the rules of deduction are entirely mechanical.   What is most surprising, perhaps, is how little formal logic is needed to solve these problems: Students make use of four logical equivalences, and four rules of deduction. For a wide range of problems, this small amount of machinery is sufficient to write formal proofs which are completely rigorous and have no logical gaps.

I generally begin the Unit with no more than a two-page summary of formal logic, which is included here as the Appendix to this article. The students learn the four basic logical connectives and their truth-tables. Then four fundamental equivalences are explained:

1. $\neg(A \wedge B) \equiv \neg A \vee \neg B$

2. $\neg(A \vee B) \equiv \neg A \wedge \neg B$

3. $\neg(A \to B) \equiv A \wedge \neg B$

4. $(A \to B) \equiv (\neg B \to \neg A)$

Finally, logical deduction is presented as an essentially mechanical procedure carried out on logical sentences by using the following four rules of inference:

From A and $A \to B$, deduce B.	*(Rule MP)*
From A and $\neg A \vee B$, deduce B.	*(Rule MP')*
From $A \wedge B$, deduce A. (Also, deduce B)	*(Rule CON)*
From A together with B, deduce $A \wedge B$.	*(Rule CON')*

Using nothing more than this simple machinery, many types of problems --- at many levels of difficulty --- can be solved in precisely the same manner as the example given above.

The strength of our approach is the simplicity of the deductive apparatus that is required. It is widely acknowledged that the limiting factor in teaching deductive or symbolic logic to students outside the sciences is the daunting complexity of the formalism, and the large number of rules of inference in the standard texts on logic. What is presented here is an unconventional logical system – a combination of four logical equivalences with four rules of inference. Such a system can be learned quickly and is very easy to use. Here is another, somewhat more difficult problem, together with its solution.

LOVE AND LOGIC.

In the problem which follows, we turn from the logic of chivalry to the logic of love. Note that these problems do not take place on the island of knights and knaves. Thus, the protagonists of the following problem are neither knights nor knaves---just folks like you and me.

Problem. I know three girls, called Marcia, Sue and Dianne, and my heart is a-flutter. My feelings for these girls may be summed up as follows:

1. I love at least one of the three girls.
2. If I love Sue but not Dianne, then I also love Marcia.
3. I either love both Dianne and Marcia, or I love neither one.
4. If I love Dianne, then I also love Sue.

Which of the girls do I love? (From Smullyan [1]).

To solve this problem symbolically, the letter M represents the proposition "I love Marcia", S stands for "I love Sue", and D represents "I love Dianne". Thus, ¬M stands for "I don't love Marcia", and so on. The four premisses are symbolized as follows:

(P1) $S \lor M \lor D$

(P2) $(S \land \neg D) \Rightarrow M$

(P3) $(D \land M) \lor (\neg D \land \neg M)$

(P4) $D \Rightarrow S$

Solution.

		Reasons
1.	Assume $\neg(D \land M)$.	
2.	$\therefore$ $\neg D \land \neg M$	Step 1, premiss (P3) and rule (MP')
3.	$\therefore$ $\neg D$ as well as $\neg M$	Step 2, rule (CON)
4.	$S \lor M$	Premiss (P1), step 3, rule (MP')
5.	S	Steps 4 and 3, rule (MP')
6.	$S \land \neg D$	Steps 3 and 5, rule (CON')
7.	M	Premiss (P2), step 6, (MP)
8.	Contradiction between Steps 3 and 7.	
	Thus, $\neg(D \land M)$ cannot be true.	
	$\therefore$ $D \land M$	
9.	D as well as M	Step 8, Rule (CON)
10.	S	Premiss (P4), step 9, rule (MP).

Our conclusions are D, M and S. As I feared, I love all three girls.

The following problem is one of my favorites.

"FROM THE FILES OF INSPECTOR CRAIG

Inspector Leslie Craig of Scotland Yard has kindly consented to release some of his case histories for the benefit of those interested in the application of logic to the solution of crimes.

Problem.   An enormous amount of loot had been stolen from a store. The criminal (or criminals) took the heist away in a car. Three well-known criminals, **a**, **b** and **c**, were brought to Scotland Yard for questioning. The following facts were ascertained:

(1) No one other than **a**, **b** and **c** was involved in the robbery.

(2) **c** never pulls a job without using **a** (and possibly others) as an accomplice.

(3) **b** does not know how to drive.

Is **a** innocent or guilty?" (Smullyan [1]).

To solve this problem in symbols, the letter A is used for the proposition "**a** is guilty", so that ¬A represents "**a** is innocent". Likewise for the other protagonists. The premisses of this problem are as follows:

(P1) $A \lor B \lor C$

(P2) $C \Rightarrow A$

(P3) $B \Rightarrow A \lor C$

Solution	**Reasons**
1. Assume ¬A	
2. $B \lor C$	Step 1, premiss (P1), rule (MP')
3. $\neg A \Rightarrow \neg C$	Premiss (P2), equivalence 4
4. ¬C	Steps 1 and 3, rule (MP)
5. B	Steps 2 and 4, rule (MP')
6. $A \lor C$	Step 5, premiss (P3), rule (MP)
7. C	Steps 1 and 6, (MP')
8. Contradiction, steps 4 and 7.	

Thus, ¬A cannot be true.

∴ A

The variety of problems like the ones given above, whose solution can be found by "computation", is virtually endless.   The intricacy and distinctiveness of problems can be increased if the logical system is extended to allow quantifiers. In that case, two additional equivalences are needed:

$$5. \quad \neg (\exists x)P(x) \;\equiv\; (\forall x)[\, \neg P(x)]$$
$$6. \quad \neg (\forall x)P(x) \;\equiv\; (\exists x)[\, \neg P(x)]$$

The last problem presented here involves the use of quantifiers. It is included here for the perusal of the hardy reader.

"The Asylum of Doctor Tarr and Professor Fether."   (Adapted from Smullyan [2]).

Inspector Craig of Scotland Yard was called over to France to investigate an insane asylum where it was suspected that something was wrong. Each inhabitant of the asylum, patient or doctor, was either sane or insane. Moreover, the sane ones were totally sane and a hundred percent accurate in all their beliefs. The insane ones were totally inaccurate in their beliefs. Everything true they believed to be false, and everything false they believed to be true.

It was known that this asylum contained either a sane patient or an insane doctor. So Inspector Craig interviewed Doctor Tarr and Professor Fether in the following words:

Craig: Tell me, Doctor Tarr, are all the doctors in this asylum sane?

Tarr: Of course they are!

Craig: What about the patients? Are they all insane?

Tarr: At least one of them is.

Craig (to Professor Fether): Dr. Tarr said that at least one patient here is insane. Is that true?

Fether: Of course it is true! All the patients in this asylum are insane.

Craig: What about the doctors? Are they all sane?

Fether: At least one of them is.

Craig: What about Dr. Tarr? Is he sane?

Fether: Of course he is! How dare you ask me such a question?

At this point Craig realized the full horror of the situation. What is it?

The statements may be translated into symbols as follows: If x is any inhabitant of the asylum, $D(x)$ is the assertion that x is a doctor, hence $\neg D(x)$ asserts that x is a patient. Also, $S(x)$ asserts that x is sane, so $\neg S(x)$ asserts that x is insane. The letter t stands for Tarr and f stands for Fether. The fact that the asylum has either a sane patient or an insane doctor is conveyed in Premiss (P1), and the remaining premisses are the two doctors' statements.

(Recall that if x is not sane, all his assertions are false. This is used in (P4) and (P7)).

(P1) $(\exists x)[\neg D(x) \wedge S(x)] \vee (\exists x)[D(x) \wedge \neg S(x)]$

(P2) $S(t) \Rightarrow (\forall x)[D(x) \Rightarrow S(x)]$

(P3) $S(t) \Rightarrow (\exists x)[\neg D(x) \wedge \neg S(x)]$

(P4) $\neg S(t) \Rightarrow (\forall x)[\neg D(x) \Rightarrow S(x)]$

(P5) $S(f) \Rightarrow (\forall x)[\neg D(x) \Rightarrow \neg S(x)]$

(P6) $S(f) \Rightarrow (\exists x)[D(x) \wedge S(x)]$

(P7) $\neg S(f) \Rightarrow (\forall x)[D(x) \Rightarrow \neg S(x)]$

(P8) $S(f) \Leftrightarrow S(t)$

Solution

	Reason
1. Assume $S(f)$	
2. Assume $S(t)$	
3. $(\forall x)[D(x) \Rightarrow S(x)]$	Step 2, premiss (P2), rule (MP)
4. $\neg(\exists x)[D(x) \wedge \neg S(x)]$	Step 3, equivalences 5 and 3.
5. $(\exists x)[\neg D(x) \wedge S(x)]$	Step 4, premise (P1), rule (MP')
6. $(\forall x)[\neg D(x) \Rightarrow \neg S(x)]$	Step 1, premiss (P5), rule (MP)
7. $\neg(\exists x)[\neg D(x) \wedge S(x)]$	Step 6, equivalences 5 and 3.

8.   Contradiction, steps 5 and 7.
     Thus, S(t) cannot be true.
     ∴ ¬S(t)   *(On condition of step 1)*
9.   ∴ ¬S(f)   *(On condition of step 1)*   Step 8, premiss (P8), rule (MP)
10. Contradiction, steps 1 and 9.
     Thus, S(f) cannot be true.
     ∴ ¬S(f)
11. ∴ ¬S(t)   Step 10, premiss (P8), rule (MP)
12. (∀x)[¬D(x) ⇒ S(x)]   Step 11, premiss (P4), rule (MP)
13. (∀x)[D(x) ⇒ ¬S(x)]   Step 10, premiss (P7), rule (MP)

Thus, it turns out that all the patients are sane and all the doctors are insane.

In the course of working these problems, students learn three important skills, and the insights which come with these skills:

1. Students come to recognize that, in certain limited domains of the English language, sentences are made up of a small number of fundamental propositions connected together by means of logical connectives.

2. Once the structure of these sentences is understood (a skill which the students learn), the sentences may be written in fully symbolic form.   In this symbolic representation the structure of the sentences is transparent.

3. Finally, students learn that deduction is essentially a mechanical procedure, and this is made especially manifest when the deduction is carried out on strings of symbols.

APPENDIX

Logic is the study of relationships between statements.
Statements in logic are also called **sentences**, or **propositions.** (Here, we'll call them sentences.)
A *sentence* is any assertion of fact, or even non-fact. The following are examples of sentences:

> *Snow is white.*
> *London is the capital of England.*
> *Snow is black.*

One of the first and most important facts you'll have to understand about logic, is that logic *disregards what it is that a sentence asserts*. To a logician, a sentence such as "snow is white" is nothing more than a blank assertion, an X. The *content* of the sentence (that is, its meaning) is totally irrelevant to logic. For that reason, there is no loss if we simply denote sentences by single letters, such as A or B. Remember: Logic is the study of the *interrelations* between sentences, and it disregards what it is that specific sentences assert.

Logic is, in many ways, like arithmetic or algebra. Arithmetic deals with numbers; it studies the properties and interrelationships between numbers. The laws of arithmetic are equations such as a + b = b + a, which are true for all numbers a and b. Logic, in comparison with arithmetic, deals with *sentences*. Logic examines how compound sentences can be formed from simple ones, and how they interrelate. The laws of logic can be expressed as identities like

$$A \text{ and } B \;\equiv\; B \text{ and } A$$

(where $\equiv$ means "is equivalent to").   Compare the above to the identity $a + b = b + a$ in arithmetic As we move on, you'll see many similarities between logic and arithmetic.

In arithmetic, you connect numbers with one another by operations such as $+, -, \times$ and $\div$. Similarly, in logic you connect sentences with one another by using the following operations:

Operation	Symbol
A and B	$A \wedge B$
A or B	$A \vee B$
If A, then B	$A \rightarrow B$
Not-A	$\neg A$

Now, every sentence is either true or false. More importantly, the truth or falsity of any compound sentence is completely determined by the truth or falsity of each of its component sentences. For example, if we are told whether A is true or false, and also whether B is true or false, then we can determine whether $A \wedge B$ is true or false, whether $A \rightarrow B$ is true or false, and so on. It is convenient to express this information in the form of tables, called *truth-tables:*

(Truth-tables for the four connectives)

Compound formulas are called *equivalent* if they have the same truth-tables. For example, the formulas $A \rightarrow B$ and $\neg A \vee B$ are equivalent, and we symbolize this fact by writing

$$A \rightarrow B \;\equiv\; \neg A \vee B$$

A few such equivalences are of great importance in logic, and are listed next. (Equivalences).

Logical Deduction By examining the truth-tables, we can also see that certain logical *rules of deduction* are valid. A logical proof consists of the following: One or two formulas (or possibly more) are given to us initially, and we are told that they are true. From these initial formulas, called premises, we logically derive conclusions. In each step of the proof, we use a rule of deduction to conclude that a statement is true. For example, we conclude that B is true, if we have already shown that A and $A \rightarrow B$ are true. The following are commonly used as rules of deduction: (Rules appear here).

BIBLIOGRAPHY

1. Smullyan, Raymond: *What is the Name of this Book*, Simon & Schuster, 1982
2. _____   *The Lady or the Tiger*, 1992, Times Books
3. _____   *To Mock a Mocking Bird*, 1980.

Subsymmetry Analysis and Synthesis of Architectural Designs

Jin-Ho Park
School of Architecture
University of Hawaii at Manoa
Honolulu, HI 96822, U.S.A.
E-mail: jinhpark@hawaii.edu

Abstract

This paper presents an analytic and synthetic method founded on the algebraic structure of symmetry groups of a regular polygon. With the method, an architectural design is analyzed to demonstrate the use of symmetry in formal composition, and then a new design is constructed with its hierarchical structure of the method.

1. Introduction

The approach of subsymmetry analysis and synthesis of architectural designs shows how various types of symmetry, or subsymmetries, are superimposed in individual designs, and illustrates how symmetry may be employed strategically in the design process. Analytically, by viewing architectural designs in this way, symmetry which is superimposed in several layers in a design and which may not be immediately recognizable become transparent. Synthetically, architects can benefit from being conscious of using group operations and spatial transformations associated with symmetry in compositional and thematic development. The advantage of operating the symmetric idea in this way is to provide architects a method for analysis and description of sophisticated designs, and inspiration for the creation of new designs.

The objective of the research resides in searching out the fundamental principles of architecture. A study of the fundamental principles of spatial forms in architecture is an essential prerequisite to the wider understanding of complex designs as well as the creation of new architectural forms. In this I stand by the Goethe's theory of metamorphosis in *The Metamorphosis of Plants*. His theory centered in the notion that there may be an ideal form with what he called *urform*, a key to understanding the development of forms. Then, based on the *urform*, a variety of new designs can be developed. An architect, Frank Lloyd Wright in his article, "In the Cause of Architecture: Composition as Method in Creation (1928)", laid emphasis on the study of the principles of composition, in his words, 'geometry at the center of every Nature-form we see'. By looking into nature and grasping the fundamental principles at work, architectural forms that are not imitative but creative are developed with the creative endeavor.

In this paper, the methodology employed in my previous papers (Park, 1996, 2000) is recounted, but a new analytic and synthetic design has been added. The study of the subsymmetries begins by characterizing three categories of process:

1. to outline the important properties of the subsymmetries of regular polygons;
2. to analyze an existing architectural design; and
3. to synthesize an abstract design with respect to the subsymmetries associated with the symmetry of the square.

2. Subsymmetry Methodology

Symmetry operations are concerned with spatial displacements which take a shape and move it in such a way that all the elements of the shape precisely overlay one another, so that, despite the displacement, the shape retains the appearance of the shape before displacement. In two dimensions, there are two symmetry groups of plane symmetry: finite group and infinite group. The finite group of plane symmetry is called the point group. Spatial transformations take place in a fixed point or line. The transformations involve rotation about the point and reflection along the lines, or the combination of both. In the point symmetry group, no translation takes place. In the infinite symmetry group, spatial transformations occur where the basic movement is either translation or a glide translation (the composite movement of a reflection and a translation). In this group, designs, which are invariant under one directional translation, are called the frieze group, and designs under two directional translations are called the wallpaper group.

Subsymmetries arise from a curtailment of some of these operations: formally, selecting subgroups from the group of symmetries. Symmetry applies to a shape as a whole, but it may also apply to components (not necessarily discrete) which make up the form, and also to the consequences of multiplying the form in some larger assembly. Symmetry may be local, or global. In Shape Grammar applications (Stiny, 1980), the local symmetry of individual shapes in a shape rule effects the number of distinct applications, or the number of colorings of the shapes. Equally, simple set grammar rules applied in parallel are sufficient to derive the standard spatial groups of global symmetry. Since most of these are infinite groups they have little relevance for material form-making: only the point groups apply to finite objects. Thus, we limit our focus on finite point group symmetry.

Let us provide an elementary account of the mathematical structure of a symmetry group, in particular, the *point groups* in two dimensions. There are two finite point groups: the *dihedral group* denoted by D_n for some integer n; and *cyclic group* denoted by C_n. The spatial transformations of the dihedral group comprise rotation and mirror reflection; yet the cyclic group contains rotation only. The point groups have no translation. The number of *elements* in a finite group is called its *order*. The symmetry group of D_n has order $2n$ elements, while C_n has order n elements. For example, the symmetry of the square which is the dihedral group D_4 of order 8 has eight distinguishable spatial transformations which define it: four quarter-turns; and four reflections, one each about the horizontal and vertical axes and the leading and trailing diagonal axes. C_4 has four spatial transformations: the four quarter-turns. In this, it may be a noticeable fact that C_n is a subgroup of D_n. Also, by computing symmetry groups of a regular polygon, it is possible to generate the entire group as well. Let us begin by examining the lattice of subsymmetries of a square.

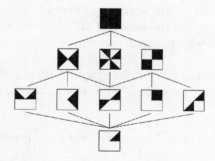

Figure 1: *The lattice of subsymmetries of the square: At the top is D_4 of order 8, below are subsymmetries of order 4, then below that again of order 2, and finally C_1 of order one, the unit element.*

The symmetry group of a square D_4 includes not only reflections in its four axes but also rotations through 0°, 90°, 180°, 270° respectively. Thus, the symmetry group of the square contains eight transformations, and these are the elements of the group. The diagram illustrates all possible subsymmetries; some with four elements, some with two, and just one, the identity or asymmetry, with one element. The structure of the diagram can be accounted for in two ways: from top to bottom, symmetries are 'subtracted' from the full symmetry of the square; and conversely, from the bottom to the top, subsymmetries are 'added' to achieve higher orders of symmetry. Starting from the top of the diagram, level 1 represents the full symmetry of the square D_4 with four rotations and four reflections. Level 2 consists of two reflexive subsymmetries D_2, one shows two orthogonal axes, and the other shows two diagonal axes at 45° to the orthogonal. Both of these subsymmetries exhibit a half-turn through 180°. The third subsymmetry shows four quarter-turns C_4, or 90° rotations. At level 3, there are five subsymmetries. Four with reflective symmetry D_1, two subsymmetries with a single reflective axis on the orthogonal, simple bilateral symmetry, and two subsymmetries with a single reflective axis on the diagonal. The fifth subsymmetry C_2 at this level has the half-turn rotation only. At the bottom level is the *unit element* or the identity of the group C_1. This element has no reflection axes, and no rotation less than the full-turn through 360°. The lattices of subsymmetries of other polygons such as an equilateral triangle, pentagon, etc. can be considered as well in its hierarchical order.

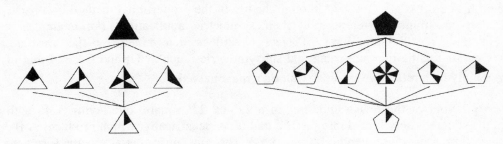

Figure 2: *The lattice of subsymmetries of an equilateral triangle, and a regular pentagon*

As with the examples of the regular polygon, above, the subgroups may be further differentiated according to axes into what we are calling here its subsymmetries. A polygon with *n* edges has at most dihedral symmetry of order $2n$, where the order of a finite group is the number of elements. The subgroups of the symmetry group of a regular *n*-gon are perhaps ordered in the lattice diagram. For instance, D_3 is the group of symmetries of an equilateral triangle, which has order 6 with its D_1, C_3 and C_1 subsymmetries. Furthermore, we can generalize the lattice diagram of the regular polygon, which shows its hierarchical order of subsymmetries.

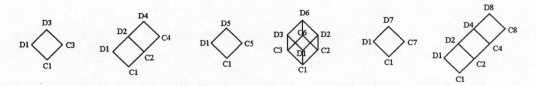

Figure 3: *The lattice of subsymmetries of regular polygons: equilateral triangle, square, pentagon, hexagon, heptagon, and octagon (from the top left)*

3. The Analytic Example

While innumerable examples could be quoted to illustrate this analysis, we take an architectural plan, which appears to use symmetry in the project, and scrutinize the plan to discover its symmetrical structure. We isolate partial elements of the design relying on its symmetrical order till the symmetry exhausts to identify its overall symmetry. Although, in most cases, the whole design is seemingly asymmetrical despite an almost obsessive concern for symmetry in the parts of the design, the analytic approach demonstrates how various symmetric transformations may be involved in each of the parts of the design, exposing the underlying structure of its spatial order. By doing so, the architect's conscious use of the symmetrical idea in the design becomes clearer.

Rudloph Michael Schindler's Popenoe House, designed in 1922 but demolished, is an example for our analysis. Schindler's debt to symmetry, particularly the hybrid use of various subsymmetries in the project, is astonishing. The plan of a single story desert cabin on a rectangular site is extremely simple, structurally and spatially; however, it is one of the most striking examples in its transparent interplay of his lifelong proportional method, "Reference Frames in Space" (Schindler, 1946), and symmetry. At large, D_4 subsymmetries, including both reflective and rotational symmetry, and his fundamental 48-inch (4') unit system guide all the major decisions of the spatial composition as well as details. The interweaving of both proportional and symmetrical ideas in his design is one of his major compositional tools throughout his career (Park, 1999). Although the use of the proportional method is consistent among his designs, there are minor differences in their symmetric application. For example, the spatial composition of the Free Public Library Project (1920) is determined by the various reflective subsymmetries without using the pure rotational subsymmetries such as C_4 and C_2, and that of the How House (1925) mainly by a reflective symmetry along a diagonal axis.

The primary layout of the house underlies in a 22' by 22' square, overlaying a 48-inch (4') unit system. Then, the square is subdivided into 6', 10', and 6', concentrically, which produces A B A rhythm. The spatial interval presents its sequential ratio to 3:5:3. The concentric spatial schema forms the absolute four-fold symmetry with the square. The *parti* is more of an underlying tool that governs spaces with regularity and shows an extreme clarity of its geometrical origin.

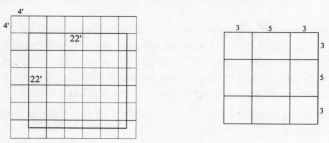

Figure 4: *Basic parti of the plan (Top left, basic parti with the unit system; Top right, the subdivision of the parti)*

Additional screened four porches, including Living, Sleeping, Kitchen, and Dining, are disposed in the pinwheel types of C_4 symmetry around the central square plan. These wrap around the primary square. The length of the porch wing increases successively, clock-wise in increments of 3', 4', 6', and 10'. It forms a spiral shape, which reinforces the rotational character. Whereas the basic composition of additional porches seems to derive from the absolute pinwheel type of C_4 symmetry, the final design is asymmetrical. If you see the architect's initial scheme as shown below right, his original idea of the composition becomes clearer. In the initial scheme, four porches, including four entrance doors and window openings, are set along the C_4 cyclic symmetry. It means that the asymmetric design derives from a disciplined understanding of the principle of rotational symmetry rather than merely being arbitrary.

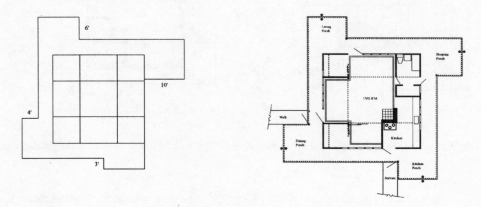

Figure 5: *Top left, the basic parti of the project; Top right, the initial scheme*

The fireplace is set along the diagonal axis of D_1 symmetry. The details of the fireplace reinforce the architect's conscious use of the diagonal symmetry as shown below right.

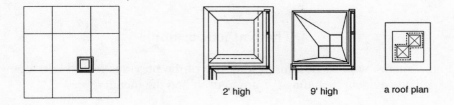

2' high   9' high   a roof plan

Figure 6: *Top left, the fireplace set along the diagonal axis; Top right, the fireplace horizontal sections cut 2' and 9' from the floor, and a roof plan*

Based on the *parti*, the major spaces of the house are juxtaposed, making its overall spatial configuration asymmetric. It breaks the strict symmetrical order for adjusting functional requirements. The living room is located in the center of the house. Each room is adjacent to the central living room. A sliding door divides the rooms, providing spatial flow as well as spatial flexibility in a minimum space.

The disposition of the raised high ceiling above part of the living room, kitchen and clothes closet is set along an orthogonal axis of D_1, allowing clerestory windows to bring light into the center of the building (below left). The symmetric juxtaposition of the ceiling gives emphasis to his ingenious use of the subsymmetries.

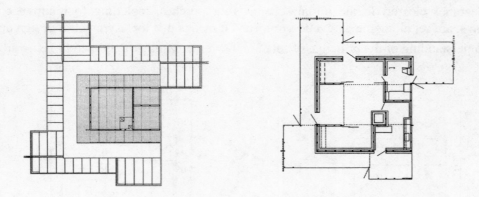

Figure 7: *Top left, D₁ subsymmetry with the ceiling design; Top right, First floor plan of the final design*

The final floor plan makes the example of the 'identity' element, which is equivalent to C_1. It can be said that in the final design, there is an abundance of symmetries within the parts while negating the strict symmetry of the whole. It is clear that the combination of the local and global symmetry is the driving force for the organization of the design. As clearly demonstrated, although the plan never used up all the possibilities of the subsymmetries of the square, various subsymmetries with rotation and reflection are superimposed into a single story design, which is an extremely rare example in architectural design. Also, it seems that Schindler sets up the symmetrical frame for the project, then he breaks it to come up with a strong asymmetrical design with functional necessity.

4. The synthetic example

Now, we build up a single building design, making use of the previous method. In a new design, all the subsymmetries are made evident including the group itself and the identity.

First of all, we need to choose a minimum building element. The main reason to choose the minimum element is to give the clearest possible picture of the design as a whole. The minimum element is composed of dots as columns with rectangles as floors. And then, as seen in the analytic example, we take the subsymmetries of the square with a grid system. The subsymmetries of the square are used as compositional tools and a grid system as an underlying *parti* to juxtapose building elements in order. Any regular polygonal grids or tessellation can be used but a rhythmic grid is implemented in our exercise. March (1981) in his paper, "a class of grids", defines and catalogues a series of grids. Among them, the permutation of the 3, 4, and 5 linear elements are chosen in this exercise. The grid itself forms D_4 symmetry.

Using the element on top of the grid, we arrange columns and floors to form a symmetric design with respect to the subsymmetries. Each floor level is designed in terms of the distinctive subsymmetry. Thus, each level defines a certain type of subsymmetry of the square. Then, we stack them up floor by floor in a standard height, creating a 3-D schematic building design.

We start at the lowest level of the symmetrical order. The first floor consists of the full symmetry of the square, which is D_4 symmetry. Columns and floors are set along the full symmetry of the square with four rotations and four reflections. The second floor illustrates the pinwheel types of rotational design of

C_4 symmetry. Column positions are the same as the previous D_4 configuration, but the floors are arranged based on the four quarter-turns C_4 symmetry. Compared to the level one, floor planes are shifted to form the pinwheel type of symmetry while preserving the column position the same as before. The third floor represents the half-turn C_2 symmetrical design. Although columns are set as D_2 symmetry, two floors are set along the half-turn rotation. The fourth floor consists of two reflexive subsymmetries of D_2. Columns and floors are set along two reflexive axes on the orthogonal. The fifth floor is composed by a single reflective axis on the orthogonal. The design looks similar to a level below, but upon a closer inspection, it shows that columns are set along the D_2 orthogonal axis, and floors are shifted from the central axis. The sixth floor illustrates the D_1 diagonal symmetry where elements and the grid are set along the single reflective diagonal axis. At the top floor is the identity, which is C_1. This is the minimum element without symmetry.

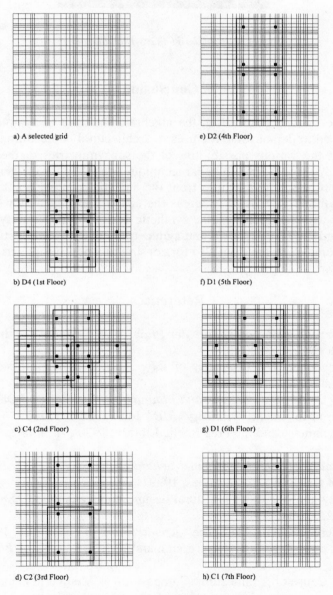

a) A selected grid

e) D2 (4th Floor)

b) D4 (1st Floor)

f) D1 (5th Floor)

c) C4 (2nd Floor)

g) D1 (6th Floor)

d) C2 (3rd Floor)

h) C1 (7th Floor)

Figure 8: *Each floor with the distinctive subsymmetries*

Figure 9: *Final design in a 3-D building form*

5. Conclusion

An analytic and synthetic method founded on the algebraic structure of symmetry groups of a regular polygon has been described to demonstrate the uses in architectural composition. In the Popenoe House, R. M. Schindler explores the various possibilities of the subsymmetries of the square where these are articulated around a single central point. A similar technique is applied for the generation of a new design. However, the synthetic process differs a little from the analytic one. Rather than a regular rectangular grid, we have used a rhythmic grid. Also, whereas the analytic example shows the superimposition of various subsymmetries in a single floor plan, the synthetic design has different types of subsymmetries in each floor plan. In the study, we have shown that symmetry is one of the effective methods not only for reading spatial order of complex designs but also for constructing new designs in architecture.

References

[1] J. W. von Goethe, *Versuch die Meamophose der Pflanzen zu Erklaren* (Gotha) translated by A Arber (1946) Chronica Botanica 10 (2) 63-126, 1790.

[2] L. March and P. Steadman, *The Geometry of Environment*, RIBA Publications Limited, London, 1971.

[3] L. March, 'A Class of Grids', *Environment and Planning B: Planning and Design* 8 325-382, 1981.

[4] L. March, 'The Modern Movement: symmetry' *RIBA* Journal 86 171, 1979.

[5] J. Park, 'Schindler, Symmetry and the Free Public Library, 1920', *Architectural Research Quarterly* 2 72-83, 1996.

[6] J. Park, *The Architecture of Rudolph Michael Schindler-the formal analysis of unbuilt work*, Ph.D. dissertation, University of California Los Angeles, 1999.

[7] J. Park, 'Subsymmetry analysis of architectural design: some examples', *Environment and Planning B: Planning and Design* 27, 2000.

[8] R. M. Schindler, 'Reference Frames in Space', *Architect and Engineer* 165, 1946.

[9] G. Stiny, 'Introduction to shape and shape grammars', *Environment and Planning B: Planning and Design* 7 343-351, 1980.

[10] F. L. Wright, 'In the Cause of Architecture: Composition as Method in Creation', unpublished essay, 1928.

Beyond the Golden Section – the Golden tip of the iceberg

John Sharp
20 The Glebe
Watford, Herts England, WD2 6LR
Winfield, KS 67156, U.S.A.
E-mail: 101503.432@compuserve.com

Abstract

The Golden Section is considered by many as the pinnacle of perfect proportion. There are many other proportions which have unusual, interesting or equally valid properties which do not seem to have been studied. They do not *appear* to have been used, but this may be because they have not been documented or because no-one has deemed to look, or because they cannot be constructed using ruler and compasses. This paper is an attempt to look at more than the tip of the iceberg in terms of proportion.

Introduction

The Golden Section is perhaps the most commonly cited proportion both in art and architecture. I believe that there are many more proportions that have just as interesting properties and that the Golden Section is just the tip of the iceberg, albeit a special one. In describing other proportions and proportional systems, I hope that eyes can be opened and possibilities explored for their use, as well as preparing the way for exploration of their existence. It is well known that the Golden Section is found everywhere because it is sought after and, in many cases, found by careful selection of data to fit aspirations rather than fact. There are a multitude of exciting possibilities for other systems of proportions which may have been used in the past but, because no one has looked for them, their use is not known. Non-mathematicians find ways around constructions that are not theoretically possible with ruler and compasses and, with the advent of computers, more is possible. Because there is so much to explore, some proportions will only be introduced rather than explored in depth. There is enough to fill a book rather than these few pages.

The Golden Section

The Golden Section is often quoted as being the perfect proportion. There are many more myths and fallacies about it than there are truths [1,16]. Some erroneous cases like the Nautilus shell [2] are perpetuated when mathematicians do not check the facts [3, 4]. The Golden Section is often found simply because it is looked for, and the facts "adjusted" to fit the hypothesis of the searcher. The truth is somewhere in the middle ground. This is not to say that it is not important and that new facts cannot be found which, strangely, are not well known. A few important and unusual constructions are as follows.

One of the most simple constructions for the Golden Section, is Odom's construction [5] with the circumcircle of an equilateral triangle which is shown in figure 1. Join the centres of two sides and produce them to the circle. Then AB/BC is in the Golden Section. This is easily proved using the intersecting chords property (Euclid's Elements XIII.8). I believe Odom found it from the properties of the icosahedron in which the Golden Section abounds. It is odd that such a simple construction was only discovered so recently.

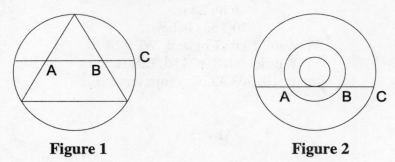

Figure 1   **Figure 2**

Sam Kutler's three circle problem shown in figure 2 is another simple construction, derived from it. The three circles are concentric; the middle circle has double the radius of the inner circle and half the radius of the outer circle and the line is tangent to the inner circle. Then AB/BC is in the Golden Section.

The Sacred Cut

This is the name given to the proportional construction shown in figure 3 by the Dane Tons Brunes [7].

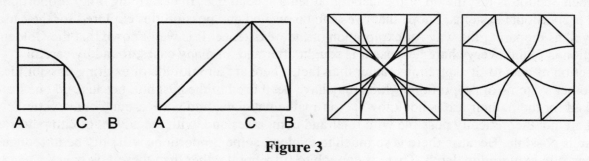

Figure 3

The two constructions shown on the left create C as the Sacred Cut of AB, giving AB/AC as √2:1. Brunes thought that ancient Greek references to the Golden Section are in fact to the Sacred Cut. The second construction has a similarity to the standard construction of the Golden Section (and the whirling squares configuration in figure 4) which uses the half diagonal of the square based on AC instead of the full diagonal. It may, in fact, be a more natural type of proportion for designs than the Golden Section since it is an easy and obvious construction. It is common in designs based on a square within a square, such as Tibetan Mandalas and is particularly common in Roman mosaics and has been found in the designs in the Laurentian Library [8] and in Roman House design and the layout of the city of Florence [9]. Paper sizes (the A series) in Europe are based on a proportion of √2:1 with A0 having an area of one square metre. Folding or cutting in half gives two pieces of the same proportion but rotated by 90°.

Dynamic symmetry and whirling squares

Although the Golden Section has been known since the time of the Ancient Greeks, and the Renaissance through Pacioli's book, it only came to real prominence at the end of the nineteenth century and the name has only been traced back to 1824 to a book by Martin Ohm (the brother of the discoverer of Ohm's law). The most famous manifestation is the whirling squares figure derived from a golden rectangle which allows the generation of an approximation to a logarithmic spiral using quadrants of circles (figure 4).

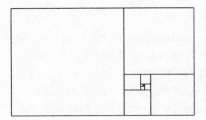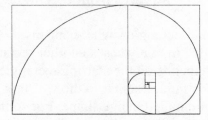

Figure 4

I have discussed the mathematics of the true logarithmic spiral fitting in the rectangle (which cannot go though the point where the square intersects the side of the rectangle without intersecting the rectangle) in another article [2]. This includes the construction of other spirals like the one shown in figure 5 which is complementary to the one in figure 4 since it uses *three* quadrants of the circle and which I have called the wobbly spiral. A number of these, combined as in the right of figure 5, show plant-like forms.

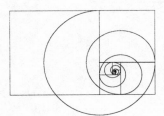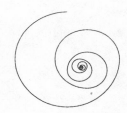

Figure 5

The equation of the wobbly spiral (not its approximation using arcs of circles) is:
$$r = (1 + 2k\sin(4\theta/3))ae^{\theta k}$$
where $k = \ln(\phi)/(3\pi/2)$ and ϕ is the Golden Section.

Such gnomonic constructions can be applied to any rectangle to yield other spirals. This is the system of Dynamic Symmetry of Jay Hambidge [10] and is achieved by constructing a perpendicular to a diagonal which gives rise to a similar rectangle as shown in figure 6, so that PBDQ has the same proportion as ABCD.

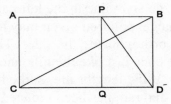

Figure 6

When the rectangle is a Golden Section one, the construction divides it into a square and another Golden Rectangle as shown here, but the system can be applied to any rectangle.

Dynamic symmetry has been very influential in the twentieth century both in art and architecture, with roots in the concept of gnomons. In looking below the tip of the iceberg, the concept can be developed in a number of ways, and as I will show later to study other systems of proportion. Edwards book on art deco design [11] is a good example of its use.

Squares from rectangles

Rather than take rectangles from rectangles, what happens if we take squares from rectangles other than the Golden Rectangle? In many cases another rectangle will result and the shape of the resulting rectangle obviously depends on the starting rectangle. If you continue to take a square off the resulting rectangle, in a similar manner to the whirling squares in figure 4, then in general the resulting rectangle will continually change shape. For some rectangles it may not be possible to remove a rectangle. There are also special cases. The most important one is the Golden Section where the resulting rectangle is the same shape as the original. A rectangle with the proportion 2:1 results in another square and you can go no further. This is the singularity, below which removal of a square yields another rectangle and above which it is not possible. One of the most interesting cases is the √2 rectangle which gives a rectangle from which two squares can be removed. Adding a square to the √2 rectangle gives a rectangle with the proportion 1+√2 and the resulting diagram can be used to construct two spirals as shown in figure 7.

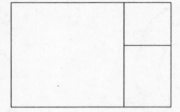 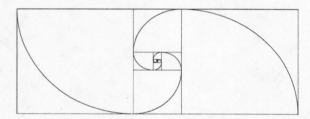

Figure 7

I can find no reference to the morphology of such systems. Many of the techniques used with Golden Section rectangles can be applied to the above proportional systems, but there is much more of the iceberg to investigate, so I will now move to some newer, unexplored possibilities.

The High-Phi division of a square

Martin Gardner, though retired, continues to write occasionally and I came across this division of a square with special properties in a student mathematics journal [12]. The puzzles originated with the New Zealand computer scientist Karl Scherer. The starting point was a puzzle to dissect a square into three similar shapes no two of which are congruent. There is an infinity of solutions, with examples in figure 8, where the slant line in the left part of this figure is inclined at various angles. Gardner thinks that the integer solution given may have the smallest values. The solution at the right is where the line is orthogonal to the sides. This cannot have integer solutions (although it can have approximate ones just like the Fibonacci numbers can be used to approximate to a Golden Section rectangle). If the lengths are marked as in the centre diagram, then x has interesting properties. (The right diagram has been reduced to a unit square.)

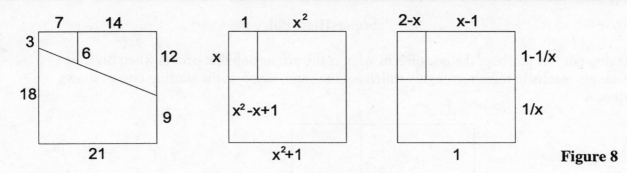

Figure 8

An interesting point about the left part of figure 8 is that the three areas are in the ratio $1^2:2^2:3^2$ and perimeters in the ratio 1:2:3.

The division yields three rectangles with the same proportions. It is a logical child of the division of the Golden Rectangle in Figure 4 and consequently Martin Gardner has given it the name High-Phi to match the often used symbol for the Golden Section which is the Greek letter phi (ϕ).

The equation for x is a cubic:
$$x^3 - 2x^2 + x - 1 = 0$$
which solves to the value 1.75487766624669276. The equation also gives rise to an associated recurrence sequence (like the Fibonacci sequence for the Golden Section) which is:
$$u_n = 2u_{n-1} - u_{n-2} + u_{n-3}$$
for which the first 20 terms of the sequence are 0, 1, 1, 1, 2, 4, 7, 12, 21, 37, 65, 114, 200, 351, 616, 1081, 1897, 3329, 5842, 10252, 17991, 31572, 55405, 97229, 170625, 299426, 525456, 922111, 1618192...

For lovers of Fibonacci curiosities, note how u_{29} is 1618192. The ratio of successive terms converge more slowly than the Fibonacci sequence, so that (if u_0 is 0) then u_{29}/u_{28} is 1.7548776665715...

The reason for Martin Gardner giving it the name High-Phi, apart from the general similarity of the diagram, is the relationship:
$$(x - 1)^2 = 1/x \quad \text{to compare with } x - 1 = 1/x \text{ for } \phi$$
There are other simple relationships derivable from this:

$$x = 1 + 1/\sqrt{x} \qquad \sqrt{x} - 1 = 1/x^2 \qquad x\sqrt{x} = x + 1/x \qquad 1 + 1/(x-1) = x + 1/x$$

The areas of the three rectangles with the notation in the centre of figure 8 are x, x^3 and x^4 whereas in the one on the right they are $1/x^4$, $1/x^2$, $1/x$ and thus giving the relationship:
$$1/x + 1/x^2 + 1/x^4 = 1$$
and a secondary equation:
$$x^4 - x^3 - x^2 - 1 = 0$$
which gives another recurrence relationship for the sequence above:
$$u_n = u_{n-1} + u_{n-2} + u_{n-4}$$
This number 1.75487766624669276 is related to the Plastic Number (see below).

Super-High-Phi

Having got this far, begs the question of what is the proportion that occurs when division of a rectangle results in three rectangles which are the same shape as the starting one as shown in figure 9.

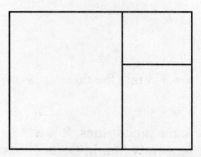

Figure 9

This leads us back to the Golden Section since the ratio of the sides of the rectangle is $\sqrt{\phi}$. The right hand side is divided in the Golden Section.

It is obviously difficult to avoid the Golden Section, but since I am trying to do so, I will move on.

The Tribonacci constant

The Golden Section is essentially two dimensional, although it is found in two of the Platonic solids, the dodecahedron and the icosahedron and in four dimensional polytopes, but not in higher dimensional ones. There is another pair of proportions hidden in the Archimedean solids. The diagonals of the snub cube and many other properties [13] are functions of the number 1.83928675521416... (which I have called η) which is the root of the equation
$$x^3 - x^2 - x - 1 = 0$$
which is the associated equation of the sequence: 0, 1, 1, 2, 4, 7, 13, 24, 44, 81, 149 ... that is the recurrence:
$$u_n = u_{n-1} + u_{n-2} + u_{n-3}$$
each number being the sum of the previous three terms rather than the previous two as is the case with the Golden Section which is often called the Tribonacci sequence.

The snub cube fits in the cube, with the square of the snub cube rotated on the face of the cube by an angle whose tangent is η.

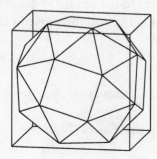

Figure 10

The snub dodecahedron also has a constant associated with it, the number 1.94315125924388817 which is the root of the equation:

$$x^3 - x^2 - x - \phi = 0$$

where ϕ is the Golden Section.

The heptagon and other polygons

Another approach is to consider regular polygons. Whereas the Golden Section is derived from the pentagon, other polygons offer numbers with much richer properties. They have not been studied in depth because of a quirk of the history of mathematics. Only a limited number of regular polygons can be constructed with the tools of the ancient Greek mathematicians, the straightedge and compasses. This is not to say that they cannot be constructed, but as the Golden Section obscures all others proportions, so construction of polygons only seems "valid" if carried out with the "approved" tools. Even the Greeks had other methods, for example using curves or, in the case of the heptagon using the so called "lost neusis" of Archimedes [14]

The pentagon only has one type of diagonal but the heptagon has two which I have labelled using Steinbach's symbols [15].

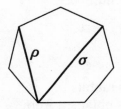 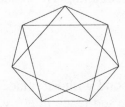 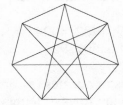    **Figure 11**

This allows two types of stellated polygons and the one at the right of figure 11 where the sides are all of the same length.

The two constants σ and ρ are roots of the equations:

$$\rho^3 - \rho^2 - 2\rho + 1 = 0 \quad \text{and} \quad \sigma^3 - 2\sigma^2 - \sigma + 1 = 0$$

giving three roots for each, but if the side of the heptagon is unity then $\sigma = 2.246979603717..$ and ρ is 1.8019377358048… with various relationships between the roots:

$$\frac{1}{\sigma} + \frac{1}{\rho} = 1 \qquad \frac{1}{\sigma} = \sigma - \rho \qquad \frac{\rho}{\sigma} = \rho - 1 \qquad \frac{\sigma}{\rho} = \sigma - 1$$

$$\rho\sigma = \rho + \sigma \qquad \sigma^2 = \rho + \sigma + 1 \qquad \rho^2 = 1 + \sigma$$

Which gives geometric interpretations like a σ rectangle and a ρ rectangle fitting together to give a square and a ρ removed from a σ rectangle resulting in another σ rectangle. There is scope for a whole series of logarithmic spirals.

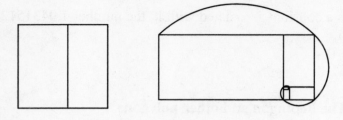

Figure 12

The familiar division of a line and fixing the position of the Golden mean also has its Heptagonal equivalent, but this time there are two points because we are dealing with a cubic construction not a quadratic one.

thus AB : CD : BC = CD : AC : BD = BC : BD : AD or writing these divisions in terms of σ and ρ, then $1 : \rho : \sigma = \rho : 1 + \sigma : \rho + \sigma = \sigma : \sigma + \rho : 1 + \rho + \sigma$ which I find more satisfying than the simple Golden Ratio perhaps because it has slightly more complexity.

From the equations for σ and ρ, there are recurrence relationships for the heptagonal equivalent of the Fibonacci numbers which I will term the σ-Heptanacci and ρ-Heptanacci numbers, namely the sequence:
0, 1, 1, 3, 6, 14, 31, 70, 157, 353, 793, 1782, 4004 ......
0, 1, 1, 3, 4, 9, 14, 28, 47, 89, 155, 286, 507, 924 ....

There is also scope for writing stories or puzzles to correspond with Fibonacci's rabbits. Consider the following. In the following rows of letters, the sequence in a row is obtained from the previous row by substitution. "A" becomes "C", "B" becomes "BC" and "C" becomes "ACB". The fractal nature of the rows is evident if you notice that each row is always the beginning of the next but one row.

A
C
ACB
CACBBC
ACBCACBBCBCACB
CACBBCACBCACBBCBCACBBCACBCACBBC
ACBCACBBCBCACBCACBBCACBCACBBC ......

If you count the number of each letter in a row, then for "A" you get 1, 0, 1, 1, 3, 6, 14, 31, 70, 157, 353, 793 .., and for "B" 0, 0 , 1, 2, 5, 11, 25, 56, 126, 283, 636, 1429 .... and for "C" 0, 1, 1, 3, 6, 14, 31, 70, 157, 353, 793, 1782.

The sequence for B arises from the expansion of the powers of σ and so there are other Heptanacci sequences also.

Another aspect is the types of triangles which have angles which are multiples of $\pi/7$ as denoted by the numbers in the following figure:

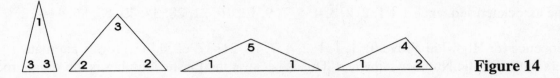

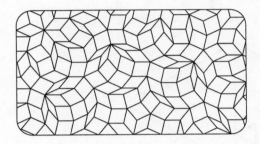

Figure 14

where the sides of triangle 133 are in the ratio 1 : σ : σ, of 223 are σ : ρ : ρ, of 115 are ρ : 1 : 1 and of 124 (which incidentally are angles in geometric progression) are 1 : ρ : σ. Such triangles in turn lead to the consideration of aperiodic tilings equivalent to the Penrose tilings as shown in figure 15.

Figure 15

These are but a few aspects of the heptagonal possibilities. As one increases the sides of the polygon so different properties appear with the increasing number of diagonals. Steinach [15] deals with some of these properties, but they are even more unexplored than the heptagon.

The Plastic Number

This is the discovery of the Dutch architect Dom Hans van der Laan (1904-1991). He started out trying to follow the work of a fellow architect who used the Golden Section, but found he could not get it to work in three dimensions. His work and how he came to discover the Plastic Number is the subject of books and three articles by his champion, the English architect Richard Padovan [17, 18, 19]. A mention of his name with some confusion of information by Ian Stewart [20] has led to the recurrence sequence being associated with Padovan.

The Golden Section was considered essentially two dimensional by van der Laan, although it obviously does have three dimensional manifestations in the dodecahedron and icosahedron. He looked for an equivalent. It is worth bearing in mind that the equation for the Golden Section is a quadratic, so the Plastic Number is the root of a cubic, derived from the equation:

$$p^3 - p - 1 = 0$$

which has the real root 1.3247179572447... and equations which are reminiscent of Golden Section related ones:

$$p = \frac{1}{p} + \frac{1}{p^2} \qquad p^2 = \frac{1}{p} + 1 \qquad p^3 = p + 1 \qquad p^4 = p^5 - 1$$

The associated recurrence relation is:

$$u_n = u_{n-2} + u_{n-3}$$

and there also others:

$$u_n = u_{n-1} + u_{n-5} \qquad \text{and} \qquad u_n = u_{n-3} + u_{n-4} + u_{n-5}$$

with the associated sequence 1,1,1, 2, 2, 3, 4, 5, 7, 9, 12, 16, 21, 28, 37, 49, 65, 86, 114.....

The sequence for High-Phi above (0, 1, 1, 1, 2, 4, 7, 12, 21, 37, 65, 114,..) forms alternate numbers in the Plastic Number sequence. This is because the Plastic Number is the square root of High-Phi. Striking out the High-Phi numbers and looking at the remainder gives another sequence which also obeys the High-Phi recurrence.

Geometrically, the Plastic Number offers both two and three dimensional possibilities. The recurrence $u_n = u_{n-1} + u_{n-5}$ leads to a set of whirling triangles to match the whirling squares of the Golden Section. These are equilateral triangles.

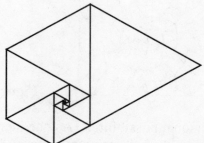

Figure 16

which in turn leads to a spiral which has special properties:

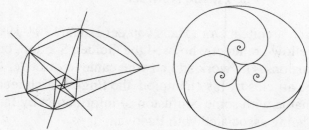

Figure 17

The spiral was discovered many years ago by Rutherford Boyd [21] when he was looking for spirals which touched a number of times. In this case, join the poles of the spirals and the lines go through the touching points.

In three dimensions, a "whirling" takes place with rectangular boxes leading to a helical spiralling of the boxes. Figure 18 shows how the boxes build up to the Plastic cuboid, using the Plastic sequence. The cuboids have dimensions 1,1,1 then 1,1,2 then 1,2,2 then 2,2,3 then 2,3,4 then 3,4,5 then 4,5,7 and so on.

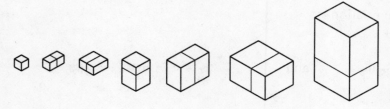

Figure 18

This sequence has been drawn in isometric projection, which is easier to draw but does not show the full beauty of the object.

In this approximation to the true Plastic cuboid, the addition each time results in a square added to one face. This can be built up to create a three dimensional spiral by combining quarter circles on the square faces of the boxes in the same way as a spiral is built in the Golden Section rectangle in figure 4.

From Theory to Practice

The above ideas are starting points, both in rich areas of mathematics but also as inspiration for art. Considering how artists have used the Golden Section by being inspired by the myths which arose in the nineteenth century, then there are many lifetimes' work here. This needs the computer in many cases since a number of these ratios cannot be constructed with standard ruler and compasses. However, square roots can and the approximations using recurrence sequences which are similar to the use many artists make of the Fibonacci sequence require no special tools. If anything, the possibilities are overwhelming. I have many examples in sketchbooks which have not gone further because too many other possibilities suggest themselves. They are the geometrical equivalent of equivalent of the Chinese curse of "may you live in interesting times". Such possibilities are easier to show in a presentation, so this paper is a theoretical background. The following examples are a taster.

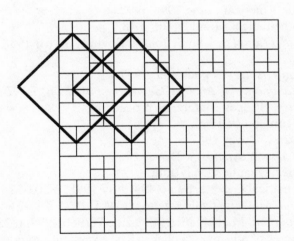 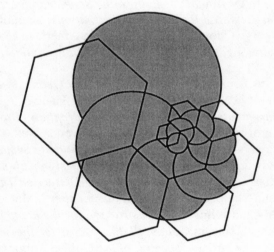

The first of the two above was designed using the heptagonal division of a square. The second comes from the whirling triangles using the Plastic number.

The mathematician Lagrange was probably the first to prove that if you take the Fibonacci sequence, and reduce it, for example, MOD 4, then you get a periodic sequence 0,1,1,2,3 … Now other sequences subjected to this transformation give seemingly random patterns. The next example shows how such patterns can be used to generate grids which, in this case might be used to create aperiodic rectangular tilings reminiscent of Mondrian. They could be also used for weaving pseudo-tartan patterns. This one uses Heptanacci numbers.

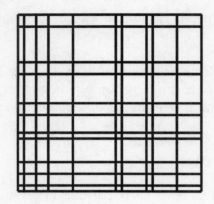
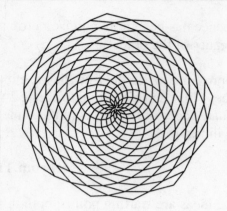

Finally, the Fibonacci numbers are not the only ones to give sunflower-like designs. This is set of spirals using the numbers 12 and 17 which are found in the continued fraction of the square root of 2 which is also used to define the special angle of the spiral.

References

1. George Markowsky, *Misconceptions about the Golden Ratio*, The College Mathematics Journal, **23** (1) Jan 1992
2. John Sharp, *Golden Section Spirals,* Mathematics in School, November 1997
3. Keith Devlin, *Mathematics, the Science of patterns*, W H Freeman 1995
4. Ian Stewart, *Mathematical Recreations*, Scientific American, June 1996, p 92
5. George Odom, Problem E3007 in American Mathematical Monthly, **90**, p 482
6. Samuel Kutler, *Brilliancies Involving Equilateral Triangles*, The St John's Review, **38** (3) 1988-9, p 17-43
7. Tons Brunes, *The Secrets of Ancient Geometry and its use*, Copenhagen 1967.
8. Ben Nicolson, Jay Kappraff, Saori Hisano, *A taxonomy of Ancient Geometry based on the Hidden Pavements of Michaelangelo's Laurentian Library*, Bridges conference proceedings 1998.
9. Carol and Donald Watts, papers given at Nexus conference 1996.
10. Jay Hambidge, *The Elements of Dynamic Symmetry*, Dover 1967
11. Edward B Edwards, *Pattern and Design with Dynamic Symmetry,* Dover 1967
12. Martin Gardner, *Six Challenging Dissection Tasks*, Quantum, May/June 1994
13. John Sharp, *Have you seen this number?*, Mathematical Gazette, **82** no 494, July 1998, p 203-214
14. Crockett Johnson, *A construction for the heptagon*, Mathematical Gazette, no 405, 1975 p 17-21
15. Peter Steinach, *Golden Fields: A case for the heptagon,* Mathematics Magazine **70** Feb 1997, p 22-31
16. Martin Gardner, *The cult of the Golden Ratio*, Notes of a Fringewatcher in The Skeptical Inquirer, **18**, Spring 1994 p 243-247.
17. Richard Padovan, *A Necessary Instrument,* The Architect, April 1986 p 54-57.
18. Richard Padovan, *Measuring and Counting,* The Architect, May 1986 p54-58.
19. Richard Padovan, *Theory and Practice,* The Architect, June 1986 p 54-58.
20. Ian Stewart, *Tales of a neglected number*, Scientific American (Mathematical Recreations page), June 1996, p 92-3
21. Rutherford Boyd, *In a Laboratory of Design*, Scripta Mathematica **XV** 1949, p 183

Towards a Methodological View on (Computer-Assisted) Music Analysis

Nico Schuler
School of Music
Central Michigan University
Mt. Pleasant, MI 48859, U.S.A.
E-mail: schuler4@pilot.msu.edu

Abstract

Music analysis is very often taught and practiced without reflecting on the method(s) used. Analysis is not possible without language and concepts. If 'pure' cognition of composition does not exist, then practicing analysis must include some reflection on its purposes. In this respect, analytical methods must be classified. This article reflects on methodological problems in music analysis. Traditional classifications of music analysis are re-visited and changes towards a more consistent system of classification are suggested with regards to the approaches taken. Finally, these methodological observations are applied to the use of computer technology in music analysis. Since methods of computer-assisted music analysis have not been classified yet, such a system is proposed here, based on detailed historical studies in this field. Each of the categories in this system of classification of computer-assisted music analysis are briefly characterized.

1. Introduction: On the Necessity of a Methodological Approach

Music analysis, including computer-assisted music analysis, is often practiced and taught without any reference to, or reflection on, the premises of the methods employed. One of the few writings acknowledging methodological concerns in the area of music analysis [7] points out that analysis is not possible without language and concepts, which means that there is nothing like 'pure cognition'. If 'pure' cognition of composition does not exist, then practicing analysis and teaching analytical methods must include some reflection on its purposes. Music analysis is not an independent discipline—and it depends on very specific theoretical systems—, nor is it an activity to be defined once and for all. Rather, music analysis is a *method*, which means it is a *way* to reach specific goals, it is a means to an end.

One way to handle these methodological problems of music analysis is to reflect on different methods; methods of music analysis need to be classified and described. While this task is partly done for 'traditional' methods of music analysis [3, 5, 6], classifications and descriptions are still lacking for new (especially computer-assisted) methods of music analysis. (An exception is, certainly, Ian Bent's monograph *Analysis* from 1987 [4].) Also, a critical evaluation is urgently needed.

So, why—in more detail—is a *methodological* approach to music analysis necessary? This question must be answered after introducing other theoretical considerations.

For Wolfgang Horn [7], analysis is, first of all, neither a doctrine nor a theory. It is not a formal-logical activity, but it has to do with the application of concepts to objects of experiences. This is the reason, why analytical activity is so hard to understand and formalize. The activity 'analyzing' is characterized by examining musical objects that are supposed to be resolved "into simpler constituent elements" [4, p. 1], and by the *manner* of resolving.

But what is 'music analysis' supposed to resolve? Ian Bent's definition continues: "Music analysis is the resolution of a musical structure into relatively simpler constituent elements, and the investigation of the functions of those elements within that structure." [4, p. 1] But the "resolution of a musical structure" into "constituent elements" is *not* the resolution of an *unknown object*, but of an *internalized experience*: Acoustical events, or their notation, function as a result of experiences and concepts [7, p. 12].

Analytical resolutions are usually communicated via language. Here, language rules need to be applied. The product, the analytical text, can be verified with the help of logic. Also important, the choice of the concepts on which the analysis is based needs to conform to the goals of the analysis. The (logical) terminological frame as well as the conceptional frame are most crucial and must be explained in the analytical text.

Wolfgang Horn distinguishes between two main approaches: The first answers the question "How is this done?"; the second answers the question "What is this?". "The results have, in both cases, only illustrative character, because the theory 'knows' concepts," and you *apply* concepts, but do so within a framework relating to a specific object. These kinds of analyses are important in historical research for getting an overview, but they are better used to catalogue compositions and put them into types. Generally, analyses are dependent on their methodological basis: "Only if the frame of an analysis is discovered, can you ask for the relevance of the analysis, and even if the question is only about the relevance to my subjective, current interest." [7, pp. 13-14]

In summation, analyzing music should not only be done "right" and "logically," but the framework of the analysis needs to be *justified*. "We should not only talk about analysis, but also, and especially, about its terms and conditions!" [7, p. 16] Reflections on the framework of music analysis, its purposes, and its goals are most important, which require *then* the application of certain methods.

2. Existing Classifications of Music Analysis

Music analysis can be classified with regard to the kind of music analyzed, the methods used, the general approach taken, etc. Any classification needs to be based on a logical framework; that means that a certain classificational level (level of abstraction) has to be on the same epistemological level. (Epistemology is the study of the methods and grounds of knowledge, especially with regard to its limits and validity. "Epistemological level" refers, here, to a level [in a system of classification] in which all 'members' have one main common characteristic, e.g. all 'members' of that level refer to *either* a method of analysis, *or* to musical categories, *or* to kinds of music, etc.)

Dieter de la Motte [5], for instance, distinguishes the following analytical categories:

❖ Large-Scale Form → Detail Structure
❖ Measure-by-Measure Analysis
❖ Analysis of Vocal Music
❖ Category Analysis
❖ Comparative Analysis
❖ Special Analysis
❖ Tendency Analysis
❖ Statistical Analysis
❖ Analytical Details
❖ Analysis with no Prerequisites

Here, different epistemological levels are mixed, such as classifying with regard to musical categories (e.g., form, structure), with regard to the kind of music (e.g., vocal music), with regard to certain methods (statistics), etc.

Ian Bent and his analytical categories offer a better example. Bent's categories of analysis are within the same epistemological level, since he only aims at specific theories. To support that notion, he mentions the author of each theory in parentheses:

❖ Fundamental Structure (Schenker)

- ❖ Thematic Process (Réti) and Functional Analysis (Keller)
- ❖ Formal Analysis
- ❖ Phrase-Structure Analysis (Riemann)
- ❖ Category and Feature Analysis (Lomax; LaRue)
- ❖ Musical Semiotics (Ruwet and Nattiez)
- ❖ Information Theory
- ❖ Set Theory

However, if Bent whishes to consider all existing, specific theories, his list is far too short and eclectic. Other theories would have to be added: different theories of harmony (e.g., Rameau, Hindemith, Perle), melody (e.g., de la Motte), rhythm, and so on.

3. Towards New Classifications of Music Analysis

For the reasons of insufficiency mentioned above, another classification of music analysis, characterized by its categories of musical elements, at which the analysis is aimed, shall be suggested here:

- ❖ Form Analysis
- ❖ Melodic Analysis
    - Thematic Analysis
    - Motivic Analysis
    - Phrase Structure Analysis
- ❖ Harmonic Analysis
- ❖ Contrapuntal Analysis
- ❖ Rhythmic Analysis
- ❖ Analysis of the Relations Between Text and Music
- ❖ Analysis of Instrumentation

Each of these categories can be sub-divided (indicated here already for "melodic analysis"). Musical categories such as range, type of motion, type of patterns, timbre, texture, sound, etc. are included.

To classify with regard to the approach used—depending on the goal of the analysis—the following categories could be distinguished:

- ❖ Schenkerian Analysis
- ❖ Transformational Grammar Analysis
- ❖ Comparative Analysis
- ❖ Measure-by-Measure Analysis
- ❖ Statistical Analysis
- ❖ Information Theoretical Analysis
- ❖ Semiotical Analysis
- ❖ Category and Feature Analysis
- ❖ Cognitive and AI Analysis
- ❖ Process Analysis

However, such a classification, based on the approach used for the analysis, cannot be complete, since new approaches are always being developed. (In this respect, the new classification of music analysis given earlier, i.e. characterized by its categories of musical elements, should be preferred. However, for certain goals of analytical research, a classification based on the approach used, will be useful.) In some of the categories shown above, specific theories are implied; however, since these are very broad and established categories of music analysis, a classification of such 'analytical approaches' seems to be justified. An additional sub-category could distinguish between the basis of the analysis: whether it is notational based or performance based (i.e. is the object to be analyzed notated music or performed music).

Another classification would be possible within the epistemological level that would refer to the "kind of presentation" of, and to the logical order within, the analytical text. (Here, de la Motte's 'Special Analysis' would fit in, which does not seek to prove something postulated in the beginning but to discover something unknown by following a specified procedure.) However, there are so many different kinds of presentation possible that a classification in this respect does not seem appropriate. The more interesting question would be if there is a classification possible with regard to *goals of analyses*, since this is the ultimate aim of any analytical work. Such a classification of analytical goals as well as the unification of the variety of classificational levels mentioned above in one system of classification, remains to be done.

4. A Classification of Methods of Computer-Assisted Music Analysis

Another methodological point needs to be made, relating to the use of technology: All analytical methods can be supported by the use of computers in music analysis. Computer-assisted music analysis provides analytical tools to help solve problems of analyzing music with traditional methods. For instance, it may clarify stylistic characterizations and questions of unclear authorship, it helps investige (historical) musical developments, it is useful for developing new theoretical systems, for research on acoustics and performance, as well as for cognitive and artificial intelligence research.

Introductory reading materials about the history of computer-assisted music analysis, such as overview articles by Bo Alphonce [1, 2] are highly selective; dozens of dissertations and numerous American and European articles are excluded. Also, most of this material fails to reflect on the subject critically. More specifically, it does not show the limits of the applications discussed. They do not show, for example, how some of the first experiments with computer-assisted music analysis are not complex enough and do not use enough musical material to support their findings.

While a critical history of computer-assisted music analysis has just been written [8], historizing computer-assisted approaches of music analysis on the one hand and classifying those on the other leads to an important epistemological problem. Even though computer-assisted music analysis has been conducted for only four and a half decades (which is very little time compared to the history of 'traditional' methods of music analysis), it has been developed under various premises, using a variety of methodologies. For that reason, it is almost impossible to talk about a real "history" of computer-assisted music analysis. Rather, approaches of computer-assisted music analysis must be initially placed within a classificational system based on their methods. On the other hand, the development of a system of classification is only possible after a thorough study of all existing approaches.

Finally, the following classificational system of computer-assisted music analysis shall be suggested here:

- ❖ Statistical and Information-Theoretical Analyses
- ❖ Set Theoretical Analyses
- ❖ Other Mathematical Analyses
- ❖ Hierarchical Analyses
- ❖ Transformational Analyses
- ❖ Schenkerian Analyses
- ❖ Spectral Analyses
- ❖ Cognitive & AI Analyses
- ❖ Combined Analyses

These categories can be divided into sub-methods. But the methods applied strongly depend on the form in which the music is analyzed; thus, "notation-based analysis" and "performance-based analysis" should create the second highest epistemological level in such a system of classification. The kind of music analyzed could be the basis for the third epistemological level.

In the following, each of the categories of computer-assisted music analysis will be briefly characterized.

Statistical and information-theoretical approaches are historically the first methods applied to computer-assisted music analysis. Even though statistical methods and information-theoretical methods are distinct from each other, in computer-assisted music analysis they are usually applied together. Statistical and information-theoretical approaches comprise frequency, mean (average), variance, standard deviation, correlation, regression, the chi square test, entropy, Markov chains, probability, redundancy, and other measurements.

For the analysis of atonal music, a number of computer programs draw on Allen Forte's set theory and on further developments of Forte's theory. Most of these programs comprise such standard procedures as calculating prime forms (most often using Forte names), interval vectors, number of occurrences, as well as (Forte's) similarity and set complex relations.

In some approaches, mathematical procedures, other than those of statistical, information-theoretical and set theoretical nature, were applied to music analysis. Structural relationships can be explicated in many mathematical ways, some of which are fractal-like descriptions, formulas for symmetrical structures or for the relationships between groups of motives (describing their characteristics) as well as formulas for calculating the inner tempo of a composition, depending on meter, metrical relationships and rhythmical structures, etc.

Hierarchical approaches to music analysis try to apply reduction procedures to music in a sense that different hierarchies of musical structures show certain dependencies as well as, on a high abstraction level, large-scale relationships (especially melodic and harmonic relationships). The basis for hierarchical approaches to music analysis is twofold: linguistic methods, especially those from the structuralistic grammar developed by Noam Chomsky, and Heinrich Schenker's concept of musical grammar. Regarding to those two main approaches, "hierarchical approaches" can be divided into "transformational analyses" and "Schenkerian analyses". Both methodological approaches comprise different abstraction levels, which can be obtained by applying certain abstraction rules.

In some cases of performance-based music analysis, spectral analysis is involved. Usually in those approaches, the sound spectrum is broken up to identify, for instance, the chord structure. While spectral analysis has been used in pure sound analysis for several decades, it became part of structural analysis of music not before the late 1980s and early 1990s.

Computer-assisted approaches of music analysis that draw on cognitive research and artificial intelligence use computer systems to simulate functions that are usually associated with human intelligence. Those functions include reasoning, learning, and self-organization (or self-improvement). Artificial intelligence approaches can exist in forms of neural network systems or expert systems. With neural networks (net-like connections of units [neurons]) as a class of dynamic systems, music theorists are trying to simulate the architecture of the human brain. Activities in single units of this network entail changes in the whole system. On the contrary, the goal of expert systems is to solve problems by drawing inferences from a knowledge base acquired by expertise; expert systems process information pertaining to a particular application and perform functions in a manner similar to that of a human who is an expert in that field.

In some applications of computer-assisted music analysis, several methodological approaches are combined in one computer system. Those systems are oriented towards interactivity, so that the user can choose which methods of music analysis to apply, depending on the goal of the specific research.

5. Conclusions

In music analysis in general, the reflection on the methods used and the awareness of how they affect the outcome and the goal of the analysis is most important. Every method of music analysis has its advantages for certain goals of the analysis. But every analytical method has also its limits. It is most crucial to know both, as well as to know when to apply which method. For methods of computer-assisted music analysis, the integration of traditional *and* computer-aided methods seems to be most crucial.

References

[1] Alphonce, Bo H. "Music Analysis by Computer - A Field for Theory Formation," *Computer Music Journal* IV/2 (1980): 26-35.

[2] Alphonce, Bo H. "Computer Applications: Analysis and Modeling," *Music Theory Spectrum* XI/1 (Spring 1989): 49-59.

[3] Beck, Hermann. *Methoden der Werkanalyse in Geschichte und Gegenwart,* 3rd edition. Wilhelmshaven: Heinrichshofen, 1981.

[4] Bent, Ian D. *Analysis*, with a glossary by William Drabkin. London: Macmillan, 1987.

[5] de la Motte, Diether. *Musikalische Analyse*. Kassel: Bärenreiter, 1987.

[6] Dunsby, Johnathan, and Arnold Whittall. *Music Analysis in Theory and Practice*. London: Faber Music, 1988.

[7] Horn, Wolfgang. "Satzlehre, Musiktheorie, Analyse. Variationen über ein ostinates Thema," *Zum Problem und zu Methoden von Musikanalyse*, ed. by Nico Schüler. Hamburg: von Bockel. 1996. pp. 11-31.

[8] Schuler, Nico. Methods of Computer-Assisted Music Analysis. History, Classification, and Evaluation. Ph.D. thesis. East Lansing: Michigan State University, 2000.

Computer Generated Islamic Star Patterns

Craig S. Kaplan
Department of Computer Science and Engineering
University of Washington
Box 352350, Seattle, WA 98195-2350 USA
csk@cs.washington.edu

Abstract

Islamic star patterns are a beautiful and highly geometric art form whose original design techniques are lost in history. We describe one procedure for constructing them based on placing radially-symmetric motifs in a formation dictated by a tiling of the plane, and show some styles in which they can be rendered. We also show some results generated with a software implementation of the technique.

1 Introduction

More than a thousand years ago, Islamic artisans began to adorn architectural surfaces with geometric patterns. As the centuries passed, this practice developed into a rich system of intricate ornamentation that followed the spread of Islamic culture into Africa, Europe, and Asia. The ornaments often took the form of a division of the plane into star-shaped regions, which we will simply call "Islamic star patterns"; a typical example appears on the right. To this day, architectural landmarks in places like Granada, Spain and Isfahan, Iran demonstrate the artistic mastery achieved by these ancient artisans.

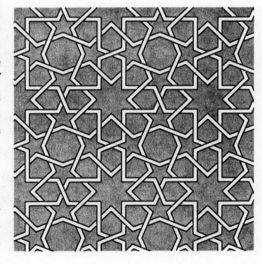

Lurking in these geometric wonders is a long-standing historical puzzle. The original designers of these figures kept their techniques a closely guarded secret. Other than the finished works themselves, little information survives about the thought process behind their star patterns.

Many attempts have been made to reinvent the design process for star patterns, resulting in a variety of successful analyses and constructions. Grünbaum and Shephard [9] decompose periodic Islamic patterns by their symmetry groups, obtaining a fundamental region they use to derive properties of the original pattern. Abas and Salman apply this decomposition process to a large collection of patterns [2]. Elsewhere, they argue for a simple approach tied to the tools available to designers of the time [1]. Dewdney proposes a method of reflecting lines off of periodically-placed circles [5]. Castera presents a technique based on the construction of networks of eightfold stars and "safts" [7].

This paper presents a technique described by Hankin [10], based on his experiences seeing partially-finished installations of Islamic art. It also incorporates the work of Lee [11], who provides simple constructions for the common features of Islamic patterns. Given a tiling of the plane containing regular polygons and irregular regions, we fill the polygons with Lee's stars and rosettes, and infer geometry for the remaining regions. We have implemented this technique as a Java applet, which was used to produce the examples in this paper. The applet is available for experimentation at http://www.cs.washington. edu/homes/csk/taprats/.

The rest of the paper is organized as follows. Section 2 presents constructions for the common features of Islamic patterns: stars and rosettes. Section 3 shows how complete designs may be built using repeated copies of those features. Techniques for creating visually appealing renderings of the designs are given in Section 4. Some results appear in Section 5. The paper concludes in Section 6 by exploring some opportunities for future work.

2   Stars and Rosettes

In our method, a regular n-gon is filled with a figure of symmetry type d_n (which has all the symmetries of the n-gon). In practice, these figures belong to a small number of families which we describe below.

For $n \geq 3$, let the unit circle be parameterized via $\gamma(t) = (\cos{(2\pi t/n)}, \sin{(2\pi t/n)})$. We construct the n-pointed star polygon (n/d) by drawing, for $0 \leq i < n$, the line segment σ_i connecting $\gamma(i)$ and $\gamma(i + d)$. Note that $d < n/2$ and that $(n/1)$ is the regular n-gon. For some values of $k \neq i$, σ_i will intersect σ_k, dividing σ_i into a number of subsegments. We often choose to draw only the first s subsegments at either end of σ_i, which we indicate with the extended notation $(n/d)s$. Figure 1 shows the different stars that are possible when $n = 8$.

Our implementation generalizes this construction, allowing d to take on any real value in $[1, n/2)$. When d is not an integer, point P is computed as the intersection of line segments $\overline{\gamma(i)\gamma(i + d)}$ and $\overline{\gamma(i + \lfloor d \rfloor - d)\gamma(i + \lfloor d \rfloor)}$, and σ_i is replaced by the two line segments $\overline{\gamma(i)P}$ and $\overline{P\gamma(i + \lfloor d \rfloor)}$. Two examples of this generalization are given in Figure 2.

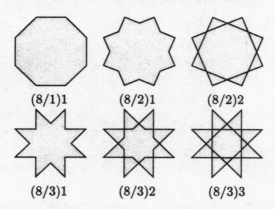

(8/1)1   (8/2)1   (8/2)2

(8/3)1   (8/3)2   (8/3)3

Figure 1 The six possible eight-pointed stars when d is an integer.

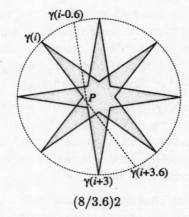

(8/3.6)2

Figure 2 An $(n/d)s$ star for non-integral d.

When sixfold stars are arranged as on the left side of Figure 3, a higher-level structure emerges: every star is surrounded by a ring of regular hexagons. The pattern can be regarded as being composed of these surrounded stars, or **rosettes**. Placing copies of the rosette in the plane will leave behind gaps, which in this case happen to be more sixfold stars.

The rosette, a central star surrounded by hexagons, appears frequently in Islamic art. They do not only appear in the sixfold variety, meaning that we must generalize the construction of the rosette to handle

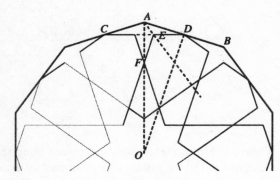

Figure 3 An arrangement of sixfold stars can be reinterpreted as rosettes. The pattern is one of the oldest in the Islamic tradition.

Figure 4 The construction of a ten-pointed rosette.

arbitrary n. The construction given by Lee [11] yields an n-fold rosette for any $n \geq 5$ while preserving most of the symmetry of the hexagons. Each hexagon has four edges not adjacent to the central star; all four edges are congruent. Moreover, the outermost edges lie on the regular n-gon joining the rosette's tips, and the two "radial" edges are parallel.

A diagram of Lee's construction process is shown in Figure 4. To begin, inscribe a regular n-gon in the unit circle and draw the n-gon whose vertices bisect its edges. Let A and B be adjacent vertices of the outer n-gon and C and D be adjacent vertices of the inner n-gon with D bisecting $\overline{AB}$. The key is to then identify point E, computed as the intersection of $\overline{CD}$ with the bisector of $\angle OAB$. Then F is the intersection of $\overline{OA}$ with the line through E parallel to $\overline{OD}$. The rest of the rosette follows through application of symmetry group d_n: edges $\overline{DE}$ and $\overline{EF}$ lead to the outer edges of the hexagons, while copies of F become the points of the inner star, which can be completed with the construction given earlier.

By sliding E along the bisector of $\angle OAB$, we can continuously vary the shape of the rosette while preserving the congruence of the four outer hexagonal edges.

Some Islamic designs feature a motif slightly more complicated than a basic rosette, where opposing limiting edges from adjacent tips of the rosette are joined up. The resulting object has the same symmetries and number of outer points as the rosette, but with an additional layer of geometry on its outside. We refer to these as "extended rosettes". A ninefold extended rosette appears on the right.

3 Filling the Plane

Equipped with a taxonomy of typically Islamic motifs that can be inscribed in regular polygons, we are now ready to create complete periodic designs. We start with a periodic tiling containing regular polygons, with irregular polygons thrown in as needed to fill gaps. For each regular n-gon with $n > 4$, we choose an n-fold star, rosette or extended rosette to place in it and replicate that motif everywhere the n-gon appears in the tiling. The motif is placed so that its points bisect the edges of the n-gon.

The result is a design like that of Figure 5(b). There are still large gaps where motifs were not placed, corresponding here to to the squares in the original tiling. Each square edge is adjacent to an edge of an octagon, and so a vertex of the chosen motif is incident to it. The presence of these vertices suggests a technique for filling the gaps in a natural way, by extending the line segments that terminate on the boundary of the region until they meet other extended segments in the region's interior. Except for degenerate cases,

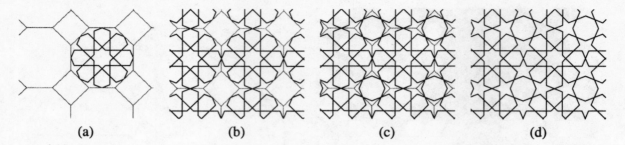

(a)   (b)   (c)   (d)

Figure 5 Given the octagon and square tiling shown in (a), we decide to place 8-fold rosettes in the octagons and let the system infer geometry for the squares. The rosette is copied to all octagons in (b), and lines from unattached tips are extended into the interstitial spaces until they meet in (c). The construction lines are removed, resulting in the final design shown in (d).

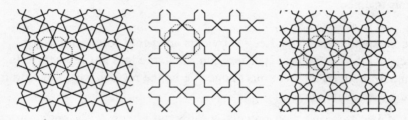

Figure 6 Some alternative patterns based on the octagon-square tiling that can be constructed by varying the motif placed in the octagons.

following this procedure guarantees that the resulting design will admit an interlacing.

Figure 5(c) shows the design with the free rosette tips extended into the gaps. Here, the natural extension creates regular octagons in the interstitial regions. To complete the construction, the original tiling is removed, resulting in the design in Figure 5(d), a well-known Islamic star pattern [3, plate 48].

Given a tiling containing regular polygons and gaps, we can now construct a wide range of different designs by choosing different motifs for the regular polygons. Even when restricted to the octagon-square tiling used above, many different designs can be created. Three alternative designs appear in Figure 6. Of course, we can expand the range of this technique in the other dimension by also encoding a large number of different tilings.

The implementation currently encodes fourteen tilings from which Islamic star patterns may be produced. Some are familiar regular or semi-regular tilings [8, Section 2.1]. Some are derived by examination of well-known Islamic patterns. The remaining tilings were discovered by experimentation and lead to novel Islamic designs shown in Section 5.

4 Rendering

The output of the construction process is a planar graph. To be sure, the graph has an intrinsic beauty that holds up when it is rendered as simple line art. Historically, however, these designs were never merely drawn as lines. Islamic star patterns are typically used as a decoration for walls and floors. The faces of the planar graph are realized as a mosaic of small terracotta tiles in a style known as "Zellij". Often, the edges are thickened and incorporated into the mosaic with narrow tiles, sometimes broken up to suggest an interlacing pattern. Islamic designs can also be found carved into wood or stone and built into trellises and latticework.

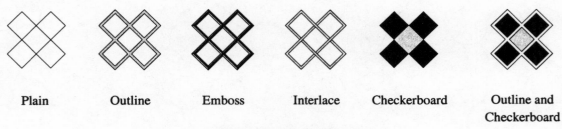

| Plain | Outline | Emboss | Interlace | Checkerboard | Outline and Checkerboard |

Figure 7 Rendering styles.

To increase the aesthetic appeal of our implementation, we provide the ability to render the planar graph in a manner reminiscent of some of these techniques (see Figure 7). The **outline** style thickens the edges of the planar graph, adding weight and character to the lines of the plain style. The **emboss** style adds a 3D effect to the outline style, simulating the appearance of a wooden trellis; the centre of each thickened edge is raised towards the viewer and the graph is rendered by specifying the direction of a fictitious light source. The **interlace** style adds line segments at each crossing to suggest an over-under relationship between the crossing edges. When every vertex in the graph has degree two or four, the crossings can always be chosen so that the graph is broken into strands that adhere to a strict alternation of over and under in their intersections with other strands. The final style, **checkerboard**, renders the faces of the graph and not the edges. When all vertices have even degree (as they must in an interlace design), it is always possible to colour the faces with only two colours in such a way that faces with the same colour never share an edge. The checkerboard style walks the graph, creating a consistent 2-colouring.

A further enhancement can be achieved by layering one of the edge-based rendering styles on top of the checkerboard style. This combination comes closest to the appearance of Zellij.

5   Results

Figures 8 and 9 present a selection of finished computer-generated drawings. The first group, Figure 8, is made up of reproductions of well-known Islamic star patterns which can be found in Bourgoin [3] or Abas and Salman [2]. Figure 9 contains designs that do not appear in either of those sources. Three of them are based on polygonal tilings that do not seem to be used by any known designs. These last three are moderately successful, though they seem to lack the harmonious balance of the well-known designs. Still, in an artform with a thousand-year tradition, any sort of novel design is certainly of interest.

6   Future Work

Our software implementation and the technique on which it is based allow access to a wide variety of designs without offering so much flexibility that it becomes overly easy to wander out of the space of recognizably Islamic patterns. There are, however, opportunities for future work that do not compromise the focus of the system.

The set of available tilings from which to form patterns is open-ended. More tilings could be implemented. Some new ones can easily be derived by inspection of patterns in Bourgoin or Abas and Salman. We could move away from periodicity by implementing aperiodic tilings with regular polygons. Castera has constructed several ingenious aperiodic Islamic star pattern based on Penrose rhombs [4]. Finally, the hyperbolic plane offers tremendous freedom in the construction of tilings with regular polygons. We hope

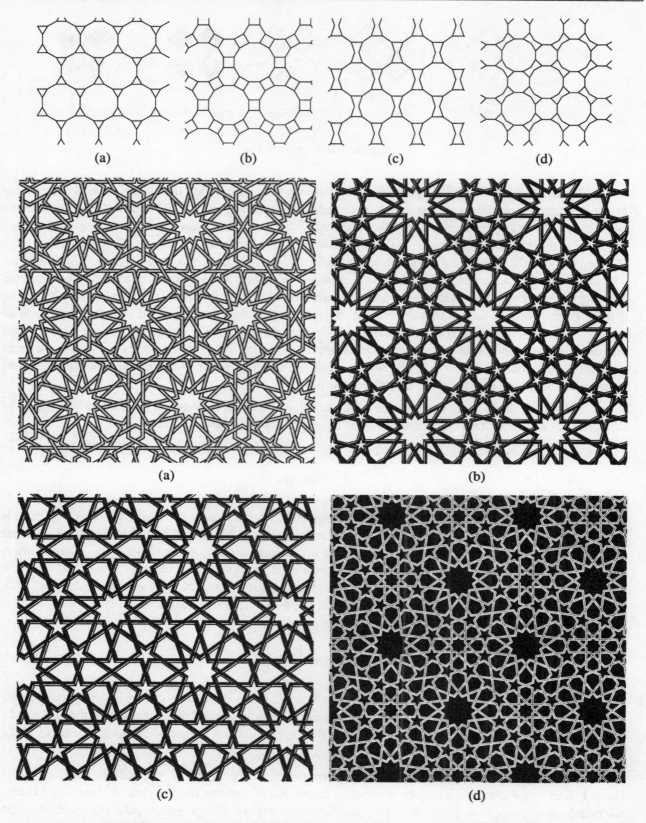

Figure 8 Some sample results based on well-known tilings from Islamic ornament. Each final design is based on the corresponding tiling in the top row.

Figure 9 Sample results not found in the literature. The pattern in (a) is similar to one found in Abas and Salman [2, p. 93], using extended rosettes instead of ordinary rosettes. The other three patterns are based on previously unused tilings.

Figure 10 A novel pattern with 7-stars. The design in the centre, resulting from the natural extension of star edges, leaves behind disproportionate octagons. The design on the right, constructed manually, corrects this by redistributing the area to new 5-stars.

to adapt the technique described in this paper to the Poincaré model of the hyperbolic plane, much the same way Dunham has done with Escher patterns [6].

One last aspect of the system we hope to improve is the naïve extension of lines into interstitial regions. Our algorithm can easily fail to produce attractive results. In Figure 10, a novel grid based on regular heptagons is turned into an Islamic pattern by placing (7/3)2 stars in the heptagons. The natural extension of star edges into the gaps leaves large, unattractive octagonal areas. With the appropriate heuristics, our inference algorithm could detect cases such as this and add some complexity to the inferred geometry in order to improve the final design.

Acknowledgments

I am indebted to Mamoun Sakkal for his guidance and helpful discussion while initiating this project in his course. Thanks also to David Salesin for his valuable input and to Reza Sarhangi for the encouragement.

References

[1] Syed Jan Abas and Amer Shaker Salman. Geometric and group-theoretic methods for computer grahpics studies of Islamic symmetric patterns. *Computer Graphics Forum*, 11(1):43–53, 1992.

[2] Syed Jan Abas and Amer Shaker Salman. *Symmetries of Islamic Geometrical Patterns*. World Scientific, 1995.

[3] J. Bourgoin. *Arabic Geometrical Pattern and Design*. Dover Publications, 1973.

[4] Jean-Marc Castera. Zellijs, muqarnas and quasicrystals. In Nathaniel Friedman and Javiar Barrallo, editors, *ISAMA 99 Proceedings*, pages 99–104, 1999.

[5] A.K. Dewdney. *The Tinkertoy Computer and Other Machinations*, pages 222–230. W. H. Freeman, 1993.

[6] Douglas Dunham. Artistic patterns in hyperbolic geometry. In Reza Sarhangi, editor, *Bridges 1999 Proceedings*, pages 139–149, 1999.

[7] Jean-Marc Castera et al. *Arabesques: Decorative Art in Morocco*. ACR Edition, 1999.

[8] Branko Grünbaum and G. C. Shephard. *Tilings and Patterns*. W. H. Freeman, 1987.

[9] Branko Grünbaum and G. C. Shephard. Interlace patterns in Islamic and moorish art. *Leonardo*, 25:331–339, 1992.

[10] E.H. Hankin. *Memoirs of the Archaeological Society of India*, volume 15. Government of India, 1925.

[11] A.J. Lee. Islamic star patterns. *Muqarnas*, 4:182–197, 1995.

The Subtle Symmetry of Golden Spirals

Alvin Swimmer
Department of Mathematics
Arizona State University
Tempe, AZ 85287-1804, USA
E-mail: aswimmer@asu.edu

Abstract

The beautiful symmetries of golden spirals are intuitively evident to everyone that sees them even for the first time. The purpose here is to show how these symmetries are manifestations of the special nature of the eyes of those spirals.

1. The Golden Spiral Circumscribing Golden Triangles

Following the ancient Greek geometers, I call every isosceles triangle whose base angles each measure twice the vertex angle a **Castor Triangle**. That is, the vertex angle measures 36 degrees or $\frac{\pi}{5}$ radians and each base angle measures 72 degrees or $\frac{2\pi}{5}$ radians. This is the triangle that occurs in the construction of a regular decagon as the central triangle determined by two radii of the circle being subdivided into 10 equal parts and a side of the decagon. It is the triangle determined by two diagonals of a regular pentagon emanating from the same vertex and the side opposite that vertex. It is also evident in the ubiquitous 5 pointed star that occurs in the flags of many countries as well as many, many other places, as each of the 5 triangles protruding from the regular pentagon that forms the interior of the star. The ratio of a side length to the base length of a Castor Triangle is the golden ratio $\phi = \frac{\sqrt{5}+1}{2}$ and that is the reason that I call the spiral it determines a Golden Spiral.

The construction of this spiral proceeds as follows. (We use $\doteq$ as an abbreviation of 'defined as'.) Let $\mathcal{C}_0 \doteq [T_0, T_1, T_2]$ denote the Castor Triangle with vertex T_0 and base $[T_1, T_2]$. We bisect the base angle with vertex T_1 and call T_3, the intersection of this bisector with the opposite side $[T_2, T_0]$. Then, because of the angle relationship $\mathcal{C}_1 \doteq [T_1, T_2, T_3]$ is another Castor Triangle. The other triangle thus formed, $\mathcal{P}_1 \doteq [T_3, T_0, T_1]$ is also isosceles. The vertex angle in $\mathcal{P}_1$ measures $\frac{3\pi}{5}$ radians and each base angle measures $\frac{\pi}{5}$ radians, the same as the vertex angle of a Castor Triangle. Still following those ancient Greek geometers that adorn my list of mathematical heroes, I call all isosceles triangles with a vertex angle measuring three times each base angle, a **Pollux Triangle**. In a Pollux Triangle the ratio of the base length to the side length is ϕ. It is the triangle formed by adjacent sides of two adjacent protruding Castor Triangles of a 5 pointed star and the line joining the two vertices. A pair of triangles, one a Castor and the other a Pollux situated so that, like $\mathcal{C}_1, \mathcal{P}_1$, they share a common side and together make a larger Castor Triangle, I call **Gemini Twins**. In this context, the ancient Greek geometers would say that the Pollux Triangle acts as a **gnomon** to transform its Castor Twin into the Mother Castor.

We can now draw the first arc of our golden spiral, from T_0 to T_1 using the vertex T_3 of $\mathcal{P}_1$ as the center and its side length as the radius.

The whole process is now repeated beginning with C_1 as the mother Castor Triangle. The angle with vertex T_2 is bisected. The endpoint of this bisector on $[T_3, T_1]$, we call T_4. Then $C_2 \doteq [T_2, T_3, T_4]$ is the new Castor Triangle and its twin is $P_2 \doteq [T_4, , T_1, T_2]$ and the arc drawn circumscribing the base of P_2 from center T_4 is the second arc of our golden spiral. Note that since the centers of the two arcs are collinear with T_1, the tangents to the two arcs at T_1 coincide. Thus the spiral curve is tangent smooth at the transition point.

The nth step in this iterative process begins with the Castor Triangle $C_n \doteq [T_n, T_{n+1}, T_{n+2}]$. The base angle at T_{n+1} is bisected thus creating a pair of Gemini Twins C_{n+1}, P_{n+1}. A tangent smooth arc of the associated Golden Spiral is then drawn circumscribing the base of P_{n+1} using its vertex as the center.

From a transformational viewpoint we map C_n into C_{n+1} by rotating the first Castor through $\frac{3\pi}{5}$ radians and simultaneously shrinking its sides by a factor of ϕ. I call this transformation the **Φ-map**. We can think of it as the point transformation that maps each vertex T_n into the 'next' vertex T_{n+1} and describe it by saying that "the Φ-map maps the vertices sequentially".

The nested sequence $\{C_n\}$ of Castor Triangles arising by the Φ-map converge to a special point E called the **eye** of the spiral. It can be located as the intersection of the medians $[T_{n+2}, L_{n+2}]$ where L_{n+2} is the midpoint of $[T_n, T_{n+1}]$. E has a number of fascinating properties. (For proofs of these properties see [1]).

(1) E divides each median $[T_{n+2}, L_{n+2}]$ in the same ratio: $\frac{[T_{n+2}E]}{[EL_{n+2}]} = \frac{\phi^2}{2}$.

(2) E is the only fixed point of the Φ-map. That is, its location in every Castor Triangle remains the same. In fact the areas of the three triangles into which C_n is subdivided when E is joined to the three vertices are in the ratio (where, in general, we use $[ABC]$ to denote the area of triangle $[A, B, C]$)

$$[ET_{n+1}T_{n+2}] : [T_nET_{n+2}] : [T_nT_{n+1}E] = 1 : 1 : \phi^2.$$

(3) The Φ-map also maps the midpoint sequence $\{L_{n+2}\}$ sequentially. Indeed each $[L_{n+2}, L_{n+3}, L_{n+4}]$ is a Castor Triangle and thus this sequence determines another golden spiral. This spiral intertwines the original spiral since it has the same eye E.

(4) It is easy to construct the image A_1 under the Φ-map of any point A_0 in the plane $\overline{T_0T_1T_2}$. Simply join A_0 to T_0 and to T_1 and then construct lines parallel to these lines passing through T_1 and T_2 respectively. The intersection of these two lines is A_1. Similarly join A_1 to T_1 and T_2 and then construct lines parallel to these passing through T_2 and T_3 respectively. This will yield A_2, the image of A_1. Of necessity then $[A_0, A_1, A_2]$ will be a Castor Triangle and thus will generate a new golden spiral intertwining the other golden spirals since its eye is also E.

Thus E is simultaneously the eye of an infinite family of golden spirals all of which can be generated from any one of them by the Φ-map.

One other fascinating property of the Castor Sequence is that in each Castor Triangle C_n, the interval $[T_n, T_{n+3}]$ on the side $[T_n, T_{n+2}]$ is divided harmonically by M_{n+1}, the midpoint of $[T_n, T_{n+2}]$ and L_{n+4}, the midpoint of $[T_{n+2}, T_{n+3}]$ (which is a side of C_{n+1}). In other words, $[T_nT_{n+3}]$ is the harmonic mean of $[T_nM_{n+1}]$ and $[T_nL_{n+4}]$. Thus the set $\{T_n, M_{n+1}, T_{n+3}, L_{n+4}\}$ is the geometric counterpart of a major musical chord.

2. The Golden Spiral Inscribed in Golden Rectangles

A golden rectangle is one for which the ratio of the length of the long side to that of the short side is ϕ. This automatically makes the long side length the geometric mean between the semiperimeter and the length of the short side.

Let $\mathcal{R}_1 \doteq [R_1, R_{-2}, R_0, R_2]$ denote a golden rectangle as in Figure 3 with vertical side lengths $[R_1 R_{-2}] = [R_0 R_2] = 1$ and horizontal side lengths $[R_{-2} R_0] = [R_2 R_1] = \phi$.

To construct the golden spiral associated with $\mathcal{R}_1$ we subdivide it by the line segment $[R_3, R_4]$ where R_3 divides $[R_{-2}, R_0]$ and R_4 divides $[R_1, R_2]$ in the golden ratio. This subdivides $\mathcal{R}_1$ into a square $\mathcal{S}_1 \doteq [R_1, R_{-2}, R_3, R_4]$ each of whose sides is of length 1 and a new rectangle $\mathcal{R}_3 \doteq [R_3, R_0, R_2, R_4]$ the ratio of whose sides is $\frac{1}{\phi-1} = \frac{1}{\frac{1}{\phi}} = \phi$ so that $\mathcal{R}_3$ is another golden rectangle. Reversing the process, we can say if to $\mathcal{R}_3$ we adjoin $\mathcal{S}_1$ we end up with $\mathcal{R}_1$ so that $\mathcal{S}_1$ is a gnomon for $\mathcal{R}_3$.

Using R_4 as a center and $[R_4, R_1]$ as the radius we inscribe in $\mathcal{S}_1$, an arc from R_1 to R_3. This is the first arc of our golden spiral.

We now iterate this process using $\mathcal{R}_3$ as the mother rectangle subdividing it into a square $\mathcal{S}_3 \doteq [R_3, R_0, R_5, R_6]$ and a new golden rectangle $\mathcal{R}_5 \doteq [R_5, R_2, R_7, R_8]$. With R_6 as the center we inscribe an arc beginning at R_3 and ending at R_5. Since R_3 is collinear with both centers R_4 and R_6 the two arcs form a tangent smooth curve at R_3.

Each step in the iteration process leads to a subdivision of a golden rectangle $\mathcal{R}_{2d-1}$ by the interval $[R_{2d+1}, R_{2(d+1)}]$ into a square $\mathcal{S}_{2d-1}$ with vertices R_n with $n = 2d-1, 2(d-2), 2(d-1), 2d$ and a new golden rectangle $\mathcal{R}_{2d+1}$ with vertices R_m where $m = 2d+1, 2(d-1), 2d, 2(d+1)$. If d is even the long sides are horizontal, otherwise they are vertical. The arc of the golden spiral associated with this square connects R_{2d-1} to R_{2d+1} and the center is $R_{2(d+1)}$.

The nested sequence of golden rectangles converge to a point E that is the eye of the associated spiral. It may be constructed as the intersection of the diagonals $[R_{-2}, R_2]$ and $[R_0, R_4]$ and it divides each of those diagonals in the ratio ϕ^2. These diagonals are perpendicular.

When R_1, the unused vertex of $\mathcal{R}_1$, is joined to E it intersects the opposite side at R_5. When R_3 the unused vertex of $\mathcal{R}_3$ is joined to E it intersects the opposite side in R_7. These two lines are also perpendicular.

Concerning these four 'spokes' emanating from E we have:

The Golden Harmonic Spokes Theorem

The pencil of four 'spokes' emanating from E (the eye of the golden spiral inscribed in a nested sequence of golden rectangles), consisting of a pair of perpendicular diagonals containing all of the even indexed vertices, and a pair of perpendicular 'non-diagonals' containing all of the odd indexed vertices, contains all of the vertices of all the golden rectangles in the sequence. Each pair bisects the other pair (so that the 8 angles they determine with vertex E are each $\frac{\pi}{4}$ radians and the two pairs together consequently form an harmonic set of lines. Thus they intersect each vertical and each horizontal side in an harmonic set of vertices.

The vertices on each of the four harmonic spokes are marvelously arranged in that any four 'consecutive' (i.e. their indices are of the form $(d, d+4, d+8, d+12)$) vertices on each of them is also an harmonic set of points.

References

[1] Swimmer, A., *On Golden Spirals: The Subtlety of Their Symmetry.* Preprint.

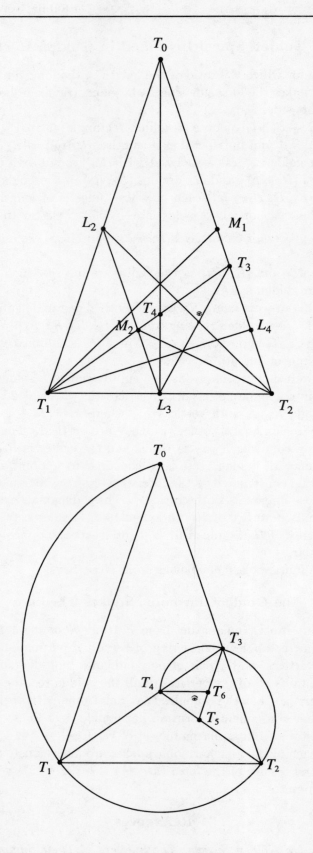

Figure 1. The Gemini Family and its Spiral

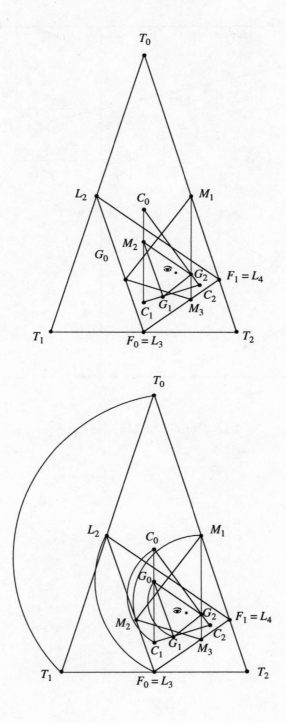

Figure 2. Five Mothers and Their Intertwining Spirals

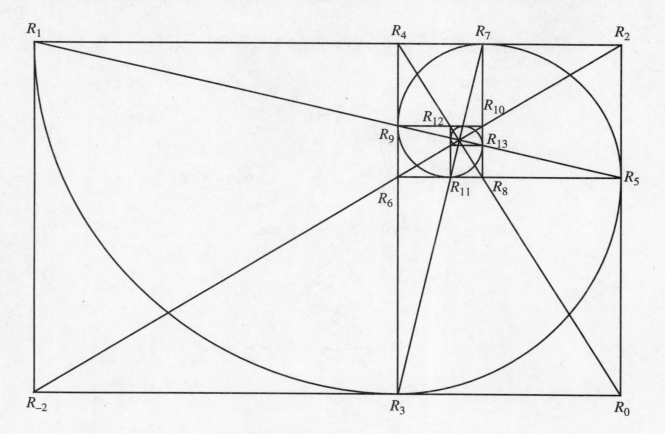

Figure 3. The Pencil of Harmonic Spokes

Nearing Convergence:
An Interactive Set Design for Dance
The Intersection of Physical Movement and Optical Form

Benigna Chilla

Department of Visual Arts

Berkshire Community College

Pittsfield, MA 01201, U.S.A.

E-mail: bchilla@cc.berkshire.org

Abstract

In this paper, I will present the development of my collaboration with choreographer, Ellen Sinopoli, and four of her dancers in creating "Nearing Convergence." My work alone consists of layered elements and often explores the interplay of geometrical forms and optical movement. In this collaboration, the challenge and the excitement for me was to add the dimension of physical movement of the dancers to a very large scale work. The challenge for the choreographers, the dancers, and myself together was to find and express the intersection of physical movement and optical form. In this lecture/slide demonstration, I will present slides documenting the conceptual development of the set design for "Nearing Convergence," and I will show video clips from the premier production of the performance.

Introduction and Background

The knowledge of geometry has been the fundamental basis of art forms throughout history. The application of geometry pervades and forms the visual world we move through; it allows us to remember and reconstruct formations and gives artists the freedom of scale and combination of different dimensions. My own work deals with hidden images: layers of planar surfaces visible from a certain angle, which change according to distance and personal interaction. Three planar surfaces, consisting of canvas and screen, are parallax or completely unrelated in their geometrical design. The surfaces, as concrete as they are alone, together convey images with constantly changing focus and appearance. One knows what is physically present but captures, and then loses the optical images in an oscillating phenomenon.

In recent years, my interest has grown to expand the scale and interaction of my pieces. I wanted to integrate actual movement into my pieces and have the viewer experience the changes, appearances, and interactions of the work. When physical movement becomes part of a piece the perception of the viewer will change. The viewer is stationary and the piece itself begins to move in front of the viewer, changing through the interactions of light, sound, and physical movement.

In 1989 I created "Four Parallax Ellipses," which measured 28 feet wide, 6 feet high, and one foot deep. While hanging and adjusting this long piece I had to move in and out between the two surfaces so for a moment I became an interactive part of the work. Ten years later the idea to bring physical movement into my work fully manifested itself and a collaboration began.

Concept and Development

I first met with choreographer Ellen Sinopoli in 1997. We spent a lot of time learning about each other's work before beginning the actual collaborative process. The choreographer, four dancers, a composer, a lighting designer, a sculptural engineer, and I got together a year later to create "Nearing Convergence" specifically for a premiere at the Egg, a performance center in Albany, NY in 1999. Composer William Harper, who was familiar with the nature and form of my work, was commissioned to compose a score for this project. Meanwhile I developed several conceptual ideas for the set design into preliminary sketches and a maquette. Ellen, the dancers, and I soon realized how many changes we had to make as our collaborative work developed. It was a work with multiple authors. We could no longer make decisions about the choreography or the set design single-handedly; together we were challenged to find and express the intersection of physical movement and optical form.

My vision and understanding of this collaboration was with the full intent of seeing such a piece as a contemporary visual art form performed for a live audience. It was about much more than just expanding the size of my work or bringing a painting to the stage. The audience could be captured for sixteen minutes to experience "Nearing Convergence." Form, motion, sound, and time made this four-dimensional piece possible.

We started with the idea of creating a prop, which consists of six linear tetrahedrons connected to each other. These forms could create a series of shapes: an open equilateral triangle in the center, a hexagon, a closed triangle, a cube, a closed triangle, a hexagon, and again the open triangle in the center. The longer diagonal tube of the tetrahedron measured 110 inches; on the opposite side the length of the tube measured 84 inches and created a 90 degree angle with the 48 inch long rotating tube.

The linear tetrahedrons were made out of PVC and aluminum tubing and are therefor very flexible. This component was the first determination of how the dancers could interact with the piece and also change its position.

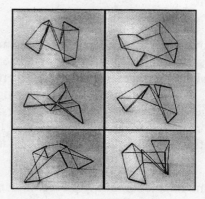

Figure 1: *Positions of the Linear Tetrahedron*

Figure 2: *Maquette (15" x 20" x 15")*

A Convergence of Design, Function, and Form

The triangle now became the basic element for design.

In "Nearing Convergence," two layers of suspended screens, one upstage and one downstage, become the physical and optical framework through which the choreography is interwoven. In a sense, the dancers become the third layer of physical and optical movement in the piece.

The set design is based on tetrahedrons (a life-sized kinetic prop) and right triangles painted on fiberglass screen which alternate between positive and negative. The configurations of triangles move in diagonals in opposite directions across the two screens measuring 30 feet high by 40 feet wide. Each of these screens comprises five panels measuring 8 feet wide and 30 feet high. The downstage screen consists of four rows of right triangles that move upwards from right to left on a 30 degree angle, whereas on the upstage screen the rows of right triangles move upwards from left to right. The two layered surfaces visually overlap and the triangles meet and cross, forming an equilateral triangle in the center.

The open linear tetrahedron was placed stage left into the space between the two hanging surfaces and was slowly moved across the stage and through different positions by the dancers.

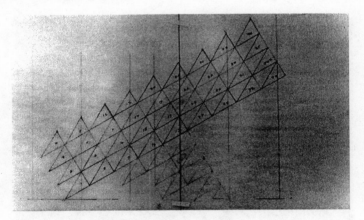

Figure 3: *Work Drawing*

Figure 4: *Moving Triangles Left*

Figure 5: *Moving Triangles Right*

The Synthesis of Movement and Form

I was forced to create a workable environment for the dancers, who began to structure improvisations as they explored the constrictions and possibilities of the space, geometrical shapes, and forms. When the four dancers enter the space, they often move in striking dissonance to their surroundings, oozing slowly over each other and making curving organic shapes. The layered effects of the screens create an intriguing confusion; it isn't always clear what dancers are on what side of each screen until they begin to interact, lifting one of the group above their heads or rolling over another's back.

The dancers enter from stage right between the scrim and the upstage screen, following the movement of the triangles, then move into the space between the upstage and downstage screens. They are now enclosed in this space, which becomes their habitat for the duration of the performance. At the very end of the piece they move through the front layer and the screens fly up.

Buddhist philosophy suggests that "space is the fundamental element of our cosmos. Its nature is emptiness and because it is empty it can contain and embrace everything...space is the precondition of all that exists." The space in "Nearing Convergence" is the space within the piece where for a limited time the empty space is activated through movement and then changed back into an empty enclosed space [1].

Figure 6: *Installation of "Nearing Convergence" with dancers*

Reflections

I did not see the piece installed until the day before the performance. Neither did the choreographer, dancers, or composer. Throughout their rehearsals and improvisations the dancers worked with the linear tetrahedron, but could only imagine the enclosed space between the two layered screens. We never had the opportunity to rehearse the dance within the complete set or with the proper lighting until the dress rehearsal in the theatre.

We were collectively faced with several constraints that limited the scope of this project. Due to our non-existing budget, we could not introduce a backdrop of a 20 feet high kinetic equilateral triangle – a concept that I had developed as part of the work in progress and executed in a variation of paintings. Similarly, we required a set that could be easily transported. Thus, the tetrahedron can be taken apart and the screens rolled up in sections and together everything fits into the back of a pickup truck.

This collaborative effort was full of challenges and new discoveries. Every step and discovery in my work is important; one piece will nourish the next one or retrieve from a previous work. Outside influences only interact when a new concept or idea is already formed. I see what I want to see, and when the eye and mind are ready to store more information, I digest and recreate. With my desire to explore the intersection of physical movement and optical form I challenged the viewer to interact with "Nearing Convergence."

Figures 7 – 9: *Dancers in rehearsal with linear tetrahedron*

References

[1] L.A. Govinda, *The Psychological Attitudes of Early Buddhist Philosophy*, Rider, London, 1961.

Photo credit: David Lee, East Chatham, New York.

Evolutionary Development of Mathematically Defined Forms

Robert J, Krawczyk
College of Architecture
Illinois Institute of Technology
Chicago, IL 60616 USA
E-mail: krawczyk@iit.edu

Abstract

With the increase use of algorithms to develop images, as well as, three-dimensional sculptural and architectural forms, additional focus should be placed on methods to teach students how to approach such a design technique. This paper reviews one such method. Architectural students are looking for inspiration for new forms to better realize their own designs and computer methods are offering them another tool to expand their investigations. This paper outlines one method used as part of a programming course developed for architectural students. The method highlights a very sequential approach to form investigation in using a common starting geometry. This approach stresses the development of rules and evaluating their results as a method to determine the next step to investigate. Equal importance is placed on the anticipated, as well as, the unexpected.

1. Introduction

As commercial Computer-Aided Drafting and Design systems have grown in power to handle both two-dimensional and three-dimensional representations, many of the basic functions for creating drawings and models have been adequately addressed. A variety of primitive shapes are available. Operations to either join or subtract them in addition to options to modify them at the vertex level are also readily available. Primitive shapes range from lines and planes to surfaces of revolution and free-form blobs. These systems rarely allow for an organized and repeatable method to create models. They all create "hand-crafted models". Some are parametrically based, which allow for dimensional relationships, such as: "dimension A changes by 10% if dimension B changes by 5%". In these systems all changes are highly regulated, and it is difficult to control a entire series of parameters or components.

Other methods to develop forms have been mathematically based. These include fractals (Yessios [10]), space curves (Krawczyk [3]), and complex surfaces (Sequin [8,9]). Still others have investigated agent based autonomous systems which are capable of collaborating, building, degenerating, and transforming, using genetic algorithms, Krause [2]. In these types of form building methods, rarely is the process explained in a sequential fashion on how to arrive at the final results. The process is very important for students to see. The drawback is that these approaches require a great amount of research and extensive mathematical knowledge, plus advanced methods in programming. The typical architectural student would find these difficult in an introductory course.

One aspect of these systems I wanted to develop was how a commercial CAD system could be used to investigate a geometrically based design concept and how a CAD programming language could be used to establish rules and procedures for such an investigation. The best way to demonstrate this idea was to have the students follow a step-by-step approach that could show them how a design concept could be developed.

These exercises were designed as "programs as pencils"; every concept, every variation was implemented by an individual program modification. The programs were meant to hold all the rules needed for the form being generated. Manual modifications were not allowed to the model itself. All ideas were to be implemented by actual program modifications. When students asked if another variation would be interesting, the usual response was "*I don't know, try it*".

Mitchell [7] took a similar approach in introducing programming to architectural students using graphic exercises. Even though the focus of his book was programming, it is a good source for how to organize work to be presented one concept at a time. Maeda [6] more recently took a similar approach in demonstrating the creation of drawings and images for those starting graphics and using programming as a method.

The incremental process of developing a computer program is a very powerful. It allows one to concentrate on a single concept at a time, make small manageable changes to the program, and most importantly, always work from success. One always begins the next step from a previously working program. This method has been shown to work with students whose primary interest is not computer science. It also introduces the idea that a program evolves and is not a static statement of a concept, but a starting point to further investigation.

Working within an established CAD environment may have some limitations, but having easy access to basic modeling primitives, operations, and visualization aids enables the student to concentrate on the design aspects of the forms.

The following section describes such a series of exercises. In developing these exercises I did not know exactly where they were going to end. I tried to incorporate in the steps some of the questions I asked myself as I developed each one. The importance of these exercises becomes the steps and decisions at each step rather than the resulting form. I wanted to show the students the process of program development and how it could be used to investigate basic geometric forms.

A few initial assumptions were made:

  a. The exercises should concentrate on the development of forms and not engineering analysis.
  b. The exercises should try to develop forms normally not available by assembling basic CAD entities.
  c. The forms generated could be somewhat anticipated. Students should be able to easily visualize each form and determine variations to it.
  d. The forms would be three-dimensional.
  e. The exercises would investigate how to develop forms based on rules easily established and modified.
  f. The primary objective was to create architecturally suggestive forms. Scale, proportion, and orientation will be considered.

To generate the forms the programming language used for the exercises was AutoLISP within Autodesk's AutoCAD R14. 3D Studio MAX was used for the renderings and animations.

2. Description of form development

Previous attempts to develop such a procedural approach to form development utilized a linear form (Krawczyk [4]). The linear form allowed for surface articulation and the development of both frame and surface models. The linear form had limitations for students to go beyond the set of exercises presented. In this series a circular form is used and then extended to the many other simple curved forms that are readily available.

Our initial class discussions included the basic geometric forms that architects have used to create built forms. Ching [1] developed an excellent description of how points, lines, planes, and volumes become the basic building blocks for form development. In addition to the basic geometric properties of each, many examples of existing architecture are shown. This discussion gave the class a context in which to work. We also covered basic computing concepts and programming language constructs.

Figure 1: *Initial circular shape*   **Figure 2:** *3D extrusion*

1: The students were given an initial form to investigate, a circle. The program accepts a center point and radius for the circle. The edges of the circle are created by a series of line segments (Figure 1).

Discussion: Input functions, looping constructs, geometry for creating a circle, and computation of the line segments. This program duplicates the circle command found in most CAD systems.

2: Add a thickness to the lines to extrude the circle edges into three-dimensions (Figure 2)

Discussion: How 2D lines can be represented as 3D planes, how 3D planes are specified, and how generally 2D shapes can be represented in 3D?

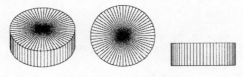

Figure 3: *Solid*   **Figure 4:** *Number of sides*

3: Add top and bottom surfaces to the edges to form a cylinder (Figure 3).

Discussion: Conversion of basic shapes to solids.

4: Add a parameter to modify the number of sides of the circle. Be able to create a series of polygons (Figure 4).

Discussion: How a circle is a multi-sided polygon.

Figure 5: *Change circle into an ellipse*   **Figure 6:** *Change center height*

5: Add option to change the values of the major and minor axis radius. Be able to turn a circle into an ellipse (Figure 5).

Discussion: The mathematical relationship between a circle and other related shapes.

6: Add a center vertex height in addition to the edge height. The center height can be less than or greater than the height of the edges (Figure 6).

Discussion: Beginning to investigate how the top surface can be articulated.

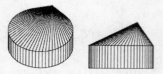

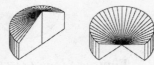

Figure 7: *Modify the center vertex offset*   **Figure 8:** *Modify the included angle*

7: Add an offset distance for the center vertex so that it can be moved is the x, y, as well as, controlled in height (Figure 7).

Discussion: Continue to investigate how the top surface can be modified.

8: Modify the included angle of the circle to be able to sweep the profile through an angle less than 360 degrees (Figure 8).

Discussion: How to control the extent of the shape. How to handle the start and end of the shape.

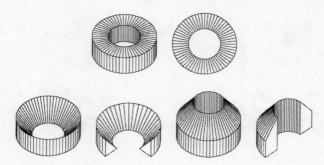

Figure 9: *Modify the inside radius*

9: Add a parameter to define an inside radius to give the circular shape an opening at the center (Figure 9).

Discussion: What are the variations in form when height and radius are considered?

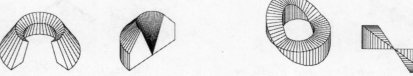

Figure 10: *Modify upper and lower radius*   **Figure 11**: *Height generated by sine function*

10: Articulate the upper and lower surfaces by giving each an individual inside and outside radius (Figure 10).

Discussion: Begin to separate the different parts of the form and study how each could be further controlled.

11: Vary the edge height by multiplying the edge height by the sine of the circular angle (Figure 11). Assume that the angle covers the same included angle as a full circle, 360 degrees.

Discussion: What type of variations can be applied to edge height? Other height functions included stepping the top surface, constant ramping from 0 to the maximum height, or ramping by increasing to a mid-point then decreasing. Use of the sine curve seemed the most promising and interesting.

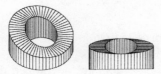

Figure 12: *Separate height for the sine function*

12: The problem with the sine function is that it generates zero and negative edge heights as seen in Figure 11. An additional height parameter is added to give the curve a positive offset, so that the edge always has a minimum height (Figure 12).

Discussion: What is the natural effect of the function selected and how can it be rectified?

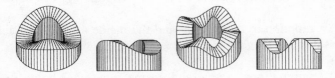

Figure 13: *Modify the angle for the sine function*

13: Add a parameter to vary the angle through which the sine function passes, which can now be different from the included angle of the circle (Figure 13). The circle will normally be drawn through 360 degrees; the top function could have any start and end angle value including multiples of 360 degrees.

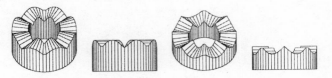

Figure 14: *Modify the sine function value*

14: Another method to modify the sine function is to compute its absolute value or its negative absolute value (Figure 14).

Discussion: How can basic functions like sine and cosine be modified to produce a variety of curves.

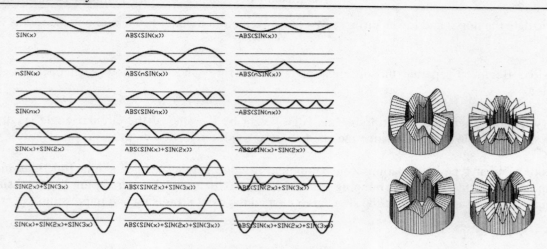

Figure 15: *Other curve functions*

15: In addition to the simple sine function, others are developed that include factors applied to it (Figure 15). In addition to these, an identical series was developed for the cosine function.

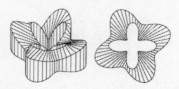

Figure 16: *Curve function on edge height*

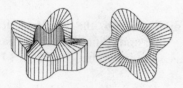

Figure 17: *Curve function on outside edge*

16: Apply a different curve function to the outside and the inside top edge. Each curve also has its own total angle, which it passes through (Figure 16).

Discussion: Given the 18 combinations of sine and cosine functions shown in Figure 15, what are the possible variations? How do we investigate each?

17: Apply a curve function to the outside edge as it extends around the circular form (Figure 17). Use the circular edges not as the edge of the form but as an axis to apply a curve function to.

Discussion: How the actual form evolved into a spine for further articulation development.

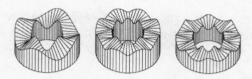

Figure 18: *Curve function on inside edge*

18: Apply a curve function to the inside edge as it extends around the circular form (Figure 18). Same as the step above using the inside edge as an axis.

Discussion: Figure 19 demonstrates some variations based on the developed parameters.

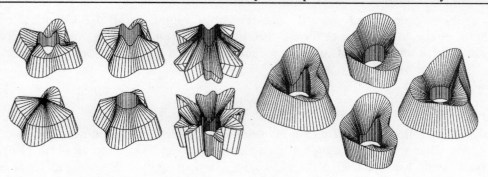

Figure 19: *Variations with all parameters*

3. Extending the basic concept

Once the students completed the previous set of exercises, the same concepts were applied to a set of more interesting curves. Each student selected a curve and produced a series of variations based on the incorporation of the parameters previously covered. The curves that were selected included the following: Bicorn, Piriform, one-half on an Eight Curve, one-half of a Leminscate of Bernoulli, Cardoid, Nephroid, Epicycloid of 3 and 4 cusps, Deltoid, Astroid, Lame Curve, and a Hippopede. These were selected for their continued use of simple combinations of sine and cosine functions and their potential to generate forms that could suggest architectural elements. Lawerence [5] was an excellent reference to develop these curves.

In addition to still renderings and drawings, one of the students created an animation by developing a series of related designs and then morphing them from one to the other. This visualization technique allows for the development of the in-between forms and gives the student more ideas on the variations to attempt. It also suggests that architectural forms could actually change in time – an interesting idea.

4. Conclusion

The process of developing these forms was the most important aspect of these exercises. The step-by-step procedure, the evaluation of each as encountered, and the trial and error required were the basic concepts discussed. The students were amazed that they could generate such forms and control their variations so easily. As we progressed through the exercises the students began to imagine what some of these variations might be. The expectations readily enabled the student to think about how to approach each succeeding modification. The strict mathematical basis for these designs focused the students to investigate the possibilities in an orderly fashion.

The discussions at the end of the course clearly indicated that the students now consider the development of programs as their own personal expression of an idea and that CAD systems could be used to investigate ideas and not only document decisions already made. They began to understand the feedback their rules created and how it could be used to clarify concepts. Many also better understood the mathematical basis for many simple forms and were more comfortable about progressing to other related areas. The ability to establish fixed and variable parameters within the model enabled the students to focus on a single design decision. Being able to duplicate results, time-after-time, also allowed for changes to be made at an earlier previously created decision point.

The greatest problem encountered was the development of values for all the individual parameters to really be able to see the entire range of form possibilities. Since enumeration was not possible, changing individual values did at times generate unexpected results – exactly what we were looking for.

5. Acknowledgments

Special thanks to the following students who developed examples and encouraged me to develop these ideas: Peng-Chien Chang, Vasavi Duvvur, Ruben Gonzalez, Ying-Chun Hsu, Hsi-Hao Hsueh, Ekkachai Mahaek, Priyadarshini Naik, Faisal Navieed, Umnt Ongoren, Alfred Sanders, and Fareeda Zayyad. Additional thanks to Ying-Chun Hsu for collecting and organizing the student work and preparing the final renderings.

Additional thanks are extended to the reviewers for their helpful comments.

References

[1] Ching, Francis. *Architecture: Form, Space & Order*. Van Nostrand Reinhold Company, 1979

[2] Krause, Jeffrey. "Agent Based Architecture". Association of Computer-Aided Design. in Architecture Conference Proceedings, 1997

[3] Krawczyk, Robert. "Hilbert's Building Blocks". Mathematics & Design Conference Proceedings, 1998

[4] Krawczyk, Robert. "Programs as Pencils: Investigating Form Generation". Association of Computer-Aided Design. in Architecture Conference Proceedings, 1997

[5] Lawrence, J. Dennis. *A Catalog of Special Plane Curves*. Dover Publications, 1972

[6] Maeda, John. *Design by Numbers*. The MIT Press, 1999

[7] Mitchell, William, Liggett, Robin, Kvan, Thomas. *The Art of Computer Graphics Programming*. Van Nostrand Reinhold Company, 1987

[8] Sequin, Carlo. "Art Math, and Computers: New Ways of Creating Pleasing Shapes". Bridges Mathematical Connection in Art, Music, and Science Conference Proceedings, 1998

[9] Sequin, Carlo. "Analogies from 2D to 3D, Exercises in Disciplined Creativity". Bridges Mathematical Connection in Art, Music, and Science Conference Proceedings, 1999

[10] Yessios, Chris. "A Fractal Studio". Association of Computer-Aided Design. in Architecture Conference Proceedings, 1987

Spiral Tilings

Paul Gailiunas
25 Hedley Terrace, Gosforth
Newcastle, NE3 1DP, England
paulg@argonet.co.uk

Abstract

In *Tilings and Patterns* [1] Grünbaum and Shephard comment that there is extremely little literature on the subject of spiral tilings. They survey what is known, and leave some unanswered questions, in particular whether there is a tile that will produce an r-armed spiral tiling for any *odd* value of $r \geqslant 5$. A new method of generating tiles from regular polygons is developed here, that allows the construction of zig-zag spirals with any desired number of arms. Spiral tilings of a type already known are considered and new tilings, related to them, are described. Further new tilings are constructed by considering some special cases.

Zig-zag Spiral Tilings

Overlapping copies of any regular n-gon will produce a shape that will tile the plane (without gaps) in several different ways, one of which is a one-armed spiral, along with various two-armed spirals [2]. If n is odd it is difficult to get much further, but if n is even then the tile that is produced can be cut into a series of rhombuses (fig.1), ensuring that each cut has the same length as a side of the polygon. Any one cut divides the original tile into two new ones, both of which will tile the plane in a variety of ways, in certain circumstances including spirals. In an n-gon it is easily verified that the angles of the rhombuses are an integer times $1/n$ turns (where 1 turn = 360°). In fact the rth rhombus has an angle of r/n turns, so it is possible to choose a cut that will produce a tile having one angle that is $1/m$ turns, if $m = n/r$.

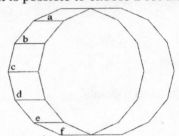

Figure 1: Overlapping copies of a 14-gon will produce a shape that will tile the plane. Tiles can be produced by cutting along any of the lines indicated. Angle a = 1/14 turn, b = 2/14 = 1/7, and so on.

Figure 2: A spiral begins when tiles are fitted around the convex edge of another tile (part of the n-gon). This can be done in two ways, although the new edge that is produced is the same It has a zig-zag profile.

By choosing a tile with an angle of $1/m$ a spiral tiling with m arms can be produced. Fitting more tiles makes the zig-zags bigger.

Figure 3:   A five-armed spiral from a tile based on a decagon. It has ten points.

Figure 4:   A five-armed spiral based on a tile from a 20-gon has twenty points. In general any zig-zag spiral based on a tile from an n-gon has n points, since the central tiles (m of them) have r edges on the convex side, each of which produces a step in the zig-zag, and $m \times r = n$.

A different *m*-armed spiral can be produced by making the cut at the (*r*-1)th rhombus and placing the tile against its mirror image, rotated by a half-turn, since the apex corresponds with angle a in figure 1, having an angle of 1/*n* turn, and the combined angle is again the *m*th part of a turn.

Figure 5: The same tile from a 20-gon can be used to make a four- and a five-armed spiral, since
1/5 + 1/20 = 1/4

The alternative way of tiling the spiral arms, indicated in the fig.2 , will only work in two cases, a three-armed spiral with a tile based on a dodecagon (fig.6) and a four armed spiral with a tile from an octagon (fig.7) . The reason is seen most easily in the three-armed example.

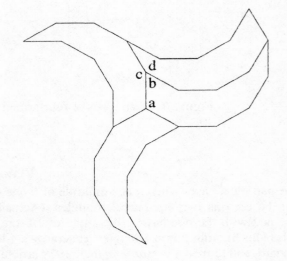

$$a = \frac{1}{m} \; turns, \; b = (\frac{1}{2} - \frac{1}{m}) \; turns$$

$$c = d = (\frac{1}{2} - \frac{1}{n}) \; turns$$

$$\frac{1}{2} - \frac{1}{m} + 2(\frac{1}{2} - \frac{1}{n}) = 1$$

$$\frac{1}{m} + \frac{2}{n} = \frac{1}{2}$$

There are four solutions, *m* = 3, *n* = 12 (fig.6); *m* = 4, *n* = 8 (fig.7); *m* = 6, *n* = 6; *m* = 10, *n* = 5, but the third is the degenerate case of a tiling of triangles, and the fourth cannot be realised, since *m* must divide *n*.

Figure 6: Two types of three-armed spiral using a tile from the dodecagon.

Figure 7: Two types of four-armed spiral using a tile from the octagon.

Versatiles

In the particular cases when *n* is a multiple of 6 one of the rhombuses will have an angle of 1/6 turn, and it can be seen as two equilateral triangles. (Actually there are two such rhombuses, but we are only concerned with the one having the angle of 1/6 on the convex side of the *n*-gon.) Making the cut that divides this rhombus into triangles generates a tile (fig.8) of the type discussed by Grünbaum and Shephard, and termed a *versatile* in their 1979 article [3].

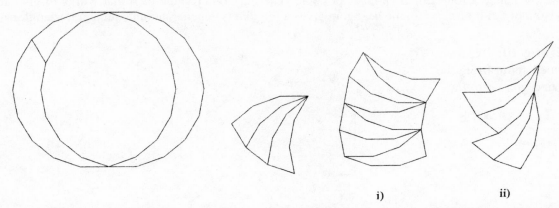

i)   ii)

Figure 8: *Versatiles* can fit together along the concave side of the *n*-gon, and along the convex side in two ways. Copies of the tile will fit into the gaps in ii) producing a profile that is the same as i). It is clear that spirals made from *versatiles* do not have zig-zags.

Grünbaum and Shephard showed that such tiles will always produce one-, two-, three- and six- armed spirals. It is not clear whether they realised that all except the three-armed case can be constructed using either tiling pattern i) or ii). They certainly never mention it, and they use type ii) in only one, a very attractive two-armed spiral (which is used on the dust-wrapper of their book).

Figure 9: A six-armed spiral having all arms of type ii), using a tile derived from a 30-gon. The spirals can been seen as starting from six tiles fitted around a 30-gon.

The Dodecagon Versatile

In the special case when *n* = 12, the dodecagon, there are further possibilities. Like any *versatile* it will produce several two-armed, a three-armed and the two types of six-armed spiral. Rice and Schattschneider [4] have also described some other ways that it will tile the plane. This particular *versatile* also generates a four-armed and an eight-armed spiral. The four-armed spiral (fig.10) is possible because the base angles of a *versatile* are 1/6 turn (making the six-armed spirals possible) and (1/3-1/*n*) turn (making the three-armed spiral possible, by combining a tile with its mirror image). When *n* = 12,

$(1/3-1/n) = 1/4$, hence a four-armed spiral is possible. The only other value of n that works to give an integer division of a full turn, $n = 6$, the hexagon, gives a tile that is degenerate (an equilateral triangle).

Figure 10: A four-armed spiral using the dodecagon *versatile*.

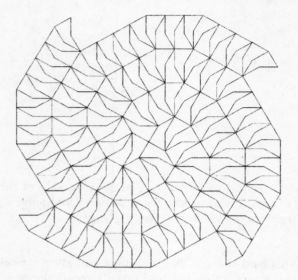

The other possibility to consider is combining the apex (of $1/n$ turns) with the 1/6 angle. The solutions of $1/n + 1/6 = 1/m$ must have $m < 6$, and $m = 3$ is the degenerate case again. The case $m = 4$ has $n = 12$, and a tiling (having eight arms) is possible (fig.11), but the tiles produced when $m = 5$ do not fit together in the same way, and there is no analogous tiling based on a 30-gon.

Figure 11: This tiling shows particularly clearly that the spiral arms, of both forms, are directed, since half are in one sense, half the other.

Rice and Schattschneider published this tiling (fig.11) with arms of type i), and again it is not clear if they were aware of the type ii) alternative. They also mention the existence of a twelve-armed spiral using this tile. There are many variants of this tiling (fig.12 shows one of them), since there are several ways to construct the central dodecagon. The next layer, which enlarges the dodecagon to twice its size, can be constructed in many different ways, and the spiral arms themselves can be of either type i) or ii).

Figure 12: One of the many variants of a twelve-armed spiral using the dodecagon *versatile*.

V-Shaped Tiles

The arms of this spiral tiling (fig.12) can be seen as originating at the mid-points of the sides of the large dodecagon, with tiles fitting around its vertices. Clearly this is only possible if the tile has no more than two concave sides, and is more or less v-shaped. The only *versatile* with this characteristic is derived from the dodecagon, but such tiles can be formed from any polygon. The one from the hexagon is big enough this time to produce tilings that are recognisable as spirals. Spiral tilings made with this tile (fig.13) are probably the simplest that have more than two arms, and it is surprising that they are not mentioned in the existing literature on the subject.

Figure 13: The six-armed spiral follows the typical construction, being formed around a hexagon of side 2. The three armed spiral is possible because of the special properties of tiles based on the hexagon.

It is always possible to cover an n-gon of double size, using a tile derived from the n-gon, when $n = 2(2r + 1)$, but only the decagon provides a v-shaped tile that does this. It generates a ten-armed spiral (fig.14), and of course in this case the arms are zig-zags.

Figure 14: A ten-armed zig-zag spiral around a decagon of side 2.

Conclusion

Tiles that are derived from regular polygons provide the most obvious possibilities for covering the plane in ways that that appear as spirals. Zig-zag spirals can be produced with any desired number of arms by following general methods, and *versatiles* provide smooth spirals, although only a limited number of arms is possible. Although many special cases have been considered, it is by no means certain that there are no more left to be found.

Also the possibility of finding spiral tilings using entirely different types of tile, not based on regular polygons, has not been eliminated. Indeed some examples are discussed in [1], and, although further development along these lines seems unlikely, a different approach might produce more fruitful possibilities.

References

[1] Grünbaum B. and Shephard G.C.,*Tilings and Patterns,* W.H.Freeman and Company, 1987.

[2] Hatch G., "Tessellations with Equilateral Reflex Polygons", *Mathematics Teaching*, No.84, Sept.1978, p.32.

[3] Grünbaum B. and Shephard G.C., "Spiral Tilings and Versatiles", *Mathematics Teaching*, No.88, Sept.1979, pp.50-1.

[4] Rice M. and Schattschneider D., "The Incredible Pentagonal Versatile", *Mathematics Teaching*, No.93, Dec.1980, pp.52–3.

BRIDGES
Mathematical Connections
in Art, Music, and Science

Musical Composition as Applied Mathematics:
Set Theory and Probability in Iannis Xenakis's *Herma*

Ronald Squibbs
Georgia State University
School of Music
P.O. Box 4097
Atlanta, GA 30302-4097
E-mail: musrjs@langate.gsu.edu

Abstract

Originally trained as an architect and engineer, Greek composer Iannis Xenakis has forged a distinctive musical style through the application of various branches of mathematics to his compositional process. Early in his career he focused on topics in probability. After his initial concentration on mathematical models of indeterminacy, he began to incorporate more deterministic methods of organization into his compositions, including set theory and group theory. *Herma* is a composition for solo piano written in 1960-61. It is representative of the coexistence of determinacy and indeterminacy in the structure of his works, and thus serves as a good introduction to the music of this remarkable composer.

1. Xenakis and the Development of Stochastic Music

1.1 Xenakis the Composer. The modern period is an era in which progressive composers have been both lauded and repudiated for their use of abstract compositional procedures. Even among his contemporaries Iannis Xenakis stands out for the direct and thorough manner in which he has applied procedures from various branches of mathematics to the composition of his music. His most important theoretical treatise, *Formalized Music: Thought and Mathematics in Music*, presents a variety of mathematical concepts along with their applications to his compositions over roughly a forty-year period [1]. In the 1950s his work was mainly concerned with the application of probability theory to musical composition, resulting in a style he dubbed "stochastic music." Beginning in the mid-1960s his work focused on formalized methods of deterministic (i.e., non-stochastic) composition including a generalized theory of musical scales. His subsequent work has been less formalized and more eclectic, including the use of freehand graphic designs in the formation of musical structures. In the 1990s, however, he has demonstrated a desire to achieve a pure form of stochastic music in the electronic domain through the creation of works in which sound waves themselves are subjected to stochastic transformations.

Xenakis came rather late in life to composition. He was born in 1922 in Romania to parents of Greek origin and received a conventional Greek education before pursuing studies in mathematics, engineering and architecture at Athens Polytechnic. His education was interrupted by the political turmoil of World War II. He involved himself in the Greek Resistance movement, was incarcerated several times and was eventually blacklisted. His father arranged for him to escape from Greece in 1947. He settled in France, where he found work in the architectural firm of Le Corbusier. Soon after his arrival in France Xenakis became interested in composition. Le Corbusier was knowledgeable in music, and through him Xenakis made contacts in the world of the French post-war avant-garde. Chief among these contacts was Olivier Messiaen, France's leading composer at the time. Messiaen accepted Xenakis as a pupil, despite his unconventional background and minimal musical training. During the early 1950s he had the opportunity to study alongside Messiaen's most famous pupils, Pierre Boulez and Karlheinz Stockhausen. Now a French citizen, Xenakis is regarded, along with Boulez, as one of France's most important senior composers.

According to his principal biographer, Xenakis turned to music partly as a form of therapy, a way of coming to terms with the traumatic events he had experienced in Greece during and after the war [2]. The expressive urge that underlies his music coexists in an uneasy alliance with his desire to create robust and enduring structures, a legacy no doubt of his architectural training. Vying with one another in Xenakis's music is the Greece of philosophers and mathematicians and the Greece of horrifying tragedies, spectacular battles, and persistent political conflicts. The result is a music that is expressive, yet shorn of all sentimentality: a music that exalts the life of the mind while stimulating the senses with uncompromising, even confrontational, intensity.

1.2 The Development of Stochastic Music. The musical world into which Xenakis entered in post-war France was a world in reconstruction, like much of European society at the time. Many composers felt a need to resume the progressive trends in the arts that had been threatened by fascist intrusions into the cultural life of Western Europe. Young composers wanted to start over again from ground zero, hoping to leave behind the retrogressive conventions that had been encroaching on modernism since the 1920s and 1930s. Stravinsky's neoclassicism was seen as a decline and a betrayal of the progressive cause, while the work of Schoenberg and Webern—which had been banned by the Nazis—seemed to point the way forward. Young composers, including Boulez and Stockhausen, began to generalize the serial organization of pitches found in Schoenberg and Webern to other aspects of musical structure, including rhythm, form, and even instrumentation. Messiaen himself did some work along these lines as well. Paradoxically, the stricter the organization of the various elements of musical sound was, the more likely the music was to be perceived as a manifestation of disorder. Xenakis saw a contradiction here between ends and means and sought a logical solution: the deliberate composition of disorder through the application of probability theory. He dubbed his new method "stochastic music."

Herma, a seven-minute composition for piano solo written in 1960-61, is the most concentrated example of Xenakis's stochastic music. Unlike Xenakis's first stochastic compositions, which were written for orchestra, chamber ensembles or electroacoustic media, *Herma* is his first stochastic work for a solo instrument. *Herma* is also a transitional work, for while the sections of the piece consist of "clouds" of pitches that are distributed according to specific randomizing functions, the selection of pitches follows a deterministic process based on set theory. *Herma* thus combines indeterminacy—manifested as stochastic processes at the local level—with determinacy—manifested as a sequence of set-theoretic operations at a global structural level. Probability also plays a role in the work's global temporal structure, which is discussed in the concluding section of this paper.

2. Set Theory as a Basis for Large-Scale Pitch Structure

The large-scale pitch structure of *Herma* is organized according to the principles of set theory. Xenakis begins by defining the 88 pitches available on the standard piano keyboard as the universal set. Using slightly non-standard terminology, he refers to this as the referential set, R. Three smaller sets—A, B, and C—were then constructed from the elements in R. Each of these sets contains about one-third of the pitches in R. The contents of sets A, B and C overlap slightly, ensuring that there will be no trivial, i.e. empty, intersections among the three sets. The standard set-theoretic operations of union, intersection and complementation form the basis for the selection of pitches in the sections of *Herma*.

Xenakis lends structure to the set-theoretic procedures by defining two sequences of operations that lead to an identical final set, F, whose contents are represented by the Venn diagram in Figure 1. Two distinct but equivalent formulas define the contents of set F:

$$ABC + \overline{A}BC + A\overline{BC} + \overline{AB}\overline{C} \qquad \text{and} \qquad (AB + \overline{AB})C + \overline{(AB + \overline{AB})\overline{C}}$$

The formula on the left is the result of a sequence of seventeen operations performed on the primary sets A, B and C. The more concise formula on the right is the result of a sequence of ten operations performed on the primary sets. Xenakis interprets the sequences of operations as competing paths toward a common goal, thus introducing an element of structural drama into the presentation of pitches as the work unfolds in time.

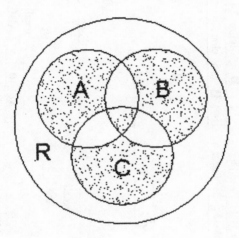

Figure 1: *Venn diagram of set F*

In his explanation of the compositional process used in *Herma*, Xenakis uses flow charts to specify the sequences of set-theoretic operations that result in set F [3]. In the finished composition, however, the order of operations in both sequences has been rearranged, some operations are omitted, and some operations appear that do not belong to either sequence. There is no mathematical justification for these modifications, but from a musical perspective the shape of the finished composition makes a good deal of sense. Figure 2 shows a structural plan for *Herma*, summarizing the labeling of sets and other information found in the musical score [4]. The work begins with an introductory section that presents the pitches in R, the referential set. This is followed by the presentation of the primary sets, each one followed directly by its complement. The presentation of the sets up to this point—at about 4', approximately three-fifths of the way through the piece—follows the musical logic of thematic presentation, in which the basic materials of a work are presented in a clearly audible sequence. The remainder of *Herma* features a musical "development" of the primary sets. Thematic development in music consists of procedures that result in the transformation of themes. In traditional music these procedures include the fragmentation of themes into smaller musical units called motives, the combination of motives from different themes, and a general increase in harmonic and rhythmic activity. In *Herma* the process of development consists of the presentation of intersections involving the primary sets and their complements. These subsets are presented in sections that are generally shorter than those in which the primary sets and their complements were first introduced. The latter part of the work also features the repetition (or "recall," to use Xenakis's term) of several of the subsets following their initial appearances.

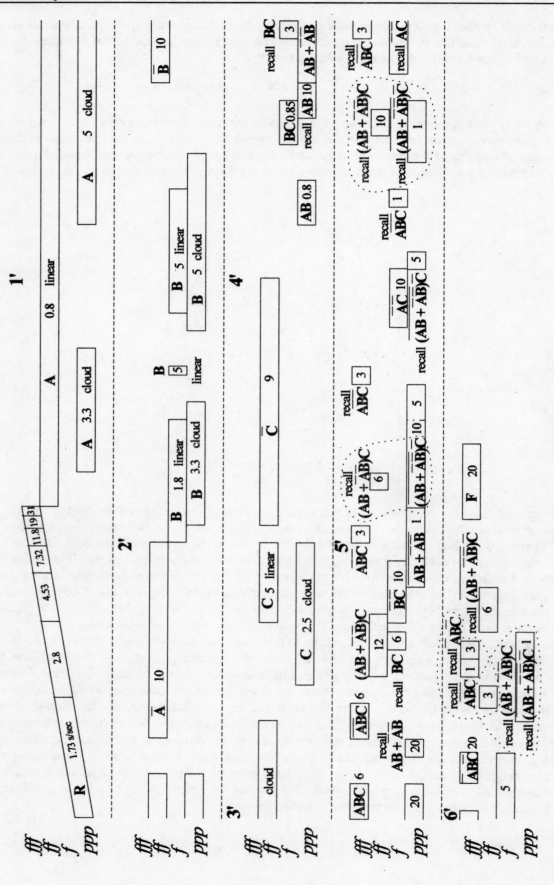

Figure 2: Structural plan for Herma

Variety is maintained on the musical surface through changes in the pitch contents of the sets and also through changes in the other characteristics that are represented in Figure 2. One of these characteristics is the loudness at which the pitches in the sets are to be played. Four levels of loudness are used in *Herma*: *pianisissimo* (*ppp*), extremely soft; *forte* (*f*), loud; *fortissimo* (*ff*), very loud; and *fortisissimo* (*fff*), extremely loud. The introduction begins at the lowest of these levels and moves progressively to the highest. The rest of the sections remain at a constant level of loudness throughout. Another characteristic shown in the chart is the density of the sections. Xenakis defines density in his stochastic as music as the average number of sounds that occur per second of music. The density increases within each consecutive subsection of the introduction while the lengths of the subsections become progressively shorter. The combination of all of these factors results in an increase in musical activity that reaches a fever pitch just before the presentation of set A. In the first part of the work most of the sections bear the descriptive terms "linear" and "cloud." In the three overlapping sections in which set A is presented, these terms appear to describe the effect of the relative density of each section: the least dense section (at 0.8 sound/sec) sounds like a linear succession of notes, while the denser sections give the impression of clouds of notes. The cloud effect is enhanced by the use of the damper pedal in these sections, which allows the resonance of the notes to continue after the keys are released. The descriptive terms do not correlate as nicely with the density or the pedaling effects in the remaining sections, however, so the reason for their inclusion in the score is not entirely clear.

The frequent overlapping of sets is a third characteristic that lends considerable variety to the musical surface. Whenever sets overlap the effect on the density of the musical surface is cumulative. Overlap also causes pitches from sets at different levels of loudness to be brought into close proximity with one another. This serves to differentiate between the contents of the sets, while the combination of different levels of loudness results in a colorful and varied musical surface. In the latter portion of the piece, the overlapping of sets allows for the gradual assemblage of the elements that combine to form set F. The contents of set F are foreshadowed several times prior to its definitive presentation at the end of the work, where it is given at the highest level of loudness and at a very high density of 20 sounds/sec. The contents of F are first foreshadowed just after 5 minutes and again just before and after 6 minutes. Each of these foreshadowings is circled with a dotted line in Figure 2. This is another example of Xenakis's adaptation of set-theoretic logic to musical purposes, for the foreshadowing of important themes is a frequent technique in classical music, especially in music written since the mid-19th century.

3. Stochastic Composition at the Local Level

The previous section presented an overview of the global pitch structure of *Herma* with respect to the three primary sets and their transformations over the course of the work. I turn my attention now to the stochastic aspects of the work's structure. This occurs at two levels: first, the composition of the pitch and rhythmic structure of the sections (represented by boxes in Figure 2); and second, the composition of the lengths of the sections and of their placement within the total time span of the work. The first of these levels will be dealt with here. The second level will be discussed in the following section.

The fundamental principle underlying Xenakis's stochastic music is the application of probability theory to the process of musical composition. The unit of structure to which probability theory is applied in this music is the interval. The interval is a musical term for the distance between two sounds with reference to some characteristic of those sounds. One may speak, for example, of an interval between two pitches, between the times at which two sounds begin, or between two degrees of loudness. The sounds of stochastic music result from the generation of pseudo-random interval successions for two or more characteristics of sound. The two most basic characteristics are pitch and the time at which a sound begins, known as its "attack time." Intervals for each characteristic are calculated independently and the endpoints of the intervals are brought together to determine the pitches and attack times of individual sounds. Each sound may therefore be thought of as a point in a multidimensional space whose coordinates are determined by pseudo-random interval successions operating independently in each dimension.

Xenakis has chosen which probability distributions to apply to which characteristics of sound by determining which distributions produce the most appropriate musical effect [5]. For the composition of the intervals between attack times he has invariably used the exponential distribution. In probability theory this distribution is used to model the waiting times between relatively rare phenomena such as automobile accidents in a city or the appearance of meteors in the night sky [6]. Xenakis has used several different distributions, however, to determine the intervals between pitches in a set. The simplest of these is the linear distribution. Since the linear distribution contains only positive values, it is useful for determining the size but not the direction of the pitch intervals. Another distribution, such as the uniform distribution, must be used in conjunction with it in order to determine whether the intervals move up or down. A situation can be set up in which there is a 50/50 chance that the interval will be given a positive sign (i.e., it will go up), or a negative sign (i.e., it will go down).

The general characteristics of the exponential, linear and uniform distributions are shown in Figure 3. All three are continuous distributions, meaning that intervals of any size may occur within the range of values defined for each distribution. Sample graphs representing the probability density functions of each distribution are shown at the top of the figure. The graph for the exponential distribution shows that, as the size of an interval x increases linearly, the probability of its occurrence decreases exponentially. The parameter λ represents the average interval of time between events in the exponential distribution. The graph of the linear distribution shows that, as the size of an interval x increases linearly, the probability of its occurrence decreases linearly. The parameter σ represents the upper limit of the interval size within the linear distribution. Finally, the graph of the uniform distribution shows that, as the size of an interval x increases linearly, the probability of its occurrence remains constant.

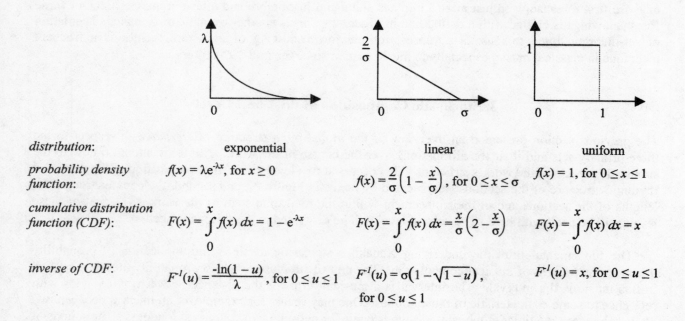

distribution:	exponential	linear	uniform
probability density function:	$f(x) = \lambda e^{-\lambda x}$, for $x \geq 0$	$f(x) = \dfrac{2}{\sigma}\left(1 - \dfrac{x}{\sigma}\right)$, for $0 \leq x \leq \sigma$	$f(x) = 1$, for $0 \leq x \leq 1$
cumulative distribution function (CDF):	$F(x) = \displaystyle\int_0^x f(x)\,dx = 1 - e^{-\lambda x}$	$F(x) = \displaystyle\int_0^x f(x)\,dx = \dfrac{x}{\sigma}\left(2 - \dfrac{x}{\sigma}\right)$	$F(x) = \displaystyle\int_0^x f(x)\,dx = x$
inverse of CDF:	$F^{-1}(u) = \dfrac{-\ln(1-u)}{\lambda}$, for $0 \leq u \leq 1$	$F^{-1}(u) = \sigma\left(1 - \sqrt{1-u}\right)$, for $0 \leq u \leq 1$	$F^{-1}(u) = x$, for $0 \leq u \leq 1$

Figure 3: *Properties of probability distributions*

The cumulative distribution function (CDF), $F(x)$, is found directly below the probability density function, $f(x)$, for each distribution in Figure 3. $F(x)$ is the integral of $f(x)$ from 0 to x within the range of possible values for x. As the integral of $f(x)$, $F(x)$ may be represented as the area beneath the curve of $f(x)$. For a given distribution, the CDF defines the probability of finding an interval whose size is somewhere between 0 and x. As x approaches its upper limit, the probability of finding an interval whose size is between 0 and x approaches 1. The probability thus varies from 0—absolutely no probability of finding such an interval—to 1—absolute certainty of finding such an interval—as the value of x increases from 0 to its upper limit.

Two methods may be used to produce successions of intervals whose statistical properties approximate the properties of specific probability distributions. In the first method, one begins by dividing the range of possible values for x into equally sized segments. In this way it becomes possible to define the probability of the occurrence of intervals whose sizes fall within specific ranges. By choosing an appropriate number of intervals from within each range and placing them end to end, one may approximate the effect of a random succession of intervals. In addition, the statistical characteristics of this interval succession will resemble those of the probability distribution with which one began. The interval succession may also be subjected to statistical tests to measure the degree of its resemblance to true randomness. Composition by this method is a slow and laborious procedure, but it was the only method available to Xenakis when he composed his first stochastic works in the early-to-mid 1950s.

The second method involves automated computation. In this method, values are drawn from a computer's pseudo-random number generator and are fed through mathematical formulas derived from probability distributions. This process results in interval successions whose statistical characteristics resemble those of the probability distributions from which the formulas were derived. This approach depends on the fact that the CDF of a probability distribution associates each interval within the range of 0 to x with a unique probability whose value lies between 0 and 1. This association works in the opposite direction also, so that one may begin with a specific probability value in order to find the unique interval size with which it is paired in the distribution's CDF. When the computer's pseudo-random number generator is used to determine the probability values, the result is a pseudo-random interval succession whose statistical properties approximate those of the desired probability distribution. The formulas used to produce interval successions that approximate the statistical properties of the exponential, linear and uniform distributions are shown in the bottom row of the table in Figure 3. The variable u in each formula represents a pseudo-random probability value between 0 and 1. For every value of u there is a corresponding value of x which is found by feeding u through the inverse of the CDF of the probability distribution, i.e. $F^{-1}(u)$.

Xenakis developed and executed a computer program in Fortran for the composition of stochastic music between 1956 and 1962. The works composed with that program were first performed beginning in 1962. Given that *Herma* was composed in 1960-61, it is possible that at least part of it may have been composed with the aid of the computer program, but it is also possible that the entire work may have been written using the original method of calculation by hand.

4. The Large-Scale Temporal Structure of *Herma*

The two methods of stochastic music outlined above may at first seem abstract, perhaps even obscure. Once one has gotten hold of the necessary formulas and understands how to implement them musically, however, both methods are actually quite straightforward and easy to use. In the descriptions of the derivation of the pitch sets and the methods of stochastic composition, my presentation has followed Xenakis's own account of these procedures quite closely. I would now like to turn the investigation of *Herma* in a more speculative direction by using the principles of stochastic composition as the basis for an analysis of the work's large-scale temporal structure. Several authors have written about *Herma*, some devoting their attention mainly to its pitch structure, but no one has yet demonstrated an understanding of how its large-scale temporal structure results from the application of probability theory [7]. The tendency to dwell on the work's pitch structure is unfortunate, since the effect of the music has much more to do with its large-scale temporal structure than with the specific contents of its sets. The composer appears to have thought so as well, for the published score remains faithful to the structural plan in Figure 2 in terms of the lengths of the sections and the densities within the sections. There are, however, irreconcilable differences between the contents of the primary pitch sets—A, B, and C—and the contents of the sets derived from them.

The first thing that one is likely notice while observing the lengths of the sections in Figure 2 is that their arrangement does not appear to be completely random. While one very short section is included in the first part of the work (set B, ca. 2'30"), the majority of the short sections are found in the work's second part, beginning with set AB (after 4'). The separation of the long sets from the short ones appears to have been done by design, for the time at which set AB begins—4'07" (247")—is approximately the golden section of the work's total duration (6'44" [404"]): 247"/404" ≈ 0.611. (The golden section ≈ 0.618.) The golden section, of course, is a classic proportional division. It forms the basis of Le Corbusier's approach to architectural design, as explained in two books on the Modulor, a proportional system of measurement that he invented [8]. Xenakis's first published essay, in fact, is the appendix to the second volume on the Modulor. In this essay, Xenakis explains how Le Corbusier's ideas regarding the structuring of architectural space were applied to the temporal structure of his first orchestral work, *Metastaseis* (1953-54). In *Herma*, the use of the golden section creates a dynamic, asymmetrical balance between the introduction of the primary sets and their complements in the first portion of the work and the transformations of these sets in the second portion.

While the division of the work into two parts according to the golden section is clear, the reason for the different lengths of the work's individual sections may not be so apparent. In point of fact, the distribution of lengths among the sections closely resembles an exponential distribution. This can be demonstrated by first removing the sections from their immediate temporal context. From this perspective, the sections may be viewed solely in terms of their duration. The section durations may be sorted by length and their distribution within the work as a whole may be compared an ideal exponential distribution. The sum of the section durations in the work is 7'22" (442"), which is 38" longer than the work's performance duration. (This discrepancy results from the overlap of several sections during the course of the work. It should also be noted that the periods of silence intervening between the sections are not taken into account in this calculation.) There are 44 sections in all. Dividing the number of sections by the sum of their durations gives an average duration of 10 seconds per section. In terms of events per second, the average section duration of 10 seconds may be expressed as 0.1 sections per second. The actual distribution of section lengths, which constitutes the observed sample, may thus be compared with an exponential distribution in which $\lambda = 0.1$. A comparison of the observed relative frequencies and their probabilities according to the exponential distribution is shown in Figure 4a. As the graph in the figure demonstrates, the two distributions are very similar.

The intervals between the start times of the sections also approximate an exponential distribution. The intervals between start times are measured with respect to the work's actual temporal structure. When the interval between the start times of two sections is smaller than the length of one of the sections, the sections overlap. An example of this is the group of sections that that present the pitches belonging to set A (see Figure 2). When the interval between start times is larger than the length of the first of the two sections, a measured period of silence results. This occurs, for example, between the presentation of set A and the presentation of its complement. The number of sections is, once again, 44, but the performance duration of the work is 6'44" (404"), which is 38" shorter than the sum of the durations of the sections. The average rate of occurrence of the sections is 44/404" ≈ 0.11 sections per second. A comparison of the

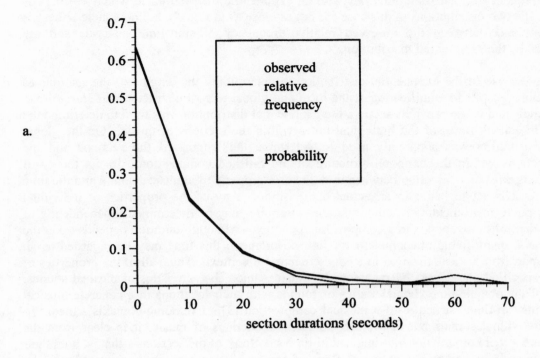

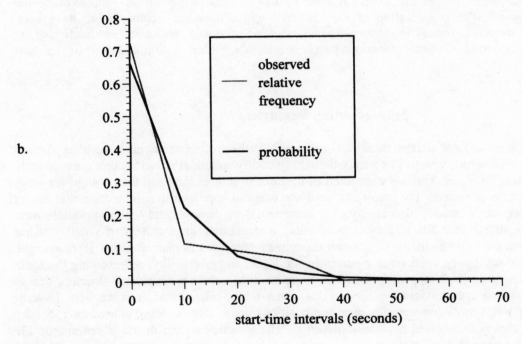

Figure 4: *Distribution of sections outside-time and in-time*

observed relative frequency of start-time intervals with an exponential distribution in which $\lambda = 0.11$ is shown in Fig. 4b. The two distributions in this case are not as close as in Fig. 4a, but nonetheless there is a noticeable resemblance between the observed relative frequency of start-time intervals and the probabilities defined by the exponential distribution.

Xenakis's apparent use of the exponential distribution in determining the lengths of the sections as well as their start times results in a unification of the local and global temporal structures in *Herma*. On the local level, as indicated in the previous section, the exponential distribution was used to determine the intervals between the attack times of the individual notes within the sections of music. On the global level, the same distribution was apparently used to determine the lengths of the sections and the arrangement of sections within the temporal structure of the finished composition. In his theory of musical time, Xenakis draws a distinction between duration as a property of a musical sound and the time at which that duration is set to begin in an actual composition. Any of the properties of individual sounds—including pitch, instrumentation, loudness, and duration—may be determined independently of the eventual placement of those sounds in a compositional context. Although duration depends upon the existence of time as a quantifiable phenomenon for its measurement, this time may be regarded as an abstract potential apart from its actualization in a concrete musical context. Thus, all of the properties of individual sounds specified above may be regarded as "outside-time" characteristics of musical sounds. The combination of the outside-time characteristics of sounds with their locations in a concrete musical context results in the "in-time" structure of a musical composition [3]. Extending Xenakis's theory to include not only individual sounds but also clearly delineated sections of music, it is clear from the illustrations in Figure 4 that both the outside-time and in-time structures of the sections—that is, both their lengths and their placement within the temporal flow of the finished composition—approximate the exponential distribution. The organization of the section lengths, however, seems not to have been randomized (as were the attack times of the individual notes within sections), but rather is indicative of the composer's will to place most of the long sections before the work's golden section and most of the short sections after it.

5. Concluding Remarks

Xenakis presents a special case of the thinking musician. Unlike other musicians, such as Arnold Schoenberg and Milton Babbitt, who used abstract thought to codify technical advances that grew directly out of previous musical tradition, Xenakis conceived of abstract structures first and then sought for ways to realize these structures in sound. The result is a bold and original approach to composition that favors explosive energy over subtle nuance, objectivity over sentimental expression, and logical principle over spontaneous improvisation. Xenakis is a builder of musical structures, an architect of volatile sound textures. His tools are the basic elements of abstract thought as expressed in mathematics. In *Herma* the specific tools used are set theory, representing deterministic logic, and probability, representing the logic of indeterminacy. Together these opposing forms of logic work to create a dialectical structure that is both dramatic and ferociously beautiful to those who can withstand the intensity of its expression. Even in this, one of the composer's most conceptually pure compositions, mathematical logic is used as a tool for the development of clearly articulated musical structures. The structure of the music is not simply the result of mathematical procedures. Rather, mathematics has been used in the service of the intended musical effect. Through the development of his unique brand of applied mathematics, Xenakis has managed to forge a distinctively original style and to infuse a renewed sense of vigor and energy into the musical avant-garde of the late twentieth century.

References

[1] Iannis Xenakis, *Formalized Music: Thought and Mathematics in Composition*, ed. S. Kanach. Stuyvesant, New York: Pendragon Press, 1992.

[2] Nouritza Matossian, *Xenakis*. London: Kahn and Averill, 1986.

[3] Xenakis, *Formalized Music*, chapter 6.

[4] Adapted from Xenakis, *Formalized Music*, Fig. VI-15, p. 177.

[5] Xenakis, *Formalized Music*, chapters 1 and 5.

[6] Douglas Kelly, *Introduction to Probability*. New York: Macmillan, 1994, p. 41.

[7] Rosemary La Grow Sward, *A Comparison of the Techniques of Stochastic and Serial Composition Based on a Study of the Theories and Selected Compositions of Iannis Xenakis and Milton Babbitt*. Ph.D. diss., Northwestern University. Ann Arbor, Michigan: UMI, 1981; Yayoi Uno, *The Roles of Compositional Aim, Syntax, and Design in the Assessment of Musical Styles: Analyses of Piano Music by Pierre Boulez, John Cage, Milton Babbitt, and Iannis Xenakis circa 1950*. Ph.D. diss., Eastman School of Music. Ann Arbor, Michigan: UMI, 1994; Eugene Montague, *The Limits of Logic: Structure and Aesthetics in Xenakis's* Herma. Master's thesis, University of Massachusetts at Amherst, 1995; Yayoi Uno and Roland Hübscher, *Temporal-Gestalt Segmentation: Polyphonic Extensions and Applications to Works by Boulez, Cage, Xenakis, Ligeti, and Babbitt*. Computers in Music Research, Vol. 5, pp. 1-37. 1995.

[8] Le Corbusier, *Modulor I and II*, tr. P. de Francia and A. Bostock. Cambridge, Massachusetts: Harvard University Press, 1980.

Number Series as an Expression Model

Elpida S. Tzafestas[*]
Intelligent Robotics and Automation Laboratory
Electrical and Computer Engineering Department
National Technical University of Athens
Zographou Campus, Athens 15773, GREECE.
brensham@softlab.ece.ntua.gr
http://www.softlab.ece.ntua.gr/~brensham

Abstract

This work is situated within the scope of the emerging complexity sciences and tries to bridge scientific and artistic intentions for understanding complexity by doing experiments. To this end, we propose the use of simple, "minimal" models coupled with expression functions that translate model results to visual elements or properties. Number series appear to be good candidates for this type of experiment, because they are simple, inflexible and infinite. Expression functions on the other hand may be arbitrary and subjective. Our case study is based on the well-known Fibonacci series and shows that, by constraining some aspect of the visual form, an expression function may translate the number series to a complex visual form. Various expression functions are investigated and their results exemplified by selected images. Several dimensions of future work are also briefly outlined.

1. Introduction

The major concern of the complexity sciences is how simple rules can give rise to complex phenomena and, inversely, how a complex phenomenon may be explained by or "encoded" in a simple rule [8][7]. On the other hand, artists have been traditionally concerned with expressing simple ideas or feelings with complex visual forms and, inversely, with understanding how complex realities can be represented with and express simple, abstract rules [6][9][10][3]. Thus, both scientists and artists often depart from the position that reality is in essence simple but possesses an infinite potential for expression and representation.

The goal of our research is to bridge the intentions of scientists and artists in understanding complexity by experimenting, rather than by using arbitrary complex models to create interesting, complex or aesthetically pleasing visual forms [1][2]. To this end, we start with "minimal" expression models and we try to explore and classify the range of visual forms that they may give rise to. While this is scientifically correct and intriguing, it is not artistically relevant unless we provide a means for artists to express themselves in unique ways. In scientific terms, we should therefore construct a framework that integrates the chosen *expression model* (the abstract function or model behind the scenes) together with an *expression function* (a translation to a visual form) that may be defined by an artist at will. This expression function will act directly on the structural elements of the visual forms and will translate the results taken with the chosen model to values for those structural elements.

Note that the term "visual form" is used in a general everyday sense that captures all features of structure, color, manner or style, etc. Those features or elements should be explicitly defined and

---

[*] Also with: Digital Art Laboratory, Athens School of Fine Arts, Peiraios 256, 18233 Agios Ioannis Rentis, GREECE.

represented in a manner with as little arbitrariness and subjectivity as possible. This is in contrast with the actual expression function from model to visual form, where every freedom is allowed and encouraged.

One minimal expression model is a **number series**, because it is simple (the whole series is usually represented with a single simple mathematical expression), inflexible (once started, it cannot be changed) and infinite (it never ends). The idea is that an arbitrary expression function can produce a wealth of interesting visual forms out of just such a series of numbers. What we seek are simple functions that can produce complex forms out of a number series. Of all available or imaginable number series, we chose the Fibonacci series, because :

- It is a series often encountered in nature and is said to possess aesthetic properties, while being mathematically intriguing as well [5].
- It has been already extensively used by at least one artist, the Italian Mario Merz [4].

Because such a series rises infinitely, it might at first glance seem uninteresting to a spectator to present consecutive values in a row (Mario Merz, "Fibonacci", 1975, and "Fountain", 1978). The next step in complexity has been to translate those numbers to a physical measure (Mario Merz, "Igloo Fibonacci", 1970, & "Fibonacci Drawing", 1977). But again, the measure (instead of the number) rises infinitely. What we would like is, on the one hand, to constrain the system so as to be able to use bounded resources (such as the area of a –finite– painting canvas) and on the other hand to produce infinite complexity despite constraints. We will see in the following sections that constraining the system allows us to do exactly this : produce potentially infinite complexity within limits. *Thus constraining ourselves allows us to become expressive.*

In a longer term perspective, we are seeking ways to express simple rules in aesthetically pleasing or innovative visual forms, so as to be abstract in essence but demonstrate expressive power and/or potential.

2. Expression 1 : Structural functions

The Fibonacci series was defined in the 13th century by Italian mathematician Leonardo Fibonacci as the series of numbers such that every number beyond the 2nd in the series is the sum of the last two : $F(n+2) = F(n+1) + F(n)$. Letting $F(1) = F(2) = 1$, we obtain the series

$$1,1,2,3,5,8,13,21,34,55,...$$

We obtain different series if we let $F(1)$ and $F(2)$ take other values, for instance we obtain the Lucas series if we let $F(1) = 1$ and $F(2) = 3$. The Fibonacci series possesses some interesting or even impressive properties, for example the ratio $F(n+1)/F(n)$ converges to the value 1.6180339... for n going to infinity.

We may bound the values given by the Fibonacci series within an interval [0,d], provided that we find a translation function $f(n):F(n) \rightarrow [0,d]$. This translation function is actually the *expression* function that expresses and represents a Fibonacci number in a different finite medium, in our case the bounded interval [0,d].

One such expression function is the *reverberation function* that translates $F(n)$ so that the goal point's distance from the start point equals $F(n)$. If the goal point falls outside the limits of [0,d], the function reverberates in the other direction, and it does so recursively in either direction (from the 0 or d end) until the goal point falls within [0,d]. The new start point is now the goal point and the expression process continues with the next number in the series $F(n+1)$. This process is illustrated in Figure 1.

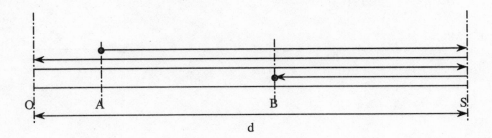

Figure 1: *Reverberation function : With A as start point, the next point is B, where (AS)+2d+(SB)=F(n), n being the current Fibonacci index.*

The reverberation function produces points on the interval [0,d] that very quickly appear to be randomly placed. Furthermore, often enough new points fall on already existing points, so that the process soon slows down in its progress. The process can be started from any start point on the interval [0,d] with any initial index within the Fibonacci series, for instance it could start from index n=7, that is with F(7)=13. A sample image is shown in Figure 3(top).

The reverberation function may also be applied to a 2-dimensional area of size [d_x,d_y], with the x- and y- values each following its own Fibonacci dynamics with its own parameters (start value and initial Fibonacci index). In this case, the process will produce points in 2D space that may be connected by line segments, and thus form a linear drawing (Figure 3 middle left). Note that if both x- and y- values start from the same value with the same initial Fibonacci index, the resulting line drawing will be a subset of the diagonal because x(n) will equal y(n), for every n.

The same reverberation function may be applied to the three color components (red, green and blue components), because their values are inherently limited (they usually range from 0 to 255). As is the case with x- and y- values of a point in space, the r-, g- and b- values will have a start value (from 0 to 255) and an initial Fibonacci index. Note that if all three r-, g- and b- values start from the same value with the same initial Fibonacci index, the resulting color will be a level of gray because r(n) will equal g(n) and b(n), for every n. Such colored drawings are shown in Figure 3 (middle right), as well as in Figures 4 to 7.

The above constitute an indication that the resulting images will be more interesting, albeit not information richer, if there is *diversity* in the values of similar parameters, for example in the initial Fibonacci indexes for the x- and y- values.

Finally, we can apply the reverberation functions as above to filled shapes instead of points or lines : in the 1D case, we will connect consecutive points with colored bands, instead of drawing colored points, while in the 2D case we will draw filled colored rectangles instead of drawing colored lines (those lines are diagonals of the corresponding rectangles). Colored bands (Figure 3 bottom) or rectangles (Figure 4) may fall on top of previously drawn ones, so that the ***drawing history*** may be hidden in the resulting image.

One can imagine many expression functions that will translate a Fibonacci number within a linear interval or a rectangular region. Another possibility, apart from the reverberation function, might be, for instance, the ***toroidal function***, which assumes that the interval or region is toroidal, i.e., one of its ends is next to the other, so that when a value exits from the right side it automatically re-enters from the left side. This process is illustrated in Figure 2, while Figure 5 presents some visual results produced by this function.

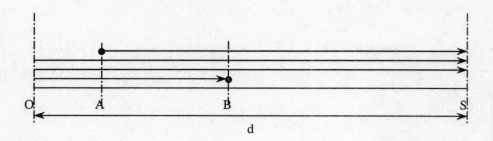

Figure 2: *Toroidal function : With A as start point, the next point is B, where (AS)+2d+(OB)=F(n), n being the current Fibonacci index.*

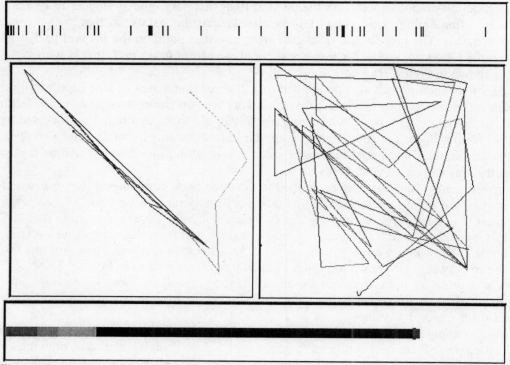

Figure 3: *Visual results of the reverberation function after n=25 or 50 steps. (top) Points in 1D space. For visualization ease, points are shown as short vertical lines. The process starts from the left end and proceeds rightward, reverberating at either end, as necessary. The current point is marked in red color. (middle left and right) Line drawings in 2D space. The x- and y- values proceed rightward and downward respectively, and reverberate as necessary. The middle right drawing uses color reverberation as well. (bottom) Colored bands in 1D space.*

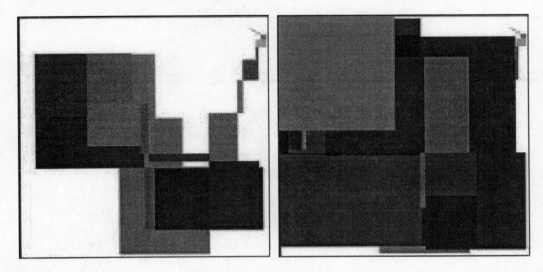

Figure 4: *Colored rectangles in 2D space. The image on the left appears at n=25, while the right one appears at n=50. Note how the small rectangles on the right of the early figure (left) are obscured by others in the late figure (right).*

3. Expression 2 : A behavioral function

Both the functions of the previous section are in essence structural functions that translate a Fibonacci number directly to a structural element of the image without intervention. A more complex case occurs if we allow the expression function to "decide" what to do with the Fibonacci value obtained. For example, in the previous case we observe that the location of the next point is unconstrained within the bounds of the given interval or region. This arrangement often produces large regions that cover older ones, so that the history of the image is lost and color or structure diversity is dramatically reduced (see for example Figure 4 right). One possible solution to this problem is to disallow points from falling far away from their predecessors, by putting some limits to their motion. This is equivalent to introducing a *behavioral* function that will act on the pure structural result (as taken in the previous section) and modify it so as to stay within limits :

```
If the next point is not within a certain range,
      then the next point is the most recently visited point that
      falls within that range,
      or, if none exists, it is the limit point in the requested
      direction.
```

As expected, the new behavioral model produces shorter line segments and smaller rectangles respectively, so that, more colored regions are allowed to coexist in the image, and more diversity is preserved (Figure 6).

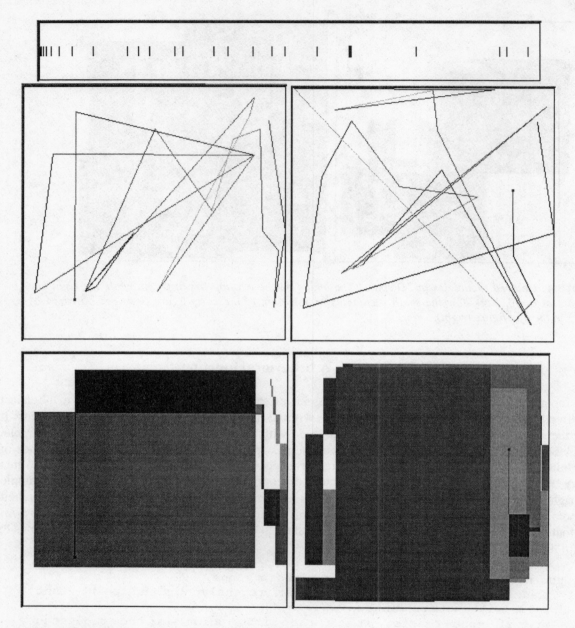

Figure 5: *Visual results of the toroidal function after n=25 steps. (top) Points in 1D space. Compare with Figure 3 (top). (middle) Comparative results in the same initial conditions of the reverberation function (middle left) with the toroidal function (middle right). (bottom) Comparative results in the same initial conditions of the reverberation function (bottom left) with the toroidal function (bottom right).*

Figure 6: *Series of colored lines or rectangles in 2D space taken with a behavioral model of range 100 (100 pixels is the maximum allowed motion) after 1000 time steps. Both images are taken with a background color other than white. Note that, despite having proceeded in time (n=1000), the images remain far richer than those of Figures 3, 4 and 5.*

Finally, all the above models may be used in the same image by switching from model to model in the course of progress. This way, noticeably more complex images may be produced. For example, the image of Figure 7 was produced by the reverberation function by first using a pure (structural) 2D colored model (rectangles) so as to fill almost the whole region with colors. We then switched to a behavioral 2D colored model (rectangles) so as to produce smaller rectangles, and we finally used the behavioral colored model (lines) so as to produce short colored line segments on top of the existing colored rectangles.

Of course, many more visual features may be defined and manipulated in the same way. Further artistic options may include, for instance, polygonal elements, the number and order of elements, the angle of rotation of geometric shapes, etc.

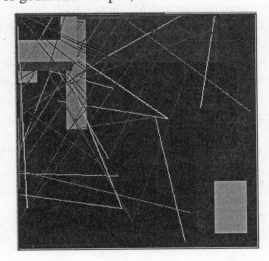

Figure 7: *Image taken by switching at will between models.*

4. Conclusion

We have briefly presented a methodology for studying complex visual forms from the bottom up by choosing a simple expression model and an arbitrary artistic expression function that translates the model's results into visual elements. The key assumption is that some aspect of the visual form has to be constrained so that the potential for infinite complexity arises. Number series are good candidates as expression models and a case study on the Fibonacci series coupled with several expression functions has been presented.

Although our demonstration focused on visual forms (drawings), the methodology may apply equally well to other media of expression. For example, application to serial music is straightforward [13]. Another fruitful direction for research is to try to express number series properties, instead of consecutive values, for example to try to express the fact that the ratio $F(n+1)/F(n)$ converges to the value 1.6180339... for n going to infinity. This line of research opens up new perspectives that match both abstract art concerns [11] and early computer art themes that were later abandoned [12].

Acknowledgment

I wish to thank painter George Harvalias for directing my attention to the work of Mario Merz.

References

[1]   *ACM SIGGRAPH Visual Proceedings* (annual conference).
[2]   *Artificial Life Video Proceedings* (bi-annual conference).
[3]   D. Batchelor, *Minimalism*, Tate Gallery Publishing, London,1997.
[4]   G. Celant (Ed.), *Mario Merz*, Guggenheim Foundation, NY, & Electa, Milan, 1989.
[5]   V. Hoggatt, *Fibonacci and Lucas numbers*, 1975.
[6]   W. Kandinsky, *On the spiritual in art*, Munich, 1912.
[7]   S.A. Kauffman, *At home in the universe : The search for laws of self-organization and complexity*, Oxford University Press, Oxford, 1995.
[8]   C.G. Langton, *Artificial Life, Proceedings of an interdisciplinary workshop on the synthesis and simulation of living systems*, Addison-Wesley, Reading, MA, 1987.
[9]   K. Malevitch, *From cubism to futurism to suprematismus*, Moscow, 1915.
[10] P. Mondrian, *The new art – the new life (Collected writings)*, Thames & Hudson, London, 1986.
[11] A. Moszynska, *Abstract art*, Thames and Hudson, London, 1990.
[12] A.M. Spalter, *The computer in the visual arts*, Addison-Wesley, Reading, MA, 1999.
[13] J.-N. von der Weid, *La musique du XXe siècle*, Pluriel, Paris, 1992.

An Iconography of Reason and Roses

Sarah Stengle
125 8th Avenue
Brooklyn, NY 11215
e-mail: sarah@stengle.com

Abstract

The author uses mathematical elements from old textbooks and engineering manuals in her artwork. This paper is a semiotic analysis of the mathematical imagery used. Semiotics, or the study of signs, is central to the concerns of twentieth century art. While most bridges between mathematics and art start by using mathematical principles to determine or augment an image, the author argues that her work uses artistic principles to augment the perception of the mathematics. Both approaches represent legitimate bridges between the disciplines, even if the former is more familiar than the latter. Many of the beautiful images and objects generated by the application of mathematics to artwork are visually compelling, but are conceptually removed from the critical dialogue associated with fine art. The author, who is an artist, believes that while the work does not break new ground mathematically, it does connect the disciplines of mathematics and art by providing an original examination of the semiotics of both disciplines.

1. Equation as Image

Mathematics is motivated by the intellect but also reflects an innately human and emotional desire for order and comprehension. The beauty of mathematics is that it offers elegant abstract constructs while providing a respite from fears and desires. Mathematics is exempt from the imperfections of biology; we are not. The author's artwork explores this theme, using pages from old mathematical textbooks and engineering manuals as raw material. The mathematical image becomes symbolic of constancy and analytical modes of thinking. Biological elements, such as leaves, roses, or bones, are then drawn or collaged into the work. These elements are chosen for their familiar connotations and symbolism. The work contrasts the analytical with the emotional and the technical with the organic, using images iconographically.

Mathematical imagery is seen through a veil of cultural assumptions that mathematics is unemotional and pure, and that its texts are stylistically neutral. The author wishes to question these assumptions, both by using mathematical imagery in her artwork, and by the semiotic analysis of the artwork presented in this paper. Hierarchical cultural values accord respect to both mathematics and fine art: the former must be logically rigorous; the latter is subject to constantly shifting interpretations. Shared, is a huge similarity in process. Both the research mathematician and the artist proceed by intuition, often aesthetically motivated, and both share a sense of discovery and achievement if and when the desired outcome is attained. Success in either endeavor feels suprapersonal and enduring.

By examining the nature and image of mathematical representation from a semiotic, rather than mathematical, viewpoint, this paper will address how art can augment the perception of mathematics. Traditionally, the interface between mathematics and art has been the mathematically augmented or determined image. Often these images are compellingly beautiful, but conceptually removed from the dialogue surrounding fine art. These images can be considered as an artistic subset of mathematics, and tend to be overlooked by museums, collectors and critics. Similarly, the work presented in this paper remains a mathematical subset of art. But a semiotic examination of mathematical imagery does break new ground in the vast grey area separating the two disciplines

In the author's work a visual vocabulary emerges as motifs develop across a variety of media, including sculpture, drawing, painting, and handmade books. Roses are feminine and beautiful, laden with sentimental associations. Leaves recall changing seasons and, by extension, the cycles of life. Bones are a reminder of death. These images serve as foils for the orderly technical information presented in the tables and equations, which also have their emotional subtext, the desire for clarity and order.

By drawing on top of the mathematical elements these motifs are placed in the unfamiliar interpretive and subjective context of art. For the mathematician, elements such as the age of the paper, the typography, and recollections of classrooms and libraries, place the mathematical element in a context that allows for interpretive analysis. Any viewer can associate the imagery with lessons fondly or sadly remembered. The connotations are emphasized over the literal content presented.

The following sections will examine a series of pieces and discuss how the work develops a visual lexicon.

2. Probability and Bones

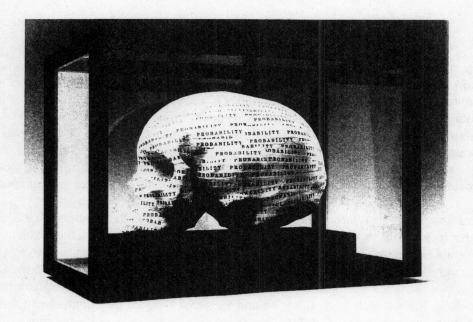

Figure 1. *"Probability", collage on skull, in a wood and glass vitrine. 1999.*

Probability is a sculpture made from a human skull, in this case a discarded nineteenth century medical specimen. It is collaged with two elements: the word "PROBABILITY" and equations from a text on probability [1]. The sculpture serves as a *memento mori*, a reminder of death. The selection of equations from probability has to do with actuarial practices, in which astonishingly accurate calculations of life expectancies of certain groups are made.

The piece began with a bisected skull. The interior was collaged with equations from a text on probability, while the exterior is entirely covered with the word "PROBABILITY" taken from the running heading of the same book. The exterior of the skull serves as a cover, with its recurring title, while the interior, which housed a human brain during life, is where the complex calculations have been placed. The skull is in a mahogany and glass vitrine to underscore its former role as specimen. The presentation of the sculptural object as a scientific specimen lends the object a more pedagogical role; it is meant to be learned from, rather than appreciated purely for its aesthetic qualities.

The use of human remains in art is problematic. In Western art there is a precedent in reliquary vessels which glorify a fragment of a deceased saint's body. These devotional containers honor people whose lives were revered and remembered. In the case of *Probability*, the skull is sadly anonymous and serves as a reminder of an unknown past life. These remains were not initially granted the respect that we would wish our own to be accorded. As a *memento mori*, the skull's anonymity reminds us that this poor soul is not only dead, but also forgotten.

3. Natural Functions

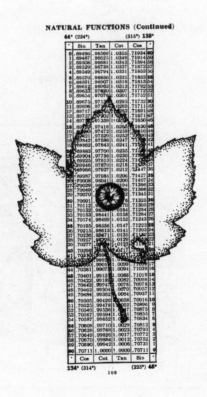

Figure 2. *"Natural Functions", collage, ink and watercolor on paper. 1998.*

The drawing, *Natural Functions*, is a collage on a table of natural functions taken from a chemist's handbook and laid down onto heavy paper. An inked and stippled outline of a decaying leaf covers the table.

The painted image of an eye looks out from the center of the leaf. The table is numerically useful, but the term "natural functions" could also apply to biological processes. The drawing supports both readings. The leaf is dying and the eye is symbolic of sight and consciousness. These elements animate the static image of the numerical table and invite a dual interpretation of the title.

This drawing is one example of an extensive series of leaf drawings over mathematical substrates over a period of four years. Rotted leaves, in life, are objects of supreme inconsequence. They are shoved into large bags and discarded. Mathematical tables, on the other hand, are implicitly confident and respected. They are predictable and correct. Unlike artistic imagery they leave little room for interpretation and opinion. By juxtaposing the delicate image of a decaying leaf with a table of natural functions, the artist quietly underscores the transient and insecure nature of life.

The leaf drawings, like the sculpture, *Probability*, are also *memento mori*. *Natural Functions* was drawn on a page taken from an old book, but many of the leaf drawings use relief prints of magic squares [2] and random numbers. The relief prints were designed by the artist and printed with a letterpress in order to do multiple drawings on a similar theme. The relief prints serve as a visual and conceptual background for the drawings. All of the leaf drawings have a symmetrical and centered composition, promoting a symbolic interpretation. The leaf does not occur in pictorial space, but instead appears as a sign centered over a field of numbers or equations. The leaf does not negate the equations, but rather brings them into a different context, where they too function visually and symbolically.

4. Reason and Roses

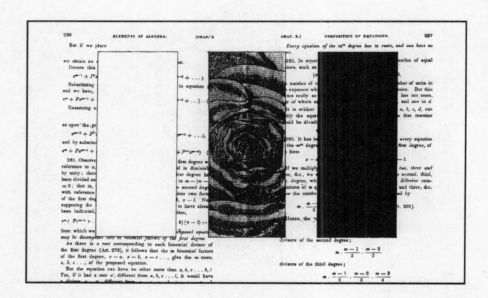

Figure 3. *"Reason and Roses", detail view, acrylic and collage on canvas. 1999.*

Reason and Roses is a painting in which the central image is a two-page spread from a nineteenth century algebra book, placed centrally on a background of large ornate roses. There are vertical rectangles on each page of the algebra lesson, one black and one white. Centered between them is a rectangle of red containing a rose identical to those on the background. The central red rose is the conceptual and visual focus of the work, framed by the geometric elements and the algebra lesson.

Reason and Roses is part of an ongoing exploration of cultural assumptions about femininity and masculinity. The locus of beauty is ambiguous; it could be in the rose, the geometry, or the mathematics. The rose has the central position, but the canvas is dominated by straightforward, symmetric geometry. Heavy black and red squares are superimposed on the decorative and feminine background of roses. This superposition is then negated by the reappearance of the rose as the central image. A tension and balance exist between the feminine and roseate imagery and the geometrical and mathematical imagery, but there are parallels as well.

Intuition and a desire for comprehension and beauty can motivate a mathematician or an artist. A mathematical proof which is entirely correct but boring is second rate, just as portrait can be an excellent likeness but artistically dull. Mathematicians and artists often use a similar vocabulary of a search led by intuition to describe their working process. Often the outcome is described in terms of discovery, meaning that the outcome was not known beforehand but seemed to exist a priori. Both often have only a sense of the outcome, rather than knowledge of it, and follow their intuition to their goals, which they recognize only when they get there. "There" is were things "feel" resolved and complete. The mathematical discovery has to withstand the rigid demands of the discipline, while the artistic discovery is subject to constant reinterpretation and debate. Both disciplines are driven by human emotions and desires, which appear explicitly in the art, and implicitly in the mathematics.

5. The Rose Album

Figure 4. *Page 127 from "Rose Album", printed text, collage and ink on paper. 1999.*

Rose Album is a handmade artist's book in which 186 drawings of roses are superimposed on a printout of the largest known prime number at that time [3]. The number was typeset and bound into an album. The length of the book was determined by the size of the number. The artist then drew into the album over the course of six months. The drawings combine images of roses taken from popular sources (such as wrapping paper) with elements from hard-edged geometric abstraction. On every page the running "text" of the prime appears as the background.

The text of the prime number is paginated like a conventional book, and presented with half and full title pages and a colophon. Is it possible to "read" such a large number? The prime number is so large that it is unintelligible. It presents information that only a supercomputer could comprehend.

The ardent effort required to generate this number is staggering. Determining a larger prime has been a classic mathematical problem for centuries. The man-years cumulatively expended seeking larger primes are evidence of a collective passion. Numbers of so many digits have few practical applications. They are unwieldy to work with and incomprehensibly large. It is "math for math's sake" and like the millionth digit of π so removed from the give and take of daily life that the number takes on almost a devotional aspect.

There are parallels in the practice of art for art's sake, which can be similarly obsessive and removed from daily life. The author chose to use visual elements from the painterly tradition of hard-edged geometric abstraction because of the hegemonic vocabulary once associated with them. Words like "pure", "timeless", "basic", "universal" were attributed to works that were free of deceptive pictorial representation. The aesthetic was reductivist, but the accompanying vocabulary was self-impressed and grandiose.

In *Rose Album*, both the mathematics and the geometric abstractions are somewhat subverted by the abundance of roses that appear. The arrangement is not meant to negate the other elements, but rather to allow for a more intricate reading of what is beautiful and what is respected. Mathematics and geometry are highbrow phenomena; roses are colorful, sensual and feminine. These lovely flowers share attributes with things that are culturally enjoyed, but intellectually marginalized.

Neither the rose nor the large prime can be explained in purely rational terms. It is obvious that no equation can define the rose's form or sweet smell. It is less obvious that the large prime exists because of an enduring collective passion that defies practical reason. Acknowledging certain applications in encryption technology and testing supercomputers, the search for a larger prime can still only be explained in terms of desire. Although the search for a larger prime occurs within the practice of mathematics, which is viewed as supremely logical, it is motivated by the desire to participate in the beauty of the infinite, the desire to achieve, and the excitement of participating in an endless process of discovery.

6. Mekanic

Mekanic is a book made from the pages of another book. Brittle, brown pages were taken from an engineer's handbook, *Des Igenieurs Taschenbuch* [4], backed with a stronger paper, and sewn into a new binding. Lines from a Shakespearean sonnet were glued into the text, along with roseate imagery taken from the paintings of Pierre Joseph Redoute [5]. The sonnet is about the transient nature of beauty; its metaphor is a rose [6]. *Mekanic* has a dual text paralleled by dual imagery. The original text and images relate to analytic geometry and appear in German. The added text and imagery relates to roses. The duality first appears the cover: the title block reads "Mekanic", but the image is a large engraving of a rose.

Roses, like a love, can be the objects of obsessive desire. Mathematicians can be obsessed with mathematics, in the way that artists can be obsessed with art.

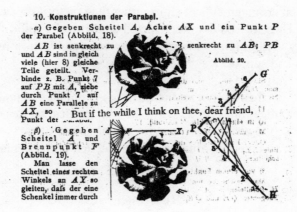

Figure 5. *"Mekanic", artist book, detail view of page 33. 1999.*

7. The Other Beauty of Mathematics

Within the community of mathematicians and scientists, there is an awareness of the conceptual beauty of mathematics. Books have been written on the art of mathematics [7]. There is another, less considered beauty within the practice of mathematics: the look of the mathematical document. Textbooks and technical publications tend to be conservatively designed, for clarity rather than style. Jan Tschichold asserts that the typographer's responsibility is "to divest themselves of all ambition for self-expression" [8]. In Tschichold's view, the most beautiful typography serves the text and does not call attention to itself. The traditional, suprapersonal aura of mathematical texts are part of their visual appeal.

Arabic numbers are highly designed symbols and, like the letters of the alphabet, their form has evolved over time. Arabic numbers have a refined beauty, particularly when they appear in classic fonts. Mathematical notation is rich in symbols, including many handsome ones taken from the Greeks. The signifiers, meaning the numbers and symbols, have their own beauty, apart from the beauty of what they signify.

Another aspect of mathematical texts is that they appear intimidating and mysterious to the uninitiated. Although there may be little comprehension, the texts "look" smart. It is somehow understood that the unfamiliar notations do not represent nonsense. Viewers with mathematical training are going to feel a familiarity with the material. For those without training, the allusion to the practice of mathematics will still be understood. In the author's artwork, the mathematical elements are used as imagery; the numbers and equations do not prove or solve anything, but instead refer to the logically rigorous practices of mathematics and science.

8. Dualities and Dichotomies

The same thematic dualities and dichotomies keep appearing throughout the artist's work. The analytic is seen with the emotional, the technical with the biological, the masculine with the feminine, and the permanent with the transient. The equations symbolize permanence, order, and clarity. The biological elements, such as the leaves, flowers, bones, and eyes, stand for the transience of life and the complexity of emotions. An iconographic reading of the images is further supported by the symmetry of the compositions, which are generally flat arrangements of signs rather than pictorial spaces. The pieces have an intentional, deliberate, quality. Even the colors are selected semantically, pink for femininity and black and white for austere rationality. Although elements of geometric abstraction recur in the work, the images retain their roles as images and support a symbolic reading.

Using images and colors for their iconographic qualities is not an unusual part of developing an artistic sensibility. Using mathematical imagery as symbolic of analytical modes of thinking is less typical. In the case of the author's work, the ambition is to draw the aesthetic of reason into the dialogue of her art and to underscore the beauty and passion implicit in the practice of mathematics.

References

[1] *Encyclopaedia Britannica*, Ninth edition, Vol. XIX, pp. 788-809. 1894.

[2] G. Greenfield and S. Stengle, *Magic Squares as Mathematical Palimpsest*, First Interdisciplinary Conference, International Society of the Arts, Mathematics and Architecture. N.Friedman and J. Barallo (eds.), The University of Basque Country, pp. 219-226. 1999.

[3] Spence, Woltman & GIMPS, $2^{2976221}-1$, downloaded in August, 1998 from http://www.utm.edu/research/primes/largest.html, a website maintained by the University of Tennessee at Martin.

[4] Academischen Verein Hutte, *Des Ingenieurs Taschenbuch*, Verlag von Wilhelm Ernst & Sons, Berlin, 1896.

[5] P.J. Redoute (1759-1860), *Redoute Roses Giftwrap*, Dover. 1990.

[6] William Shakespeare, *Sonnets*, edited by H. E. Rollins, Harvard University. Appleton-Century-Crofts, Inc. New York, Sonnet 54, pg. 27, 1951.

[7] D. Davis, *The Nature and Power of Mathematics*, University of Princeton Press. 1993.

[8] J. Tschichold, *The Form of the Book*, Hartely and Marks, pp. 11, 1991.

The End of the Well-Tempered Clavichord?

Prof. W. Douglas Maurer
Department of Electrical Engineering and Computer Science
The George Washington University
Washington, DC 20052, USA
E-mail: maurer@seas.gwu.edu

Abstract

The scale used by all pianos today is based on the idea of the well-tempered clavichord, an instrument for which Bach wrote a famous collection of pieces showing its versatility. Our contention here is that, in an era of electronic instruments, the well-tempered scale has the potential to be completely outmoded, and replaced by an adaptive scale which uses natural harmonics whenever possible.

Introduction

Johann Sebastian Bach, in 1744, wrote a collection of 24 keyboard pieces entitled "The Well-Tempered Clavichord." Today these are usually played on the piano, but Bach wrote them for the clavichord, an instrument which, at the time, represented a technological advance. For the first time, a single musical instrument could be played, without being obviously out of tune, in each of the twelve major keys (C, Db, D, Eb, E, F, F#, G, Ab, A, Bb, and B) and each of the 12 minor keys (C, C#, D, Eb, E, F, F#, G, G#, A, Bb, and B). Bach celebrated this by writing 24 pieces, one for each of the 12 major and 12 minor keys. An instrument that is out of tune can be said to be "ill-tempered," so Bach's clavichord, by contrast, was "well-tempered."

Our concern here is with the fact that the design of the well-tempered clavichord involved a compromise. Despite appearances, there is a sense in which the well-tempered clavichord and its successor, the piano, are still out of tune. They are merely so little out of tune that the ordinary ear does not hear the difference. This compromise was necessary so long as the tuning of musical instruments was a painstaking process, performed once at the start of each concert or occasionally at the start of each piece within a concert. Today, however, computer-driven musical instruments can be tuned and retuned in less than a millisecond; and it is this that suggests to us a new tuning system in which compromise, although it is still there, is a fraction of what it was before.

Natural Production of Sound

To understand the compromise, it is necessary to review the mathematics of sound production in a tunable, non-computerized instrument. There is always something that is vibrating, whether it be a string, as with a violin or piano, or a column of air, as with a flute or trumpet. This string, or column of air, has a certain length L, and can be made to vibrate along its entire length or along parts of its length of equal size. If we play the E string of violin without touching it, we get the note E, a little more than an octave above middle C. If we play this same string while touching it very lightly at the midpoint of its length, we get a note one octave higher, also called E. This arises from the fact that the string is now vibrating in two equal parts.

Suppose now that we touch the string very lightly, not at the midpoint of its length, but at a point representing one-third of its length. The string will now vibrate in three equal parts, and the note will be B, almost three octaves above middle C. The string can also be touched at a point representing one-fourth, or one-fifth, or one-sixth of its length, causing it to vibrate in four, five, or six equal parts. The notes produced in this way are called harmonics of the original note. In the same way, the column of air inside a trumpet can be made to vibrate in anywhere from two through six equal parts, and sometimes more, producing harmonics of the *fundamental note*. This is the note that would be produced if the air column vibrated along its entire length (although actually doing that is rare in a trumpet).

This is natural music, without compromise. (It is also a bit of an oversimplification; see the section on Resonance below.) We can hear it when we hear the notes of a bugle before a horse race. Try to play a bugle call on the piano, and there is something mysteriously missing. What is missing is that the bugle is in tune, in a way that the piano is not. The bugle plays natural harmonics, while the piano plays only an approximation of harmonics, a result of the compromise which is necessary for the piano to play in 24 different keys. (Note that the bugle can play in only one key.) This effect is in addition to the fact that the bugle has a different timbre than the piano.

Natural Frequencies

To understand the compromise further, we will have to look at frequencies. The musical notation A-440, for example, means than anything which vibrates 440 times per second will produce the note A (just above middle C). There is an A string on the violin which vibrates that often; that is its frequency. If we touch the A string at the midpoint of its length, it will vibrate in two equal parts, and twice as frequently, or 880 times per second. If we make the string vibrate in N equal parts, its frequency will be 440 times N. If we make the air column in a trumpet vibrate in N equal parts, its frequency, similarly, will be N times the frequency of the fundamental note.

Now let us look, specifically, at the harmonics corresponding to four, five, six, and eight times the fundamental note. This is the so-called major chord; in the key of C, it is C, E, G, and C, while in the key of A, it would be A, C#, E, and A. If the fundamental note has frequency 110, then these four notes have frequencies 440, 550, 660, and 880. We can also start with the lowest note of the major chord, in this case A, and calculate the other frequencies from that. If we start at 440, then to get to 550, we multiply by $5/4$; to get to 660, we multiply by $6/4$, or $3/2$. To get to 880, which is an octave higher, we multiply by 2; and indeed multiplying the frequency by 2 always produces the next octave up. The same thing happens in the key of C; if we start at C, with frequency N, then E will have frequency $5/4$N, G will have frequency $3/2$N, and C, an octave up, will have frequency 2N.

These are the frequencies of the natural harmonics. They are not the frequencies of the well-tempered clavichord. To understand why not, let us start at C, with frequency N; go up to E, with frequency $5/4$N; and then *start* at E and go up to the next note in the E major chord, which is G#. Now the frequency is $5/4$ times $5/4$N, or $25/16$N. But on the piano, G# is the same as Ab, and we can now go up to the next note on the Ab major chord, which is C again. This time the frequency is $5/4$ times $25/16$N, or $125/64$N. The trouble is that we are now an octave higher than we were before, and therefore the frequency should be 2N. Instead, it is $125/64$N, which is a little smaller than 2N (which would be $128/64$N, since 64 times 2 is 128).

This is the bane of natural harmonics: *they don't come out even.* We can't play the same tunable instrument in even the three keys of C major, E major, and Ab major without some form of compromise. Once you think of it, the compromise is obvious. We multiplied N by $5/4$, three times, and we wanted to end up with 2N. This didn't work, so we need another number, not $5/4$ but very close to $5/4$. Instead of 1.25 (or $5/4$), the number is approximately 1.25992, the cube root of 2. If we multiply

N by this three times, we clearly get 2N, because the definition of the cube root of 2 (call it X) is that $X^3 = 2$.

The Well-Tempered Scale

The basic principle of the well-tempered clavichord is an extension of this idea from three multiplications to twelve. Instead of a number X with $X^3 = 2$, we now want a number R with $R^{12} = 2$, also known as the 12th root of 2, or approximately 1.05946. If the frequency for C is N, then the frequency for C# is N times R; the frequency for D is N times R times R, and so on up the chromatic scale. In particular, the frequency for G, which in natural harmonics would be $3/2$N, or N times 1.5, is N times about 1.49831, on the well-tempered scale. (Other bases than 12 may be used; with base 24, for example, we get the so-called quarter-tone scale. All such scales are sometimes referred to as "just scales," where "just" means "fair," since all 24 keys are treated fairly and equally.)

More generally, let us look at the notes of the major scale: C, D, E, F, G, A, B, C. This is made up of the major chord (C, E, G, C), sometimes called the *tonic chord*, together with the first three notes of the *subdominant chord* (F, A, C) and the *dominant chord* (G, B, D, but with the D an octave down). The natural frequencies for these are:

C	D	E	F	G	A	B	C
N	$9/8$N	$5/4$N	$4/3$N	$3/2$N	$5/3$N	$15/8$N	2N

It is easy to see that, going up the major chord that starts with F ($4/3$N), you get A ($5/4$ times $4/3$N, or $5/3$N) and C ($3/2$ times $4/3$N, or $4/2$N, or 2N). Going up the major chord that starts with G ($3/2$N), you get B ($5/4$ times $3/2$N, or $15/8$N) and D ($3/2$ times $3/2$N, or $9/4$N), and one octave down from there would be half of that, or $9/8$N. Now let us see the extent of the compromise, when we use the well-tempered scale. Our factors, to be multiplied by N in each case, are:

Note	C	D	E	F	G	A	B	C
Natural harmonics	1	1.125	1.25	1.333	1.5	1.667	1.875	2
Well-tempered scale	1	1.12246	1.25992	1.33484	1.49831	1.68179	1.88775	2
Percentage off by	0%	−0.3%	+0.8%	+0.1%	−0.1%	+0.9%	+0.7%	0%

The reason that the well-tempered scale works as well as it does is that none of these notes is off by as much as one percent. Nevertheless, they are off, and today, with computers, we can do better.

The Computer-Driven Grand Piano

One might object at this point that computers are used only in synthesizers and electric pianos, and not in serious musical instruments. It is true that violins and trumpets have nothing to do with computers, although trumpets already use natural harmonics, while violinists can use an infinite gradation of frequencies by pressing their fingers firmly on the strings to change the length of that part of the string which vibrates. Meanwhile, there is the electronic grand piano, which produces a quality of sound equivalent to that of most conventional grand pianos.

We will need some basic information on computer production of sound. The earliest sounds produced by computer involved producing clicks at a given frequency. If the frequency is 440, then, after each click, the computer would wait for 1/440 of a second before producing the next click. This is feasible because 1/440 of a second is 2,273 microseconds — a long time, for a computer. However, the sound produced in such a way is monotonous.

Today's computers calculate the sound wave amplitude y associated with a particular sound. This time, if the frequency is 440, the wave equation is $y = sin\ (440x/2\pi)$, where x represents time in seconds. The value of this can now be calculated (for example) once every microsecond, and transmitted to the sound-producing device. The point is that this device is now capable of accepting, not just a yes-or-no signal, but an amplitude. The result is that waveforms of any kind may be generated. Furthermore, the wave equation is the result of a numerical calculation in which frequencies appear; and changing the frequencies in the equation can be done in less than one microsecond.

Fixed Harmonic Scales

We start by considering an idea which is inferior, although well adapted to computers. This would require the player, before playing a piece, to set the major key; and then all the notes in all the major scales of that key would be set to the natural harmonics of that major key. Whatever the frequency N of the tonic note, the frequencies of the next six notes would be found by multiplying N, respectively, by $9/8$, $5/4$, $4/3$, $3/2$, $5/3$, and $15/8$, as indicated earlier. These are then the frequencies that would be played.

The problem with this is that a piece in C major (for example) is not *entirely* in C major. A typical piece in C major will go into F, G, and D major at times; and this is only the beginning. If the harmonic scale is fixed, then, as soon as you get out of the major key, some of the notes will be even worse off than they were before.

What needs to be done is that the computer should sense, at all times, what key is being played in, and adjust the frequencies to the needs of that particular key. What we need is a way of translating a chord C into the key K in which that chord naturally occurs. The question, however, is how this is to be done, given the wide variety of possible chords C that can be played.

The Universal Adaptive Scale

The solution to the above problem arises from our observation that, in considering a chord, one only has to consider twelve notes, not 88. That is to say, for the purposes of determining K from C, it is not significant what octave a note is in. Thus, for example, C-E-G is in C major, regardless of whether the G is above the C, on the piano, or below the C, in a different octave. In fact, each of the 12 notes is either in the chord, or it is not; and thus the total number of possible chords is 2^{12}, or 4096. A table that size is well within the capabilities of today's memory chips, each of which is at least 1,000,000 bytes in size.

We thus envision a table of 4,096 positions, which, for each note combination, gives the key, major or minor. Whatever chord is being played, the frequencies of all notes will be adjusted according to the chord, and it is those frequencies that will be played. Some cases will be complete dissonances, such as C-C#-D-Eb; and some cases will be ambiguous, such as C-E-G-A, which could be either C major 6th or A minor 7th. It is important to note, however, that *most cases will not be ambiguous*.This is what we call the Universal Adaptive Scale.

Notes Other Than The Major Scale

In our treatment of natural harmonics for the notes of the scale (C, D, E, F, G, A, B, C) we left out the five additional notes (the black notes, in the scale of C). The natural harmonics here can be chosen in various ways, of which the following are the simplest:

C	Db	D	Eb	E	F	F#	G	Ab	A	Bb	B	C
N	$^{16}/_{15}$N	$^{9}/_{8}$N	$^{6}/_{5}$N	$^{5}/_{4}$N	$^{4}/_{3}$N	$^{25}/_{18}$N	$^{3}/_{2}$N	$^{8}/_{5}$N	$^{5}/_{3}$N	$^{9}/_{5}$N	$^{15}/_{8}$N	2N

These are justified as follows:

- Eb is $^{6}/_{5}$N because the Eb major scale (Eb, G, Bb) includes G, at $^{5}/_{4}$ times $^{6}/_{5}$N or $^{6}/_{4}$N (or $^{3}/_{2}$N), as specified.
- Ab is $^{8}/_{5}$N because the Ab major scale (Ab, C, Eb) includes C, at $^{8}/_{5}$ times $^{5}/_{4}$N or 2N.
- Bb is $^{9}/_{5}$N because the Eb major scale includes Bb, at $^{3}/_{2}$ times $^{6}/_{5}$N or $^{18}/_{10}$N (or $^{9}/_{5}$N). This is simpler than tying Bb to F through the Bb major scale, which would result in $^{16}/_{9}$N (since $^{16}/_{9}$ times $^{3}/_{2}$ is $^{8}/_{3}$, and then $^{4}/_{3}$ is an octave below that).
- Db is $^{16}/_{15}$N because the Db major scale (Db, F, Ab) includes F, at $^{5}/_{4}$ times $^{16}/_{15}$N or $^{4}/_{3}$N, as specified. The fraction $^{16}/_{15}$ may seem complex, but it is $1^{1}/_{15}$; any other close fraction would be just as complex.
- F# is $^{25}/_{18}$N because going up from F# to A is like going up from C to Eb: you multiply by $^{6}/_{5}$, producing $^{6}/_{5}$ times $^{25}/_{18}$ or $^{5}/_{3}$, as specified. This gives a simpler fraction than any of the obvious alternatives.

Monophonic Passages

A passage, or small part of a piece of music, is monophonic if only one note is being played at a time. Obviously the key in which to play such a passage cannot be inferred from the passage itself. It can be inferred from the preceding chord in most cases (as, for example, when a chromatic scale is being played), but not in all. To improve the way that such a passage sounds, a computerized keyboard might have a mode in which the left hand plays one or two notes (which do not sound), indicating the key, while the right hand plays a monophonic passage.

Relations Among Scales

We mentioned that, when a piece in C major goes into G major, some of the frequencies change. Actually, not as many of them change as one might think. It is still necessary, using the universal adaptive scale, for the computer to know what the tonic is (in this case C major), because the G major scale, for example, will now be based on G as it appears in the C scale, not in any other scale (including the well-tempered scale). Let us look only at the two major scales involved, one based on N and the other on M = $^{3}/_{2}$N:

	C Scale:							
	C	D	E	F	G	A	B	C
	N	$^{9}/_{8}$N	$^{5}/_{4}$N	$^{4}/_{3}$N	$^{3}/_{2}$N	$^{5}/_{3}$N	$^{15}/_{8}$N	2N

	G	A	B	C	D	E	F#	G
G Scale:								
(in terms of M)	M	$^{9}/_{8}$M	$^{5}/_{4}$M	$^{4}/_{3}$M	$^{3}/_{2}$M	$^{5}/_{3}$M	$^{15}/_{8}$M	2M
(in terms of N)	$^{3}/_{2}$N	$^{27}/_{16}$N	$^{15}/_{8}$N	2N	$^{9}/_{4}$N	$^{5}/_{2}$N	$^{45}/_{16}$N	3N

Lowering some notes by an octave (that is, dividing the frequency by two), and rearranging individual notes so that they correspond, we obtain the following comparison:

Note:	C	D	E	F	F#	G	A	B	C
C Scale:	N	$^{9}/_{8}$N	$^{5}/_{4}$N	$^{4}/_{3}$N		$^{3}/_{2}$N	$^{5}/_{3}$N	$^{15}/_{8}$N	2N
G Scale:	N	$^{9}/_{8}$N	$^{5}/_{4}$N		$^{45}/_{32}$N	$^{3}/_{2}$N	$^{27}/_{16}$N	$^{15}/_{8}$N	2N

Clearly F# is missing in the scale of C, while F is missing in the scale of G. Aside from these two, however, the only note with a different frequency in the two scales is A, which is 1.3% higher in the G scale than in the C scale.

This brings up an important issue. Every so often, the universal adaptive scale is going to present the listener with two notes, one after the other, which "ought to be the same" and yet are noticeably different. The Blue Danube Waltz is a good example of this; if it is played in C, there are three measures in which the melody is

B / B D A / A

where the first two bars are in G major (actually G7) and the last bar is in C major. According to the correspondence above, the last two notes of this, even though they are both A, have different frequencies, and an astute musical ear will recognize this. Furthermore, this phenomenon does not occur on the piano or on any other instrument using the well-tempered scale.

Will the listeners, being aware of this, criticize it and prefer, on that account, the well-tempered scale to the universal adaptive scale? Our contention (admittedly debatable) is that they will not. The reason is that the universal adaptive scale, or any scale involving natural harmonics, exhibits the phenomenon of resonance in a way that the well-tempered scale never can. We will now explain this.

Resonance

To understand resonance, we have to go back to the study of vibrations. Whenever a string or an air column vibrates, even if it is vibrating on the fundamental note, it is also, to a very small extent, vibrating on one or more of the harmonics as well. These are called *resonant harmonics*; their amplitudes differ from one musical instrument to another, and it is the relative amplitudes of the various resonant harmonics which determine, to a great extent, the timbre of a particular musical instrument.

Now suppose that we are playing a chord which involves several natural harmonics. As a specific example, let us assume that three trumpets are playing Bb, D, and F, while trombones and tuba are playing low Bb at the same time. The natural harmonics being played by the trumpets will also, in a way imperceptible to the human ear, be coming from the trombones and tuba. If all these instruments have been properly tuned, then the lower instruments will be adding sound to the higher instruments, on the exact notes that the higher instruments are playing.

But they will be doing more than that. Resonance, as a phenomenon in physics, has to do with the effect of transmitting a certain frequency in an environment in which there are other transmitters for which that is a natural frequency. In particular, resonance can occur when one instrument naturally resonates with the frequency, or a multiple of the frequency, being played by another instrument nearby.

In electronics, resonance can be inappropriate. The annoying low-pitched hum that comes from speakers when they are not properly adjusted, for example, is due to resonance in the speakers caused by ordinary 60-cycle current (it is always on a B, two octaves below middle C, whose frequency is approximately 60). In sound production, however, resonance is desirable. In particular, a low Bb played by trombones and tuba will resonate in nearby trumpets, even if no one is playing the trumpets.

All this, however, depends on natural harmonics, and doesn't work when the well-tempered scale is used, except for octaves. If a chord involving a fundamental note and one or more of its harmonics is played on the piano, or on an old-fashioned clavichord, the only resonance will be between octaves. There the intervals are exact, even in the well-tempered scale. Indeed, if the damper pedal is down, a low F#, for example, will resonate through all the unplayed F# strings, up and down the piano, and similarly for any other low note.

Other harmonics don't work here on pianos or clavichords because the natural harmonic frequencies are not exact. A low Bb doesn't resonate in a higher D, even though D would be one of the harmonics, because the D is tuned to the Bb times 1.25992 times 4 (the 4 is for the two-octave difference). This is the Bb times 5.03968, which is not the same as the Bb times 5, for the fifth (or as it is sometimes called, the fourth) harmonic. Indeed, when two notes that are almost, but not quite, the same are played at the same time, the result can be harsh and grating, as can be heard when listening to a band which is not properly in tune with itself.

Our thesis here is that the universal adaptive scale will bring resonance back to the electronic grand piano and similar instruments. The compromise that was necessary in the design of the well-tempered scale will now be necessary only in a few situations, such as chromatic scales and atonal music. With the new possibilities brought about by computers, we can look forward to a new and resonating musical future.

Other Studies of Computer Music

Computer music has a long and rich history, which is admirably summarized in the 1,234-page reference work by Roads et al. [1]. This covers sound analysis and synthesis, music languages and editors, mixing and signal processing, psychoacoustics, performance software, and algorithmic composition systems, among other subjects. It does not, however (if its 24-page index is to be trusted), address such subjects as the clavichord, well-tempered scales, just scales, or quarter tones.

Reference

[1] C. Roads, et al., *The Computer Music Tutorial*, MIT Press. 1996.

* * * * * * * * * * * *

The Generation of the Cube and the Cube as Generator

María Antonia Frías Sagardoy
Ana Belén de Isla Gómez
E.T.S.Arquitectura, Universidad de Navarra
31080 Pamplona (Navarra) Spain
E-mail: mafrias@unav.es

Abstract

In this study we submit for consideration how, in parallel with the different ways of conceiving the generation of a geometric shape, often accompanied by certain arguments, we can encounter similar artistic developments which lead to original results and show the fecundity that scientific advances have for art. For now we will exclusively consider the cube, which is, perhaps after the sphere, one of the most productive shapes. Without intending to exhaust the subject, we shall limit ourselves to highlighting one of the most representative episodes.

Plato in the *Timaeus [1]*

In the Timaeus, Plato refers to the moment "when an essay was being made to order the universe," when "everything... were then first by the creator fashioned forth with forms and numbers." On the basis that "God formed them to the most fair and perfect," he goes about rationalizing "the disposition of each and their generation." As is well known, Plato ends by assigning to each of the elements: fire, air, water and earth (which, by observation, must be distinct from one another) the shapes of regular polyhedrons, since these must be the most beautiful and perfect forms. This is thus a preconceived idea.

Moreover Plato tries in his plausible reasoning to assume the notion that some elements are "capable of being generated out of each other by their dissolution." And this could in fact occur with the three polyhedrons whose faces are equilateral triangles, as long as these forms are considered from the point of view of the surface. But the two remaining polyhedrons have squares or pentagons as their faces. Because of its similarity to the sphere, the dodecahedron ends up being the most apt for the Universe, and the cube can better respond to the stability of the Earth. The choice therefore seems clear, and the cube will be selected to configure this element.

Thus we find that Plato, in order to explain the birth of the fourth element in such a way that it might have something in common with the other three, finds himself forced to subdivide the faces of all the polyhedrons chosen into right triangles. And so upon these he raises his argument in the following manner: "In the first place, that fire and earth and water and air are material bodies is evident to all. Every form of body has depth; and depth must be bounded by plane surfaces. Now every rectilinear plane is composed of triangles."

A triangle that contains a right angle is, in Plato's mind, more perfect than a triangle composed of any other sort of angles. And among these there are two classes: those of equal sides (isosceles) and those of unequal sides, which are infinite. However, the most perfect of the latter group should be those "which two conjoined form an equilateral triangle." And thereby the first type of triangle, the isosceles, shall originate the cube; and from this triangle the faces of the rest of the chosen polyhedrons shall come forth. As we can see, the force of the argument, which seeks to find a common process of formation, just as the idea that what is most regular is always most beautiful, is that which brings Plato to imagine a process of the cube's formation, which, seen from the today's perspective, seems surprising. Consequently, the complete description of the cube will be the following: "The isosceles triangle generated the fourth, combined in sets of four, with the right angles meeting at the centre, thus forming a single square. Six of these squares joined together formed eight solid angles, each produced by three plane right angles: and the shape of the body thus formed was cubical; having six squares planes for its surfaces." The reason for the number four in the triangles that form each face should perhaps be found in that this number harmonizes best with the definition of the square. For according to Plato, the issue that interests him is "what is the form in which each has been created, and by the combination of what numbers." In the cube, according to its description, we will find the numbers four, six, eight and three: four isosceles triangles, six equilateral rectangular figures, and eight solid angles of which each is made up of three flat angles.

In summary, the conception of the cube and the other polyhedrons is for Plato not volumetric but superficial, in this case the isosceles triangles being the generators of the squares that constitute its faces. With respect to the meaning that it has acquired in this process, the cube ends up being formed by the most beautiful of all possible triangles (right isosceles triangles). Furthermore it is the one which enjoys the greatest stability: "To earth let us give the cubical form; for earth is least mobile of the four and most plastic of bodies."[1] The cube has that stability, Plato comments, as much because the quadrilateral surfaces that form it are more stable than triangles, as because of its origin, since isosceles triangles of equal sides are naturally more stable than triangles of unequal sides. This is again a superficial conception and not a volumetric one, although the stability that this volume has with respect to gravity seems to us more evident when it rests on a flat surface. In effect, this is what seemingly takes place: "When earth meets with fire and is dissolved by keenness of it, it would drift about, whether it were dissolved in fire itself, or in being some mass of air or water, until the parts of it meeting and again being united became earth once more; for it never could pass into any other kind." [1] It could not, because there is no other element composed of right isosceles triangles.

Regarding the squares that form its faces and their right angles, we have a significant reference from Proclus: "The Pythagoreans thought that the square more than any other four-sided figure carries the image of the divine nature. It is their favourite figure, indicating immaculate worth, for the rightness of the angles imitates integrity, and the quality of the sides abiding power."[2-3]

After the Timaeus of Plato, the theme of those forms since labeled the "platonic solids" –which include the cube– continued to be treated by philosophers, geometricians, astronomers, artists, etc. throughout the centuries. Among them Archimedes, Piero della Francesca, Luca Paccioli, Durero and Daniel Barbaro are indispensable. By way of again presenting a signification by cosmic attribution, we will only cite the contribution of Kepler (1571-1630). He, in moving from Plato's microscopic scale to the astronomic, attributes the regular polyhedrons to the planets, also providing himself with a rationalization.

Juan de Herrera in his *Discourse on the Cubic Figure [4]*

Based on his familiarity with Euclid, Juan de Herrera writes about the cube with the intent to illustrate the teachings of Ramon Llull. He considers that the cubic figure opens the door to "great and high mysteries and secrets which are difficult to comprehend," being the "root and foundation of so-called Llullian art." What interests the architect about the cube is in this case a process, an operation which is carried out in it, thereby having a dynamic conception of the figure, which, being natural, can also accompany rational discourse. As for the geometric figure, Herrera defines the cube, again superficially, as formed by six square surfaces, which he draws on a plane in the form of a Latin cross. He describes how these surfaces go about raising themselves perpendicularly or at right angles until the final surface comes into place, closing the cube(fig.1).

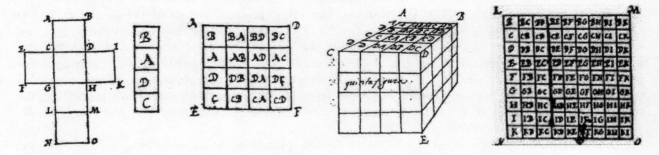

Figure 1: *Herreras' drawings*

But this is not the process which will serve him in his discourse, rather another, derived from the Greek conception of numbers: lineal, superficial or planar, and solid. The lineals have no other origin than themselves. Superficial numbers are those which proceed from two multiplied numbers, while the solid numbers come from three numbers multiplied together. If the two numbers that produce a superficial number are equal, a squared number is the result. If the three numbers that produce a solid number are equal, a cubed number results. Thus, Herrera, being attentive to Euclid's definitions, affirms: "The cubic number is that one which is formed by three equal numbers."

The unit number is imagined as a tiny corpuscle, in this case cubic. So the number 4, lineal, may be formed by four tiny cubes aligned. The superficial –square– 16, originating from two 4's multiplied together, should be formed by 16 tiny cubes arranged within a square area of 4x4 cubes. And the solid number –cubic– 64, originating from three 4's multiplied together, must be formed by 64 cubes arranged according to a macrocube of 4 cubes per edge (fig.1). This is a way of understanding the operation physically, as if the line were displaced as it is multiplied, forming the surface, and as though both the line and the surface did this in the other dimension, forming the volumetric body.

Herrera combines this operation with the Llullian concepts, and he explains it to us this way: "because the cube results from three dimensions and two operations: the first operation is of the *tivum* or the line itself from whose operation comes the *bile* or surface; the second operation, which is the *Are*, is jointly of the *tivum* or line and of the *bile* or surface. And the *are* together with the *tivum* and the *bile* are the cube, or the total plenitude of being and operating..." The cube is the plenitude of being and operating, because it contains the three essential dimensions: the *agente*, the *agible* and the *agere* (for example: the visual potential, the visible and vision), as Herrera says; and this is intrinsically as well as extrinsically. The line is active, the surface passive when engendered by the line, and the two together are again active when they produce the cube. Just as none of these three dimensions can be absent in the cube, neither can they be absent from being and operating.

All of this applies to the four elements (earth, fire, air and water), considering a cube of four units per side, to form all of the possible combinations among them, which make all bodies possible. And in the same way, considering a cube of nine per side, it applies to the nine absolute principles and to Ramon Llull's nine relative principles, which will make way for all of the relationships of these principles in everything that exists. Later he will add Aristotle's accidents, which also number nine if considered in their three dimensions. Definitively, the cube comes to be "a type which is the rule and measure of the rest, treating all things with perfection." Going even further, Juan de Herrera comes to state that the ternary relation, which engenders the cube from the union of the line with the surface, is the greatest vestige of the Most Holy Trinity within creatures. Along with this, the signification assigned to the cube by the operations which it necessarily carries in all created beings becomes transcendent.

Following these discourses, Herrera will depict the four elements as a line of four tiny cubes, over each of which figures a letter (A,B,C,D). Next, this line will engender a surface of 4x4 tiny cubes, and in each one of the 12 remaining he will figure in all of the combinations of four letters taken two by two, and their variations, which are produced upon successively adding the other three letters to each letter of those that form the line. The precedence of one of them would imply its being predominant or active, as opposed to the other, passive element, which makes one note that there are no repetitions. When he explains that this combination of line and surface will engender at the same time the cubic figure, completing the relationship of all with all, he draws the corresponding cube of 4 small cubes per edge (fig.1).

Herrera follows the same process to combine the nine principles, with tiny cubes designated with the 9 letters of the Llullian alphabet. Thus he draws the surface of 9x9 cubes placing the 9 isolated principles on one line and over the remaining 72 he places the combinations of letters taken two by two with their variations (fig.1). And he again indicates the next process which results in the cube of 9 units per side, making a general reference to the combinations of all with all. It seems that in this way the 9 isolated relationships would have been obtained, the 72 combinations of these 9 taken two by two, and 648 possible combinations would remain (in the case of the accidents –which also number nine– Herrera mentions a total of 729 solid cubes). However, this number does not seem to coincide with any type of combinations of nine nor with their sum, which, including variations, would be 986,409.

We can observe through present understanding that if in fact the n elements taken in isolation (the combinations of n over 1) plus the combinations of n over 2 with their variations of sequence, always add up to n squared $[n+n(n-1)]$. However, continuing the sum of n over 3, n over 4, etc. until arriving at n over n with its variations, it does not result that the sum is n cubed, as Herrera, who never manages to prove it, would have wished. The sum would be instead equal –at its limit– to Euler's number multiplied by n! (n! being the product: $n(n-1)$ $(n-2)…3 \times 2 \times 1$). When n is the number four we have the only exception that turns out to be truthful: so this total is equal to four cubed; and this exception is the one which permits our author to establish in a general sense the cubic simile as "plenitude and the absolute end of differences in plenitude, with neither lack nor excess, and the totality of mixtures and qualities."

In this way one arrives at the composition of all creatures that have in themselves the four elements and all of the substantial and accidental principles. And since they are composed "thus as many cubes that work together with one another, what will result from this multiplication of cubes with other cubes will necessarily be a cube," and the result of this is the elemental cube or individual.

What results from all of this is, therefore, a dynamic vision of the cube, fruit of the linear engendering element of the surface and of both together as generators of the volume. And at the same time, a sense of plenitude has been attached to this cubic body, as the bearer of the three dimensions necessary for being and operating; the cube also provides a combinatory mechanism that equally achieves the formative plenitude of the complex totalities, which pertain to every limited being created. In this way, the cube is at the base of every individual, of everything in existence, as an explanation of its being and operating, and beneath all reality as an explanation of its diversity in the specific and its plenitude in the combination. According to these explanations, it is no wonder that in the paintings of Luca Cambiaso on the ceiling of the Escorial, the great architectonic work of Herrera, should appear the figure of Our Lord Jesus Christ, supreme referent of all Creation, seated upon the cubic figure in the context of the Trinity.

After Descartes

The passage into modernity introduced a higher degree of abstraction, which while it distances us from the sensory permits us to overcome the material limitations which the sensory imposes. As Dan Pedoe states: "An enormous simplification came into being with the introduction of sensed magnitudes, which could be either positive or negative, and the introduction of co-ordinates by René Descartes in 1637 changed the face of geometry for all time."[3] The conception of geometry advances even beyond everything imaginable for man, giving rise to recognizably differentiated methods. Each way of operating opens its own possibilities, and mutual reference is always possible: generally, science expands its horizons and leaves the sum of prior understanding as a particular case of the latest conception developed.

Dan Pedoe make us take note again: "At about the same time that Descartes was producing his epoch-creating ideas, Girard Desargues (1593-1662), an architect from Lyons, was engaged in an extension of geometry which can be regarded as being more fundamental than that created by Descartes. We call this projective geometry...in which there is no reference to measurement." And along with many other scholars he came to demonstrate that "Euclidean geometry is merely projective geometry with reference to a special pair of points" (the complex points I and J on a line in infinity). The set theories would end up again orienting these studies in another direction that is even more universal: By beginning with a set of points within which the lines would be determined subsets, and by establishing certain axioms, one would find a geometry the deductions of which could be extended to other sets of objects that fulfill those determined conditions.[3]

The scientific development outlined by Pedoe arrives at the possibility of considering more than three dimensions for geometry, and even doing so without denying a representation of three dimensions in space of what is defined in four. This representation would be an intersection of such a four-dimensional space in a space of three dimensions, in the same way that we can consider the intersection of a plane in an object defined three-dimensionally. He notes that this fourth dimension might be time, and he offers the example of how the inhabitants of a two-dimensional world might interpret as a variation in time the figure that would be traced in a plane by the intersection of a three-dimensional sphere which, moving through three-dimensional space, would pass through that plane; the first point would be produced, then a series of circles, increasing until one reaches the diameter of the sphere, and then diminishing until once again ending in a point just before disappearing. The concept of the hypercube is assumed in terms of mathematics. Although we may not be able to draw them nor even imagine them, it is possible to speak of bodies equivalent to a cube in three dimensions, defined in four, five or more dimensions. The hypercube may be a body in a space of n dimensions that has equal sides. In this case, in the definition of the hypercube, what holds a singular importance is the identical dimension of its sides.

In speaking about fractal geometry, Mandelbrot characterizes Euclidean geometry as that of the "sets for which all the useful dimensions coincide," that is, "dimensionally concordant sets," while fractal geometry observes that "the different dimensions of the sets fail to coincide; these sets are dimensionally discordant."[5] He refers to the topological dimension (which is always a whole number) and to the Hausdorff-Besicovitch or fractal dimension; in fractals the latter dimension is greater than the former, while in Euclidean geometry the two remain equal. In the new fractal geometry, also, the latter dimension is typically not a whole number, in contrast to Euclidean geometry in which it is always a whole number. Indeed, a special form of the fractal dimension is the self-similarity dimension. If we divide a cube into an equal number of small cubes (in a way similar to the constructions of Herrera), we find that the number of resultant cubes is equal to 1 divided by the factor of reduction (the relationship between the edge of the small cube with respect to that of the initial cube) raised to the dimension of self-similarity. This relationship fulfills any case considered. Whatever the factor of reduction that we consider, the dimension remains constant, and in the cube it is equal to its topological dimension, that is, 3.[6]

The cube as generator of art forms

This immense panorama opened by science can do no less than to influence the imagination of those artists who work with plastic elements: painters, sculptors and above all, architects. It is commonly said that in their art, consciously or unconsciously, they metaphorically render many of these argumentations, they symbolize them or even come to predict them, in the most unusual occasions. Although the complete development of these ideas would require more space, we would like to outline at least some examples.

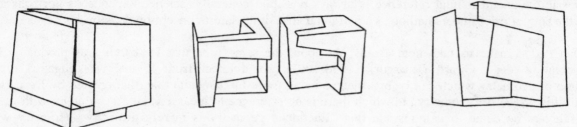

Figure 2: *Jorge Oteiza's sculptures*

The generation of the cube in a superficial manner, which we see mostly in the case of the Platonic argument, could be tied to the experiments with the cube carried out in some of Jorge Oteiza's sculptures. Oteiza is interested in the empty space that these planes define in their interior, which is shown to the viewer by opening up the surface that encloses it, and in this way said space is "deoccupied," according to the expression used by the artist. The sculpture is defined by materializing those surfaces in steel sheets of very little thickness, and by showing this extreme to the viewer (fig.2). In *The Metaphysical Box of Fra Angelico. Conjunction of Two Trihedrons[7]*, from 1958, we are permitted to access the interior void by slightly separating the two trihedrons that form the cube, as its name indicates; along with this, six of the edges are duplicated and the total virtual volume increases. In *Empty Box*, from 1958, each surface appears cut back by the sculptor's removing a trapezoidal shape from each face; although without a uniform law, the edges appear affected, partially disappearing, while the volume of the virtual cube remains identical.[8]

Figure 3: *Schröder's house*

The generation of the cube as a line that moves, engendering a flat surface, so that both (line and plane) are displaced, engendering the cube, could be tied to the neoplasticist constructions, especially Rietveld's. In these, the lines and planes, which remain as if frozen in their distinct positions, are emphasized, always avoiding the form of dihedrons –and that of trihedrons even more so– which would give the impression of a closed volume. Especially representative is Schröder's house (fig.3), where, as Theodore M. Brown already had already affirmed in 1958 [9] the architectural elements extend beyond the points of their intersection, showing lines and separated planes in their complete integrity. The void and the glass act as separators between the superficial and linear elements. Also, the different colors reinforce the singularity of the superficial and linear elements. And above all, the lines of the elements in I (being extended) act as axes of a coordinated three-dimensional system. The reference to infinite space where this architectonic body is situated and defined is therefore evident in this case. We should point out, however, some peculiarities. The generating planes are fragmentary and are established in the three directions simultaneously: these planes are vertical as well as horizontal. Among the horizontal ones, which extend between the eaves and the floor at the ground level, we should include the fragmented flooring of the first level, because of the distinct color that it takes on.

The generation of the cube as orthogonal lines that extend in the three Cartesian directions with identical measures is present in the preference of Le Corbusier for the module of 2.26 x 2.26 x 2.26 meters (the measure of a man with his arm raised upward, the origin of his Modulor [10]), which gives way to the proposition of honeycomb structure in which to develop human dwellings. In this structure, according to its author, one discovers " 'un contenent d´hommes', affirmation d´un élément volumétrique capable de mettre de l´ordre, de transformer les règlements et d´aider l´architecture des temps modernes dans sa lourde tâche de créer les logis de la civilisation machiniste"; also adding at another moment: "l´exactitude est encore ici une source de confort physique et intellectuel." The author himself carries out the application of this structural network, according to his patent of December 15, 1950, in such studies as "ROQ" y "ROB" in Cap Martin [11] on the Azure Coast, such as the "Maison de l'Homme" (Le Corbusier Center) in Zurich, where he uses it in the structure enclosing the interior space[12]. In order to better understand the meaning that this cubic structure had for Le Corbusier, we can recall what he left us written on the Modulor. As he states, Mediterranean architectural art was *the spirit placed beneath the sign of the square,* while *the spirit placed beneath the sign of the triangle and of the convex pentagon or star, and its volumetric consequences: the icosahedron and the dodecahedron,* characterized the Germanic. Masculinity, referring to architecture, pertained to the former, while femininity, as a subjective and abstract symbolism, characterized the latter. *The man of the ruler* would be the first of these artifices, *the man of the compass* the second, the one which appears to have been predominant in those times; however, our author considers that today (in his time), the ruler is necessary and the compass dangerous. [13] And in accord with these ideas he draws and writes from 1947 to 1953 his "Poem of the Right Angle." [10]

The imaginary generation of a four-dimensional cube that leaves three-dimensional traces interpreted as temporal variations could be placed in relation with the experiments of the Eisenman's houses, in which the deep formal structures introduced, or the transformations suffered by the initial solid during the process of creation, leave their impressions on the final result. Those transformations or deep formal structures are independent of both the function and the construction, because of which they never stop being simple geometric play that enriches the shape and the final space.

We can say that the law of generation itself, more than its result, is the object that minimalist works attempt to represent; so it happens in many sculptures, especially the works of Sol LeWitt, which consider the cube as a point of departure. It is a law of generation that, in contrast with the cases considered up to this point, is presented physically, that is, with a presence simultaneous with all the intermediate steps (instead of being deducible by the trace that the process utilized leaves in the final body). Each process is a work and each work is a process. Examples of this may be "Serial project (ABCD)" of 1966 or "3 part set" of 1968. The self-similarity also seems inherent in these works, apparently conceived, that try to capture the attention of the viewer and focus it onto themselves, impeding other associations or analogies that might transcend that which is immediately present. In some works by Donald Judd we find this self-similarity in the way in which identical cubes are repeated; in other works by Sol LeWitt such as "Modular Cube" of 1962, we find it as cubes obtained from another cube by subdivision (the application of a factor of reduction) or if we consider the inverse, by aggregation.

A segment of the most current architecture seems once again to be inspired by this poetry that minimalist architecture emphasizes. In this case, its signification is reinforced by the complexity that its greatest scale introduces, in terms of construction, perception and use. But even a passing glance a the most emblematic cases would still require a more extensive study that merits an exclusive development.

In the pictorial representation of the cube in two dimensions, the possibilities increase, since one can draw even that which is impossible to construct in three dimensions, such as tricks of false perspective show us. Escher is an emblematic example of the utilization of these resources, as observed already by mathematicians, in the most figurative manner. The empty cube that changes the connection with the base of the anterior and posterior vertical edges is found represented in a drawing and in a model that is held in the hand of a character in Escher's work "Belvedere" of 1958, and at the same time the principal motif of the drawing produces this same effect in an architectural belvedere of two floors that is as impossible as the cube, occupied by characters who, like those who operate the ladder, contribute even more to make us note its impossibility. In his "Metamorphosis II," xylograph from 1939-40, Escher himself attends more to the superficial aspect that the full cube presents, by means of drawings such as the well-known floor of rhombuses (white, black and gray) that simulate relief in cubes with three faces visible. It is not without significance that in this work the cubes are subdivided producing architectonic conglomerations of little Mediterranean houses.

All of the productions of this artist acquire a disquieting effect, nearly surreal, that traps the viewer, making him/her participate in these impressions. Meditation on the fragility of the real meaning of perception itself, which captures only appearances, and the suspicion about the instability of the perceived world which could correspond to that perception, is supported paradoxically on properties arising from the commonly considered aseptic and firm world of science.

The examples mentioned here may give an idea of the stimulus that, for the artist, means emphasizing the different conceptions of the cube which have been given to it through the various geometries. We have seen moreover that artists and architects tend to find associated meanings for the cube that transcend the mere physical conception of the figures. The fecundity that this very basic and apparently very simple shape has had and continues to have in the art of our time leads us to consider in this sense the expression "infinity is a cube without vertices," a statement which we can make by paraphrasing the one who wrote that infinity is a square without corners. [14]

References

[1] Plato *The Timaeus of Plato* Ayer Publishing, Salem, 1988

[2] Glen R. Morrow (ed.), *Proclus' Commentary on the First Book of Euclid´s Elements*

[3] Dan Pedoe: *Geometry and the Liberal Arts*, Dover Publications, New York, 1976

[4] Juan de Herrera: *Sobre la figura cúbica*. Preliminary study Manuel Arrate Peña. Santander: Universidad de Cantabria. S. Publicaciones; Camargo: Ayuntamiento, 1998 (English translation ours)

[5] Benoît Mandelbrot: *The Fractal Geometry of Nature*. WH Freeman and Company, New York, 1983

[6] Peitgen, Heinz-Otto, Jürgens, Hartmut, Saupe, Dietmar: *Chaos and Fractals. new frontiers of science*. Springer, New York, 1992

[7] Pedro Manterola: "Cinco pasos en torno a la Pasión de Jorge Oteiza", en *Oteiza-Moneo*, AAVV, Pabellón de Navarra Exposición universal de Sevilla 1992

[8] Roberto Ercilla: "Desocupación espacial del cubo" in *Less is more. Minimalismo en arquitectura y otras artes*, Vittorio E. Savi & Josep M. Montaner, C. O. Arquitectos de Cataluña y Actar, Barcelona 1996

[9] Theodore M. Brown *The Work of G. Rietveld, architect* A.W. Bruna & Zoon, Utrech, 1958.

[10] Le Corbusier: *Modulor 2 1955 (la parole est aux usagers) suite de "Le Modulor 1948"*, Editions de L´architecture d´aujourd´hui, 5 rue Bartholdi-Boulogne (Seine)

[11] Willy Boesiger: *Le Corbusier. Oeuvre complète 1946-1952*. Vol 5. Editions d´Architecture Zurich,1953

[12] Willy Boesiger: *Le Corbusier*. Vol 8. Editions d´Architecture Artemis Zurich,1970

[13] Le Corbusier *Le Modulor. Essai sur une mesure harmonique a l´echelle humaine applicable universellement a l´architecture et a la mécanique.*

[14] Bruno Munari: *La scoperta del quadrato*. Nicola Zanichelli Editore S.p.A., Bolonia, 1978

Applications of Fractal Geometry to the Player Piano Music of Conlon Nancarrow

Julie Scrivener
1721 Sunnyside Drive
Kalamazoo, MI 49001
E-mail: julie.scrivener@wmich.edu

Abstract

The relationship between music and geometry goes back thousands of years to the Greek quadrivium. Fractal structures have been explored in music and sound since at least 1978 (Gardner) and this work has recently been extended to specifically explore fractal structures in melodies (Mason and Saffle, 1994; Chesnut, 1996) and in musical forms and phrase structures (Solomon, 1998). Among the fractals that have been identified in musical structures are Sierpinski's triangle, Peano curves, and the Koch snowflake.

This paper is an effort to apply fractal observations to the player piano studies of American-Mexican composer Conlon Nancarrow. The most clearly mathematically-oriented of Nancarrow's Studies are the canons that explore mathematical relationships as simple as two voices in the relationship 3:4 or as complex as twelve voices proportional to the pitches of the justly-tuned chromatic scale. In particular, those of the canons which are also "acceleration canons"—that is, using carefully controlled rates of acceleration and deceleration among the voices—offer compelling possibilities for study of fractal properties. Among the studies which will be examined here are Nos. 14 (two voices) and 19 and 27 (three voices).

Introduction

The relationship between music and geometry goes back thousands of years to the Greek quadrivium. Fractal geometry is a relatively newly-described branch of mathematics based on the 1977 work of Benoît Mandelbrot [1] in which elements of self-iteration and scaling are recognized in a variety of naturally-occurring objects as widely diverse as coastlines and bodily structures such as the brain and bronchial lobes. Mandelbrot's theories began to be applied to music and sound beginning in 1978, with fractal structures being identified in the nature of sound itself [2, 3], in melodies [4, 5], and in musical forms and phrase structures [6].

Review of the Literature: Fractals, Music, and Sound

In studying fractal qualities in music, the properties of self-iteration, scaling, and space-filling have been the focus of study. According to Solomon [6], "Perhaps the most important defining property of fractals is self similarity on many different scales; i.e., they have self-iterating geometric structures that repeat in different sizes." Solomon uses the beautiful example of a fern frond, a natural object in which the same leafy shape is iterated on a number of different scales.

The *space-filling* property of fractals is also important in music. Consider Figure 1, a simple line fractal known as a Peano curve. With each new iteration of the generating shape, the space is more tightly filled and the length of the line drawing the curve increases. The number of iterations and the length of the line can reach infinity, moving toward filling the space but never completely doing so. In music, the property of space-filling takes place in the time dimension when a pattern is reiterated in proportionally shorter time values.

Let us now apply the same ideas of self-iteration, scaling, and space-filling to music. Figure 2 shows a short "generating motif" that can be compared to the largest triangle of a Koch snowflake. Notice how, with the second iteration of the melody, the same melodic shape is reiterated (self-iteration), in proportional time values (scaling), and more sonic space is filled. These are not new concepts to musicians, who recognize self-iteration in melodic imitation, scaling in rhythmic augmentation/ diminution, and space-filling (in a sonic sense) in the application of these procedures together.

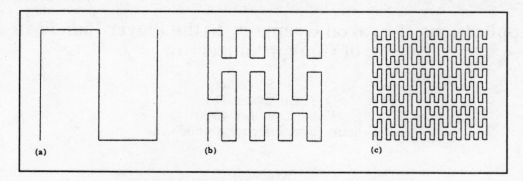

Figure 1: *A Peano curve ([4], p. 31), which illustrates the space-filling ability of fractals. With each further iteration of the curve, the length of the line drawing the curve approaches infinity.*

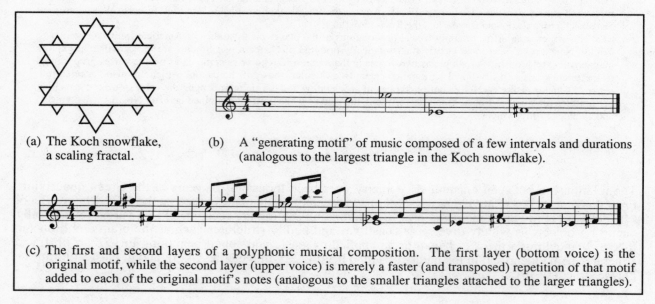

(a) The Koch snowflake, a scaling fractal.

(b) A "generating motif" of music composed of a few intervals and durations (analogous to the largest triangle in the Koch snowflake).

(c) The first and second layers of a polyphonic musical composition. The first layer (bottom voice) is the original motif, while the second layer (upper voice) is merely a faster (and transposed) repetition of that motif added to each of the original motif's notes (analogous to the smaller triangles attached to the larger triangles).

Figure 2: *Properties of self-iteration, scaling, and space-filling in a musical segment ([7], p. 190).*

Fractals can be observed in music in other ways. In 1978, Gardner [2] wrote of the work of Richard Voss, in which the nature of sound itself was revealed to have fractal properties. Mandelbrot and Voss discovered a special class of sounds in which the property of scaling is actually present in the waveform itself. These sounds, which Mandelbrot terms "scaling noises," have the fascinating property that the sound—including its pitch—does not change if the sound is played at a different speed.

Voss's work also focused on how the nature of sound relates to the construction of pleasing melodies. Gardner [2] describes how Voss identifies the frequency spectra for three types of "noise"—white, $1/f$ ("pink"), and Brownian—and demonstrates how the properties of these different waveforms could be interpreted as melodies. It turns out that $1/f$ ("pink") noise exhibits fractal self-similarity whereas white and Brownian noise do not, and it is the melodies based on $1/f$ noise that most people in a test audience found most appealing, based on the melodies' effective balance between complete randomness (surprise) and extreme correlation (expectation).

Fractal properties of melodic structures have been further studied by Mason and Saffle [4], who showed how right-angle drawings known as Lindenmayer (L-system) curves could be used to create melodies—albeit of questionable musical value. Melodies are created from the curves by interpreting horizontal line segments as durations and vertical line segments as pitches. Mason and Saffle also assert that many existing melodies can be shown to have strong correlations with L-system curves, although their work in 1996 is very preliminary. They did, however, identify L-system curves that "generate tunes that are similar or even identical to hundreds of existing melodies by classical and popular composers" (p. 35).

Solomon [6] relates the ternary divisons and forms commonly found in music to the fractal known as Sierpinski's triangle (Figure 3), which he compares to a ternary (ABA) scherzo form. The ternary relation is obvious, but Solomon also notes that within the larger divisions one often finds binary divisions (such as a rounded binary form in the first A section), and these would relate to the binary division of the triangle's sides that create further iterations of the Sierpinski triangle.

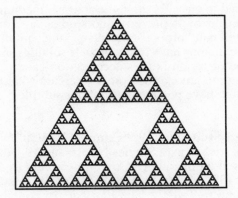

Figure 3: *An example of Sierpinski's triangle [6].*

And, of course, it should be mentioned here that the Schenkerian system of analysis has, as a primary goal, the identification of self-iterating melodic and harmonic patterns that are represented in both surface details and structural components—thus confirming both the properties of self-iteration and scaling as being inherent in many tonal compositions.

Finally, several writers have made fascinating speculations on a basic connection between fractal properties in nature, in the nature of sound itself, and our perception of musical beauty. Mason and Saffle's work [4] led them to conjecture about a fundamental relationship between our perception of beauty in melody and musical form and the presence of fractal qualities in music. As they state:

> Aspects of certain theories about the origins and fundamental structures of melodies suggest that much—perhaps all— beautiful music is, in some essential sense, fractal in its melodic material and internal self-similarity. (p. 35)

Later, they say:

> Is there something universally appealing about music—something that transcends individual cultures and tastes? We believe the human mind may use one or more models of perception in order to determine whether a given melody or musical structure is ugly or beautiful. (p. 36)

Gardner [2] notes that Mandelbrot, too, has raised similar questions in regard to abstract art:

> Is it possible, Mandelbrot asked himself many years ago, that even completely nonobjective art, when it is pleasing, reflects fractal patterns of nature? Mandelbrot has some unpublished speculations along these lines. He is fond of abstract art, and maintains that there is a sharp distinction between such art that has a fractal base and such art that does not, and that the former type is widely considered the more beautiful. (p. 24)

Nancarrow and His Player Piano Studies

Conlon Nancarrow began writing his remarkable player piano studies in the late 1940s, and by the time of his death in 1997 this body of work consisted of about 50 pieces. Nancarrow became interested in writing music based on "temporal dissonance," or multiple and often conflicting tempos, after reading Henry Cowell's book *New Musical Resources* in 1939. At this time Nancarrow was preparing to flee the United States for Mexico as a result of his Socialist party affiliations. He spent the rest of his life in Mexico, working in virtual isolation until his music began to attract notice in his later years and he began to be regarded as an eccentric genius (confirmed in 1982 when Nancarrow was among the first class of recipients of the MacArthur Foundation's "genius" grants).

Nancarrow's interest in the player piano was a practical one: as a composer in the late 1930s, he was intrigued by Cowell's radical ideas about rhythm yet frustrated by the limitations of human performers in interpreting complex tempo relationships. He found the player piano to be the best means available at the time for realizing total control over performance. It is fortunate for us that Nancarrow reached this nexus in his compositional career when he did—he commented in later years that, had he not found the player piano, he would surely have turned in the 1950s to the electronic medium for his compositions.

More than two dozen of Nancarrow's player piano studies are canons, which are basically of two types. In the first type, what I will call "proportion canons," unchanging ratios of tempos are established among the canonic voices. These canons have ratios in their subtitles, for example "Canon 12/15/20" and "Canon √2/2." The second type of canon is the "acceleration canon," which uses carefully controlled rates of acceleration and deceleration among the voices to create what Gann [8] calls a "sense of curved time" (p. 146). These canons often have percentages in their subtitles, such as "Canon 5%/6%/8%/11%" (Study No. 27).

Study No. 14, "Canon 4/5" and Study No. 19, "Canon 12/15/20". We can begin our exploration of fractal properties in Nancarrow's studies with Studies No. 14 and 19, which are both part of a group of six canons (Nos. 14-19) that are based on the additive rhythm formula (n-1, n, n+1, n). In both canons, the total length of the canonic part (337 eighth note beats) is derived from three versions of the additive formula: 3+4+5+4, 5+6+7+6, and 6+7+8+7. The first pattern contains a total of 16 eighth notes, the second 24, and the third 28; the smallest common multiple of these patterns is 336, and to make the patterns converge at the end a final note is added for a total of 337 beats (see Figure 4). A fourth voice declaims the pattern 4+5+6+5, although this pattern is varied among the six pieces so that they do not all have the same rhythm. In Study No. 14 [10] the four patterns are collapsed into a resultant rhythm; the

Formula = **n-1, n, n+1, n**
Pattern 1: 3+4+5+4 = 16 eighth-note beats
Pattern 2: 5+6+7+6 = 24 eighth-note beats
Pattern 3: 6+7+8+7 = 28 eighth-note beats
Smallest common multiple = 336

Figure 4: *Derivation of additive rhythm formulas in Studies 14-19. A fourth pattern, 4+5+6+5, provides rhythmic variety among the six pieces. The total beat length of Studies No. 14 and 19 is 337 beats (i.e., beginning and ending with a convergence of the three patterns).*

resulting rhythmic attacks are shown in Figure 5. A glance at the rhythmic structure reveals another additive formula that occurs among the four voices: beginning with the top voice, additive patterns of 3+4+5+6, 4+5+6+7, and 5+6+7+8 (identified in the shaded boxes) can be traced descending toward the right, with each pattern lengthening in time. The patterns soon become obscured as they overlap, but before this happens I believe it is possible to perceive an arithmetical deceleration effect as the note values increase from the second measure ($\frac{4}{8}$) to the third ($\frac{6}{8}$). Whether they can be heard or not, the three additive patterns can be traced to the end of the piece, and represent self-similarity and scaling on a large scale.

Study No. 14, subtitled "Canon 4/5," is a two-voiced "proportional" canon in which the top voice states the canon 20% faster than the bottom voice. The bottom voice begins the canon, and the top voice enters at exactly the point that allows the two voices to converge in the center—at which point, the "follower" voice becomes the "leader" and vice versa. The interval of imitation is 2 octaves plus a fifth (see Figure 6). The bottom voice ($\downarrow$=88) states the first 33.7 (337 x 20% = 67.4 / 2 = 33.7) beats of the canon before the top voice enters at $\downarrow$=110 (a tempo relationship of 4 to 5). Once the canon is underway, the top voice—going 20% faster—states 30 beats in each system to the bottom voice's 24 (in

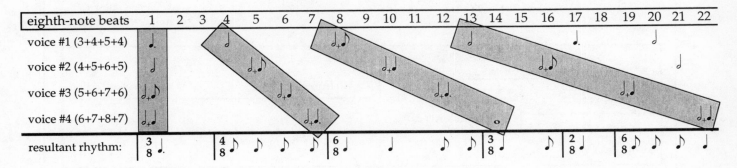

Figure 5: *Derivation of rhythmic attacks in Nancarrow's Study No. 14.*

Nancarrow's hand-written scores, the amount of space between notes is proportional to elapsed time). The two voices converge at the midpoint (beat 169), at which point the top voice becomes the leader; the piece concludes with the bottom voice stating its final 33.7 beats alone. Carlsen [9] calls this an "arch-shaped canon" (p. 18).

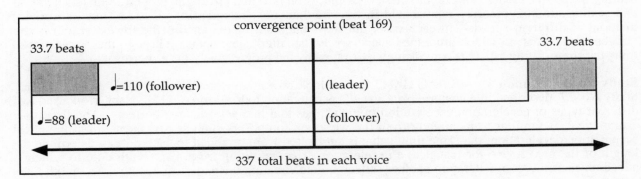

Figure 6: *Structure of Study No. 14. The top voice states the same musical material as the lower voice but at a higher pitch and a faster rate of speed. Canonic voices converge in the middle to form an arch.*

Study No. 19 [11], subtitled "Canon 12/15/20,"[1] uses three canonic voices and is constructed so that the point where all three voices converge is the very last note of the piece. In Study No. 19, the four additive rhythm formulas are clearly delineated into four distinct registers, opening with a 4-note chord that spans four octaves. Both Gann [8] and Carlsen [9] note that each of the four voices declaims basically the same melody, with the slower voices occasionally dropping a note in order to keep up with the faster voices. Carlsen [9] calls attention to this as a sort of Chinese-nested-boxes fractal relationship.

Like Study No. 14, No. 19 begins with the lowest (and slowest) voice ($\downarrow$=144); it then adds a faster middle voice ($\downarrow$=180) and a faster-still top voice ($\downarrow$=240). In order for the convergence point to occur at the end, the first voice must state 67.4 eighth-note beats before the second voice enters. The interval of imitation between the voices is an eleventh, and the range of each voice is four octaves. Since the range of Nancarrow's piano is narrower than a standard piano by two keys on the bottom and three keys on the top, the entire keyboard is used, with the voices intentionally arranged so that the middle voice is symmetrical about the piano's middle note (E_4). The basic structure of Study No. 19 is shown in Figure 7.

---

[1]I can find no reason why Nancarrow specifically chooses the ratio 12:15:20 instead of the superparticular ratio 3:4:5 to which it reduces, despite the claim of Tenney [12] that 12:15:20 "incorporates" the three ratios 3:4, 4:5, and 3:5 used in this series (p. 51). This canon and No. 17 (which uses the same ratio, 12:15:20) are the only ones which use a non-reduced ratio in the subtitle.

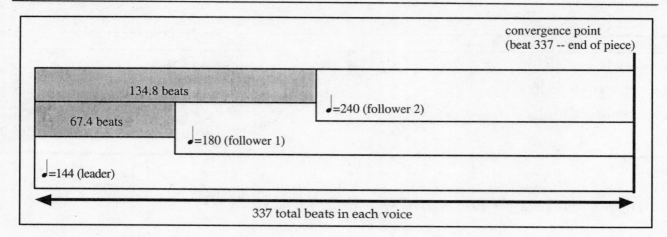

Figure 7: *Structure of Study No. 19. The musical material in the lowest voice is restated twice at progressively higher pitch levels and faster rates of speed. The voices converge on the very last note.*

Perhaps the best way to perceive the fractal nature of the overall forms of these two studies is by hearing them—the reader can listen to these brief pieces on [13] and [14]. The properties of self-iteration and scaling are convincingly represented in the simultaneous statement of identical (but transposed) material at different speeds—in each case, the first statement of the canon (the lowest voice) is the "largest" statement, to which are added successively "smaller" statements. Through the dimension of time, the space-filling aspect is convincingly portrayed by the progressively faster voices.

Study No. 27, "Canon 5%/6%/8%/11%." Let us look now at an example of an "acceleration canon": Study No. 27, described by Gann as "Nancarrow's acceleration tour-de-force" [15]. Nancarrow used two different types of acceleration/deceleration in his studies that had very different effects. The first type, arithmetical, is familiar to us already from the works of composers such as Messiaen. In arithmetical acceleration/deceleration, the same time value is subtracted from or added to each note to determine the length of the next note—for instance, a deceleration effect resulting from a sixteenth note to which is added on each successive iteration another sixteenth note, creating the pattern sixteenth note, eighth note, dotted eighth note, quarter note, etc. The resulting effect is not a smooth continuum, but a constantly increasing rate of change.

In geometric acceleration/deceleration, on the other hand, the rate of change is kept constant. The resulting rhythm is not a chain of standard note values and is too unwieldy to notate conventionally. Nancarrow found arithmetical acceleration to be adequate for small-scale effects, but geometric acceleration is far superior for the long, smooth acceleration and deceleration effects that could have structural significance. Study No. 27 [11] is one of eight studies Nancarrow wrote using geometric acceleration and deceleration; the percentages in the title indicate the four rates of acceleration and deceleration that are used in the piece. The piece also features a "clock" line in the middle of the texture that repeatedly states the same four pitches and forms a constant throughout the piece, creating a frame of reference against which the tempo changes can be heard.

Although in Studies No. 14 and 19 there is a clear mathematical basis for the rhythm of the voices, such does not appear to be the case for Study No. 27. Because there is no meter and the geometric acceleration technique requires spatial notation in the score, even in the clock line it would be difficult to discern a definite pattern. Whether or not any fractal structures will emerge in the rhythm is an area for further study.

Unlike the simpler forms of Studies No. 14 and 19, Study No. 27 is actually constructed of a series of 11 different canons. Its texture is also more complex because of the greater number of voices: there are four canonic voices plus the clock line in the middle. Within this structure, however, are smaller structures that are similar to the simpler canons in their fractal nature. Near the end of the piece, the structure of the ninth canon overlaps the four voices so that they are symmetrical about their centers in the same way that Study No. 14 was (see Figure 8). The voices progressively enter from highest to lowest in descending 1/2-step increments: the first voice on D, the next on C#, then C, and finally B.[2] This chromatic progression happens to be a transposed version of the four chromatically adjacent pitches which comprise the clock line, whose notes are D#, E, F, and G♭.

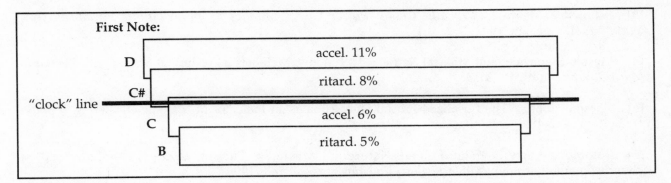

Figure 8: *Diagram of the structure of the ninth canon in Study No. 27. Each voice states the same musical material at a different pitch level and at a different rate of acceleration or deceleration (ritard.=ritardando).*

In this piece, the coincidences of the numerous canonic structures can be quite striking to the ear when they emerge from the contrapuntal texture. The very end of the piece is an excellent example—as in Study No. 19, Nancarrow sets up the four voices so that they converge on the last note in an audible fractal structure, with the same melodic pattern in all the voices but at different scales (rates of speed). On the final ascending scalar pattern from A_3 to G_4, each voice is assigned a rate of acceleration that becomes progressively faster in relation to the registral placement of the voice: the lowest voice accelerates at 5%, the next highest at 6%, the next highest at 8%, and the highest voice at 11%. The smoothness of the geometric acceleration as the voices race progressively faster to the final note is a stunning effect.

The reader is encouraged to hear the piece in its entirety on [14].

Conclusion

Nancarrow's interest in temporal relationships, realized through the structure of the canon and the medium of the player piano, provided him the means to create what we can now recognize as fractals in sound. The pieces discussed here represent only a small portion of Nancarrow's ingenuity in manipulating musical resources such as tempo and form to convey the fractal qualities of self-iteration, scaling, and space-filling. The space-filling quality, in particular, finds new expression in Nancarrow's canons through the dimension of time as expressed in tempo.

It is my hope that this brief outline of the study of fractals in music over the past twenty-plus years, and the overview of Nancarrow's canons and possible fractal applications for their study, will reveal exciting new possibilities for the further study of both. Scholarship in both areas is still emerging and dates back only a few years. There is undoubtedly a great deal more to be revealed on the fractal nature of music, and the player piano music of Conlon Nancarrow offers a unique and exciting field of study for identifying fractal structures in music.

References

[1]   Mandelbrot, Benoît B. The Fractal Geometry of Nature [revised ed., first published 1977]. San

---

[2]There is an error in the score at the end of this canon. The second voice from the top (ritard. 8%) should end on a G octave in order for the voices to end chromatically on F, G♭, G, and A♭. Instead, the final two octave chords in this voice are in error, stating E and A instead of the required D and G.

Francisco: W. H. Freeman, 1983.

[2]   Gardner, Martin. "Mathematical Games: White and Brown Music, Fractal Curves and One-over-*f* Fluctuations." *Scientific American* 238/4 (1978): 16-32.

[3]   Schroeder, Manfred. *Fractals, Chaos, Power Laws: Minutes From an Infinite Paradise.* New York: W. H. Freeman, 1991.

[4]   Mason, Stephanie, and Michael Saffle. "L-Systems, Melodies and Musical Structure." *Leonardo Music Journal* 4 (1994): 31-38.

[5]   Chesnut, John. "The Localized Fractal Dimension as an Indicator of Melodic Clustering." *Sonus* 16/2 (1996): 32-48.

[6]   Solomon, Larry. "The Fractal Nature of Music." Tucson, AZ: 1998 [accessed at <http://community.cc.pima.edu/users/larry/fracmus.htm> on 5/28/1999].

[7]   Dodge, Charles, and Curtis R. Bahn. "Musical Fractals." *Byte* 11/6 (June 1996): 185-196.

[8]   Gann, Kyle. *The Music of Conlon Nancarrow.* Cambridge: Cambridge University Press, 1995.

[9]   Carlsen, Philip. *The Player Piano Music of Conlon Nancarrow: An Analysis of Selected Studies.* I.S.A.M. Monographs, no. 26. Brooklyn, N.Y.: Institute for Studies in American Music, 1988.

[10] Nancarrow, Conlon. Collected Studies for Player Piano, Vol. 5: Studies No. 2, 6, 7, 14, 20, 21, 24, 26, and 33 for Player Piano. Santa Fe: Soundings Press, 1984.

[11] Nancarrow, Conlon. Conlon Nancarrow: Selected Studies for Player Piano [With Critical Material by Gordon Mumma, Charles Amirkhanian, John Cage, Roger Reynolds, and James Tenney], edited by Peter Garland. Soundings, Book 4. Berkeley, CA: Soundings Press, Spring-Summer 1977. [Studies No. 19 and 27]

[12] Tenney, James. "Conlon Nancarrow's STUDIES for Player Piano." In *Conlon Nancarrow: Selected Studies for Player Piano*, edited by Peter Garland, pp. 41-64. Soundings, Book 4. Berkeley, CA: Soundings Press, Spring-Summer 1977.

[13] Nancarrow, Conlon. *Studies for Player Piano*, Vols. I/II [Studies No. 3a, 3b, 3c, 3d, 3e, 4, 5, 6, 14, 20, 22, 26, 31, 32, 35, 37, 40a, 40b, 41a, 41b, 41c, 44; Tango?]. Compact discs. Wergo: WER 6168-2/6169-2, 1987.

[14] Nancarrow, Conlon. *Studies for Player Piano*, Vols. III/IV [Studies No. 1, 2a, 2b, 7, 8, 9, 10, 11, 12, 13, 15, 16, 17, 18, 19, 21, 23, 24, 25, 27, 28, 29, 33, 34, 36, 43, 46, 47, and 50]. Compact discs. Wergo: WER 60166/67-50, 1987.

[15] Gann, Kyle. "Conlon Nancarrow: List of Works." <http://home.earthlink.net/~kgann/cnworks.html> [accessed on 9/11/1998].

BRIDGES
Mathematical Connections
in Art, Music, and Science

Maximally Even Sets

A discovery in mathematical music theory is found to apply in physics

Richard Krantz
Department of Physics
Metropolitan State College of
Denver
Denver, CO 80217
e-mail: krantzr@mscd.edu

Jack Douthett
Department of Arts and Sciences
TVI Community College
Albuquerque, NM 87106
email: jdouthett@tvi.cc.nm.us

John Clough
Department of Music
SUNY Buffalo
Buffalo, NY 14209
e-mail:
clough@acsu.buffalo.edu

1. Introduction

We wish to seat eight dinner guests--four women and four men--about a round table so that the guests of each gender are distributed as evenly as possible. The obvious solution is to seat men and women alternately. Now suppose we have five women and three men. Ignoring rotations and reflections, there are essentially five ways to seat them (Fig. 1). Which of these is the most (maximally) even distribution? (We will see later that the optimum distribution for one gender guarantees the same for the other.)

On the basis of the informal "most even" criterion, the best choice seems to be Fig. 1e (which happens to be the only arrangement that avoids seating three or more women together). This is a relatively simple case, but "dinner table" cases with larger numbers also have a unique best solution. How can we formalize our intuition about evenness for all such cases?

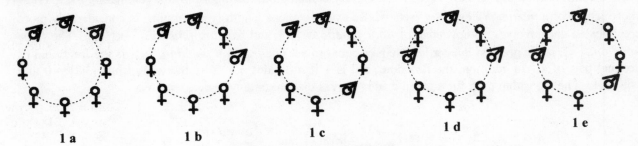

Fig. 1: *Seating Arrangements*

2. Defining Maximal Evenness

There are many ways to define maximal evenness consistent with the dinner guest problem above. We will give three very different but equivalent ways of defining such an arrangement.

2.1 The Measurement. Let us put this problem in a more abstract setting. Given c equally spaced points around the circumference of a unit circle, we wish to select d of these points to form a *maximally even distribution* that accords with intuition. One way is to choose a distribution that maximizes the average chord length between pairs of selected points when compared with all possible distributions of d points [1]. In this way the selected points are, on average, pushed as far apart as possible. For the unit circumference circle with c points, the length of the chord that connects points that are n points apart is given by

$$\text{chord} = 2\sin\left(\frac{\pi \cdot n}{c}\right).$$

(1)

The average chord length for men in Fig. 1 is worked out in Fig. 2, where the selected points are shown as filled circles. These configurations are listed from least to greatest average chord lengths and range from a *minimally even set* in Fig. 2a with an average of 0.98 to a *maximally even set* in Fig. 2e with an average of 1.70. Up to rotation and reflection, this figure lists all possible ways of selecting three out of eight points around the unit circle.

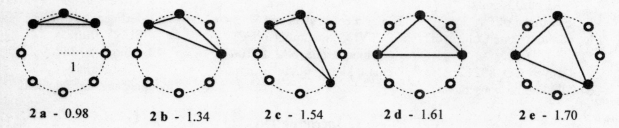

2 a - 0.98   **2 b - 1.34**   **2 c - 1.54**   **2 d - 1.61**   **2 e - 1.70**

Fig. 2: *Average Chord Length*

Minimally even sets (sets whose average chord length is minimum) have comparatively simple structures; all the filled circles--and hence, all the open circles--are clustered together. Using Eq. 1 it is possible to construct an algorithm that, for a given c and d, gives the minimum average chord length:

$$\text{Ave}_{min} = \frac{4}{d(d-1)} \sum_{k=1}^{d-1} (d-k) \sin\left(\frac{\pi \cdot k}{c}\right).$$

For the configuration in Fig. 2a, $\text{Ave}_{min} = 0.98$. By comparison, the algorithm for calculating the average chord length of a *maximally even set* (sets whose average chord length is maximum) is considerably more complicated and involves a knowledge of *floor functions*, $\lfloor x \rfloor$, and *fraction functions*, $\{x\}$ [2]. The *floor function*, $\lfloor x \rfloor$, is the greatest integer less than or equal to x. The *fraction function*, $\{x\}$, is the fractional or decimal part of x. In addition, the function $[d|n]$ is 1 if d divides n and 0 otherwise; thus $[3|8] = 0$ and $[4|8] = 1$. The algorithm that, for a given c and d, gives the maximum average length is

$$\text{Ave}_{max} = \frac{2}{(d-1)} \sum_{k=1}^{d-1} \left(2\left(1 - \left\{\frac{c \cdot k}{d}\right\}\right) - [d|c \cdot k] \right) \sin\left(\frac{\pi \cdot \lfloor c \cdot k/d \rfloor}{c}\right). \tag{2}$$

For the configuration in Fig. 2e, $\text{Ave}_{max} = 1.70$. The configuration of filled circles in Fig. 3b, below, is also a maximally even set. Using the Pythagorean Theorem, it is not difficult to observe that the average chord length for this figure is $(2 + 2\sqrt{2})/3$. Using the algorithm in Eq. 2, $\text{Ave}_{max} = (2 + 2\sqrt{2})/3$ as well.

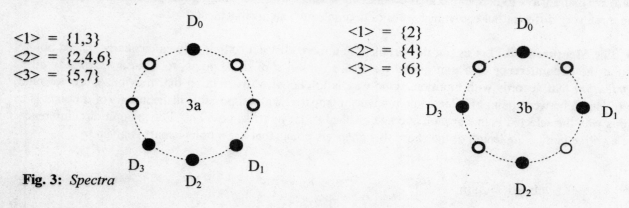

$\langle 1 \rangle = \{1,3\}$
$\langle 2 \rangle = \{2,4,6\}$
$\langle 3 \rangle = \{5,7\}$

$\langle 1 \rangle = \{2\}$
$\langle 2 \rangle = \{4\}$
$\langle 3 \rangle = \{6\}$

Fig. 3: *Spectra*

2.2 The Spectra. A particularly important property of maximally even distributions as defined above (important both to music and physics) involves what some music theorists call a *spectrum* [3,4]. Count, always moving clockwise, the number of points from one selected point (i.e., filled circle) to another in two different ways--by counting only the <u>selected</u> points and by counting <u>all</u> of the points between. The first of these measurements is called the *generic length* and the second the *specific length*. The *generic length* from a selected point to the next selected point is 1; if there is a selected point between them, the generic length is 2; if there are two selected points between them, the generic distance is 3; etc. For example, if the filled circles marked D_0, D_1, D_2, and D_3, in Fig. 3a are the selected points, the generic lengths from D_0 to D_1, from D_1 to D_2, from D_2 to D_3, and from D_3 to D_0 are all 1; from D_0 to D_2, from D_1 to D_3, from D_2 to D_0, etc. are 2, and from D_0 to D_3, from D_1 to D_0, etc. are 3. The *specific length* from one point to another is simply the total number of points between them (technically, if the endpoints are included, the length is one less than the number of points; if the endpoints are not included, the length is one more than the number of points). That is, the specific length from D_0 to D_1 is 3, from D_0 to D_2 is 4, from D_1 to D_2 is 1, from D_0 to D_3 is 5, etc.

The *spectrum of a generic length* is the set of specific lengths associated with that generic length. In Fig. 3a the set of specific lengths associated with the generic length of 1 is {1, 3}, since the specific lengths from D_0 to D_1 and from D_3 to D_0 are 3 and the specific length from D_1 to D_2 and from D_2 to D_3 is 1. This is written <1> = {1, 3}, where the number inside the angled-bracket represents the generic length and the numbers inside the curly-bracket represent the associated specific lengths. There are 3 specific lengths associated with a generic length of 2; <2> = {2, 4, 6}. Similarly <3> = {5, 7}.

In these terms, it has been shown that the average chord length is maximized precisely when each generic length is associated with either one or two consecutive specific lengths [1]. The arrangement in Fig. 3a does not qualify for two reasons; the spectra do not contain a single or consecutive integers, and <2> contains more than two integers. Fig. 3b, however, does qualify since each spectrum consists of a single integer. The 3 in 8 arrangements of Figs. 1 and 2 are reproduced in Fig. 4 with the spectra listed. The only arrangement for which each spectrum consists of one or two consecutive integers (i.e., each generic length corresponds to one or two consecutive specific lengths) is that shown in Fig. 4e, which is the same arrangement that maximizes the average chord length.

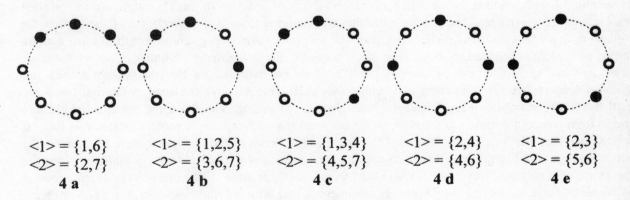

<1> = {1,6}	<1> = {1,2,5}	<1> = {1,3,4}	<1> = {2,4}	<1> = {2,3}
<2> = {2,7}	<2> = {3,6,7}	<2> = {4,5,7}	<2> = {4,6}	<2> = {5,6}
4 a	**4 b**	**4 c**	**4 d**	**4 e**

Fig. 4: *"Seating" Spectra*

2.3 The ME Algorithm. If we label c sites evenly spaced on the circumference of a circle consecutively with integers 0 through $c - 1$ then the maximally even (ME) algorithm provides a convenient way of selecting d points distributed in a maximally even way [5]

Let i be a fixed integer such that $0 \le i \le c - 1$, and assign the selected points to sites with labels $\lfloor (ck+i)/d \rfloor$ where $k = 0, 1, ..., d - 1$ and $\lfloor x \rfloor$ is the floor function.

Clough and Douthett [5] have shown that, for any given c and d and for each i, $0 \le i \le c$ - 1, this algorithm generates all and only the distributions for which each spectrum consists of one or two consecutive integers. These authors have also shown that all such sets are equivalent under rotation and that i, called the index, determines the rotation. In view of the discussion above, concerning the connection between average chord length and sets whose spectra consist of one or two consecutive integers, there are now three equivalent ways to define a maximally even set:

1. A maximally even set with parameters c and d is a set that, when compared with all other d-point distributions on c points, has a maximum average chord length.

2. A maximally even set is a set in which every spectrum consists of one or two consecutive integers.

3. A maximally even set with parameters c and d is a set whose integer elements are $\lfloor (ck+i)/d \rfloor$ where $k = 0, 1, ..., d$ - 1 and i is a fixed integer such that $0 \le i \le c$ - 1.

If we reproduce the arrangement of Fig. 4e and label the sites with integers 0 through 7 (see Fig. 5) then the set of labels of the selected points is {0, 3, 6}. This corresponds to the maximally even set with $c = 8$, $d = 3$, and $i = 2$. Other values of i produce rotations of the distribution shown. For the maximally even set in Fig. 3b, $c = 8$, $d = 4$, and $i = 0$.

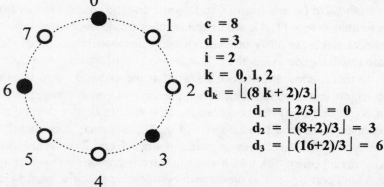

Fig. 5: *Selected Points*

$$c = 8$$
$$d = 3$$
$$i = 2$$
$$k = 0, 1, 2$$
$$d_k = \lfloor (8k + 2)/3 \rfloor$$
$$d_1 = \lfloor 2/3 \rfloor = 0$$
$$d_2 = \lfloor (8+2)/3 \rfloor = 3$$
$$d_3 = \lfloor (16+2)/3 \rfloor = 6$$

2.4 Summary. The above provides three equivalent definitions for maximal evenness; one in terms of average chord length, another in terms of spectra, and the third in terms of the algorithm. In the original work on maximally even sets, the spectra definition was adopted [5]. In that work it was shown that the complement of a ME set is also a ME set. Since ME sets maximize average chord length, it must also be true that the average chord length of the open-circle pairs in Fig. 2e is greater than that of any of the other distributions in Fig. 2. This can be seen in Table 1 where the averages for the *filled-circle chords* and *open-circle chords* (chords connecting filled and open circles, respectively) are listed in the first two rows for all the configurations in Figs. 2 and 4. Note that as the average chord lengths for filled-circle sets increase from one configuration to the next, so do the averages of their complementary open-circle sets; it can be shown that this is generally the case [2]. The filled-circle set of the configuration in Fig. 2e has the greatest average chord length, and hence, so does its complementary open-circle set. In addition (since the complement of a maximally even set is maximally even), each spectrum of the open-circle distributions in Figs. 3b and 4e consists of one or two consecutive integers, and for $c = 8$ and $d = 5$ there exists an integer i, $0 \le i \le 7$, such that the open-circle labels in Fig. 5 can be computed via the ME algorithm. The ME algorithm can also be used to compute the open-circle set in Fig. 3b. It is left to the reader to verify that this is in fact the case.

Table 1 - Average Chord Lengths

	Fig. 2a	Fig. 2b	Fig. 2c	Fig. 2d	Fig. 2e
Filled -Circle Chords	0.98	1.34	1.54	1.61	1.70
Open-circle Chords	1.30	1.41	1.47	1.49	1.52
Mixed Chords	1.62	1.47	1.40	1.37	1.33

Finally, it can be shown that if the average distance between men (and hence, the average distance between women), in our dinner party example, is maximum then the average distance between members of the opposite gender is minimum [2,6]. Thus, if our dinner party is meant to be a "mixer" the optimal seating is also a maximally even distribution. This too can be seen in Table 1 in which the last row lists the average mixed chord (a chord connecting a filled circle and an open circle) length for the configurations in Figs. 2, and 4. Thus, maximum average chord length for filled-circle chords, maximum average chord length for open-circle chords, and minimum average chord length for mixed chords occur simultaneously and occur precisely when one--and hence, both--of the sets are maximally even.

3. Even(ness) in Physics

The simplest description of the pairwise interaction of "spins" on a lattice is given by the Ising model [7]. The "spins" may represent angular momentum, intrinsic electron spin, or any two-state variable. In this model, spins at each lattice site, represented by the arrows in Figs. 6, 7, and 9 may take on the values ± 1 depending on whether the spin orientation is up (+1) or down (-1). The spins interact pairwise according to some convex distance dependent interaction energy. An exponentially decreasing function that depends only on distance between pairs of spins is an example of such a convex distance dependent interaction [e.g., $J(|i - j|) \propto \exp(-|x_i - x_j|)$]. Of particular interest are systems in which the interaction favors the antiparallel arrangement of spin pairs (i.e., spins that preferentially line up opposite to one another due to their pairwise interaction). Such a system is said to be "antiferromagnetic."

The so-called configurational energy, U, of an arrangement of N spins interacting antiferromagnetically by a pairwise distance dependent interaction energy $J(|i - j|)$ is attained by adding the contributed energy of all distinct pairs:

$$U = \sum_{\substack{i,j=0 \\ i \neq j}}^{N-1} J\left(|i - j|\right)\sigma_i\sigma_j \tag{3}$$

where i and j represent the lattice points 0 through N - 1. The sum is over all distinct pairs of spins. The argument of J is written as an absolute value to indicate that the interaction depends only on the distance between spin pairs. The σ's take on the values +1 or -1 depending on whether the spin at a particular lattice site is up (+1) or down (-1).

Consider the arrangement of spins shown in shown Fig. 6. For any given arrangement of up and down spins, all spin pairs cannot be antiparallel. Therefore, some distribution of up and down spins must result. This arrangement of up and down spins must minimize the energy defined in Eq. 3. In physics, analysis of a one-dimensional system as shown in Fig. 6 is simplified by considering the system shown in Fig. 7.

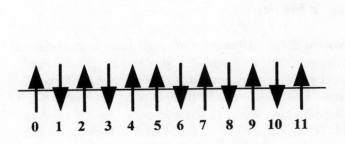

Fig. 6: *Spins on a one-dimensional lattice*

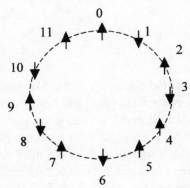

Fig. 7: *Periodic Boundary Conditions*

In other words, a one-dimensional lattice of up and down spins can be thought of as a cycle of lattice sites, some of which are occupied by up-spins and the rest are occupied by down-spins; technically, this is referred to as "invoking periodic boundary conditions." This simplification is, in fact, an approximation that introduces some error into the calculation of the configurational energy for small values of N. It can be shown that as N gets large, the error in the configurational energy introduced by invoking periodic boundary conditions can be neglected.

Comparison of Figs. 4 and 7 suggests that the problem of distributing up and down spins on a lattice is analogous to the "seating" arrangement problem analyzed above. It turns out that the solution to our spin distribution problem is also analogous to the solution of the dinner seating problem [6].

If the lattice consists of N lattice sites (c in our seating problem) $N+$ of which are occupied by up-spins (d in our seating problem) then the up-spins occupy sites (up to rotation and inversion) defined by the ME algorithm. In other words, the distribution of up-spins (or equivalently, down-spins), in which all pairwise interactions are antiferromagnetic, is maximally even.

Even when our one-dimensional antiferromagnetic spin system is put in an external magnetic field (H) the spin distribution that minimizes the energy is still maximally even. In an external magnetic field a spin acquires an additional energy proportional to $\sigma \cdot H$ due to its interaction with the field. For our Ising spins ($\sigma = \pm 1$) the direction of the external magnetic field defines the plus-direction and the energy becomes:

$$(4)$$

$$U = \sum_{\substack{i,j=0 \\ i \neq j}}^{N-1} J\left(|i-j|\right)\sigma_i \sigma_j - \sum_{i=0}^{N-1} \sigma_i H.$$

The last sum accounts for the energy of all the individual spins due to the interaction of each one with the external field.

Notice that spins aligned along the field ($\sigma = +1$) reduce the energy in Eq. 4, while spins aligned opposite to the field ($\sigma = -1$) increase the energy due to the minus sign on the last sum. At the same time, in the first sum, pairs of spins aligned in the same direction increase the energy while spins aligned in the opposite direction decrease the energy. Therefore, there is a competition between the field contribution to the energy trying to align spins in the plus-direction and the pairwise interaction energy trying to align spins in opposite directions. Equilibrium is established when the total energy, Eq. 4, is a minimum. Remarkably, Krantz, Douthett, and Doty [8] have shown that a maximally even distribution of up and down spins minimizes this total energy even when an external field is applied. Specifically, when the external magnetic field is small, one-half of the spins are up and one-half are down. They are distributed, in a maximally even way, with alternate spins up and down. As the external magnetic field is increased, the number of up spins ($N+$) increases relative to the number of down spins ($N-$), but the distribution is still maximally even. As the external field is increased further, eventually, all the spins align along the field. The field at which this occurs is the, so-called, "critical magnetic field."

4. Even(ness) in Music

As cited above, the discovery of the ME algorithm occurred in the mathematics of music theory. In music, a *pitch class* is the class of all pitches named "C," including "middle C" and all higher and lower C's. For Western music in general, we recognize just 12 such pitch-classes, corresponding to the pattern of 12 piano keys that repeats several times to form the full keyboard. This system can be thought of as the hours on a clock, but with 0 replacing 12. We need only map the 12 pitch-classes to the integers 0, 1, ... , 11--for example all the C's to pitch-class 0, all the $C^{\#}/D^{b}$'s to 1, D's to 2, ..., B's to 11 (see Fig. 8). Then the set of white keys maps to the set $\{0, 2, 4, 5, 7, 9, 11\}$ and the black keys to the set $\{1, 3, 6, 8, 10\}$.

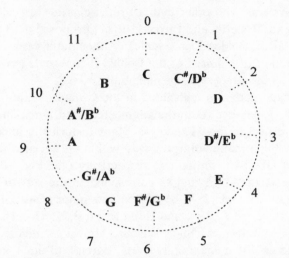

Fig. 8: *Pitch Class*

The connection between the Ising model and the 12 pitch-classes can be described as follows; if 7 out of every 12 spins are up (and, 5 out of every 12 are down) and the lattice energy is minimum, then the spin configuration of the lattice is the same as the white and black key configuration on the piano keyboard (see Fig. 9).

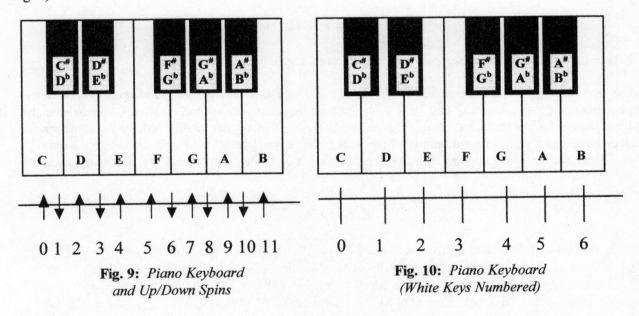

Fig. 9: *Piano Keyboard and Up/Down Spins*

Fig. 10: *Piano Keyboard (White Keys Numbered)*

Thus, the set of white key pitch-classes is a maximally even set as is the set of black key pitch-classes. In addition, many other important musical scales and chords are maximally even in the universe of 12 pitch-classes: the *augmented triads* are characterized by three equally spaced pitch-classes in the twelve pitch-class universe (e.g., {0, 4, 8} representing the notes C, E, and G$^\#$). Similarly, *diminished seventh chords* have four equally spaced pitch-classes (e.g., {0, 3, 6, 9} representing the notes C, E^b, F$^\#$ and A). These pitch-class collections are commonly used in the music of the 18th and 19th centuries. Debussy, a well known composer who lived at the turn of the last century, used the *whole-tone scale*, which is a collection of six equally spaced pitch-classes (e.g., {0, 2, 4, 6, 8, 10} representing the notes C, D, E, F$^\#$, G$^\#$, and B^b). In the 20th century many composers, including Stravinsky, used the *octatonic scale*, which is characterized by pairs of consecutive pitch-classes with skips between them (e.g., {0, 1, 3, 4, 6, 7, 9, 10} representing the notes C, C$^\#$, E^b, E, F$^\#$, G, A, and B^b). All are maximally even sets! Many interesting scale structures in microtonal systems (systems with other than 12 pitch-classes) are, also, maximally even. Most notably, Bohlen [9] and Mathews, Roberts, and Pierce [10, 11] proposed a 13 pitch-class system with a scale of 9 pitch-classes (the, so called, Bohlen-Pierce scale). Although the concept of maximally even sets had not yet been advanced, the properties required of the Bohlen-Pierce scale (most importantly, generalization of the cycle of fifths) forced its construction to be maximally even.

Whereas no three-note sets embedded in the C-major scale (the white keys) are maximally even within the 12 pitch-class universe (only the augmented triad is maximally even), some three-note sets are maximally even with respect to the white keys. Many historically important structures in music consist of three white keys spaced in a maximally even way within the seven white key universe. Suppose the white keys C through B are labeled 0 through 6, respectively (see Fig. 10). Then, for those with a musical background, the major triads C, F, and G correspond to the sets of labels $\{0, 2, 4\}$ ($i = 0$ in the ME algorithm), $\{0, 3, 5\}$ ($i = 2$), and $\{1, 4, 6\}$ ($i = 5$), the minor triads Dm, Em, and Am to $\{1, 3, 5\}$ ($i = 3$), $\{2, 4, 6\}$ ($i = 6$), and $\{0, 2, 5\}$ ($i = 1$), and Bdim to $\{1, 3, 6\}$ ($i = 4$). All these are maximally even with respect to the white key universe. Such sets are known as iterated ME sets [12] since they are the maximally even "children" of a maximally even "parent" (3 in 7 in 12). In addition to these triads, historically important four note collections (seventh-chords in musical parlance) are also iterated ME sets (4 in 7 in 12). Iterations may even go deeper (maximally even "grand children"). Such an iterated substructure of triads can be observed in the Bohlen-Pierce 13-pitch-class system (3 in 4 in 9 in 13) [11, 13].

5. Coming Full Circle

We conclude with yet another example of maximally even ordering:

Let c and d be positive integers such that $d < c$. Place d white points equidistantly around the circumference of one circle, and place $c - d$ black points equidistantly around another. Assume that the radii of the circles are the same. Superimpose the circles so that no two points occupy the same space. Pick any point as a starting point, and label the points consecutively clockwise with the integers 0 through $c - 1$. Then, the set of black point labels (and hence the set of white point labels) is a maximally even set [2,5].

References

[1] S. Block and J. Douthett, Journal of Music Theory, **37**, 21, (1993).
[2] J. Douthett, Ph.D. Dissertation, University of New Mexico, (1999).
[3] J. Clough and G. Myerson, Am. Math. Mon. **93**(9), 695, (1986).
[4] J. Clough and G. Myerson, Journal of Music Theory, **29**, 249, (1985).
[5] J. Clough and J. Douthett, Journal of Music Theory, **35**, 93, (1991).
[6] J. Douthett and R. Krantz, J. Math. Phys. **37**(7), 3334, (1996).
[7] J. Simon, The Statistical Mechanics of Lattice Gases (Princeton University Press, 1993).
[8] R. Krantz , J. Douthett, and S. Doty, J. Math. Phys. **39**(9), 4675, (1998).
[9] H. Bohlen, Acoustica, **39**, 76, (1978).
[10] M. V. Mathews, L. A. Roberts, and J. R. Pierce, J. Acoust. Soc. Am. **75**, S10 (1984).
[11] M. Mathews, J. Pierce, and L. Roberts, "Harmony and new scales," in Harmony and Tonality
     (J. Sundberg - ed.), Royal Swedish Academy of Music (Pub. No. 54), (Stockholm 1987).
[12] J. Clough, J. Cuciurean, and J. Douthett, Journal of Music Theory, **41**(1), 67, (1997).
[13] M. Mathews, J. R. Pierce, A. Reeves, and L. A. Roberts, J. Acoust. Soc. Am. **84**, 1214, (1988).

On Musical Space and Combinatorics:
Historical and Conceptual Perspectives in Music Theory [*]

Catherine Nolan
University of Western Ontario
Faculty of Music
Department of Theory and Composition
London, ON N6A 3K7, Canada
E-mail: cnolan@julian.uwo.ca

Abstract

Music theory has enjoyed a ubiquitous association with mathematics from its earliest beginnings. Among the abundant mathematical models that have inspired theories of music, combinatorics has played a continuous role since the seventeenth century, articulating musical spaces in which relations between elements of a discrete system can be articulated and quantified. This paper explores some representative theories of musical relations expressed in spatial terms using combinatorial techniques, revealing the abstract and profound ways in which mathematics, specifically combinatorics, informs music theory.

1. Introduction

"Nowhere do mathematics, natural sciences and philosophy permeate one another so intimately as in the problem of space" [1]. In this statement, Hermann Weyl depicts space as the primary locus of interconnection among three enormous domains of intellectual and scientific knowledge and endeavor. Weyl's designation of space as a problem reflects the difficulty of articulating the ontology of space, since the concept of space embraces both concrete and abstract or metaphorical meanings. Space may be broadly defined as an extent in which objects or phenomena exist in relative positions, relations, or distances from one another. This expansive definition applies both to concrete spaces in which physical objects or beings reside, as well as to abstract or metaphorical spaces constructed in the human imagination, in which reside concepts, ideas, and abstractions of physical objects or phenomena.

While concrete space is profoundly important to the study of musical acoustics and cognition, which bear close ties with music theory, it is primarily in the realm of compositional and speculative music theory that metaphorical space as a component or product of rationalism emerges as intrinsic to the discipline. This paper explores the pervasiveness of abstract or metaphorical space in music theory through the frame of the mathematical subfield of combinatorics, the study of enumeration, groupings, and arrangements of elements in discrete systems. This brief study begins at the time of the incursion of mathematical combinatorics into music theory in the seventeenth century, almost immediately upon its debut in mathematical discourse, and surveys combinatorial techniques and attitudes in some seventeenth- and eighteenth-century writings on musical composition. From there the discussion turns to two little known theorists of the late nineteenth and early twentieth century, whose theories inductively develop algorithms from combinatorial techniques within the discrete, modular system of twelve pitch classes. These writings demonstrate an important shift in music-theoretical thought, in which combinatorial techniques, applied in the seventeenth and eighteenth centuries to surface musical relations and

---
[*] I wish to express my gratitude to the Social Sciences and Humanities Research Council of Canada for its generous support in carrying out much of the research for this paper.

configurations, later become hidden from view, and applied not to surface configurations, but to the elements from which those configurations—melodies and chords—are made. The paper concludes with some reflections on the place of combinatorics and the significance of metaphorical space in contemporary music theory.

2. Marin Mersenne's *Harmonie universelle*

Numerical models and rational systems of classification formed the infrastructure of speculative music theory from its earliest Pythagorean and Platonic sources to the Renaissance, reflecting the status of music as one of the four sciences of the medieval quadrivium (along with arithmetic, geometry, and astronomy). In the seventeenth century, the Scientific Revolution and the efflorescence of new branches of mathematics, especially probability theory and analytic geometry, irrevocably altered the nature of the interface between mathematics and music theory. Probably the earliest conspicuous manifestation of the new mathematical models in music theory was the adoption by renowned mathematician and music theorist Marin Mersenne (1588-1648) of mathematical processes informed by combinatorics in his landmark treatise, *Harmonie universelle* (1636-37). In the section on melody ("Livre second de chants" or "Book Two on Melodies"), he applied now classic combinatorial formulas to perform such operations as tabulating all 720 (= 6!) permutations of a hexachord (a collection of six notes). A portion of Mersenne's complete table, which occupies twelve pages of text, is shown in Figure 1. Mersenne also recorded the number of permutations of from 1 to 22 discrete pitches in the three-octave range of the diatonic gamut of his time (the last resulting in a colossal number of 22 digits), and performed more complex calculations such as the partitions of 22 or the multisets within the 22-pitch gamut comprising melodic figures that included note repetitions [2].

Mersenne's adoption of combinatorial methods reflected his mathematical expertise, and was motivated by his desire to demonstrate objectively the vast, yet computable number of possibilities for melodic construction. As a composer and musician, he found combinatorial means of explaining the rejection of some permutational possibilities on aesthetic grounds; for example, the melodic interval of the major sixth was regarded as improper, both ascending and descending, so among the 720 permutations of the major hexachord, he calculated that 1/3 (= 240/720) were syntactically incorrect, because they included the notes *ut* (C) and *la* (A) in succession. That is to say, using combinatorics, he found a means to quantify with precision fundamental principles of melodic syntax.

Figure 1: tabulation of the 720 permutations of 6 notes, excerpt
(Mersenne, *Harmonie universelle*, 1636-37)

3. *Ars combinatoria*

The view of musical materials as a finite set of elements from which combinations are selected inspired a number of treatises and practical manuals on rational methods of musical composition in the later seventeenth and eighteenth centuries. Figure 2 illustrates an example of the 24 (=4!) permutations of a four-note melodic-rhythmic figure from Joseph Riepel's 1755 treatise *Grundregeln zur Tonordnung* [3]. Allusions to or explicit techniques derived from mathematical *ars combinatoria* became a familiar feature in eighteenth-century composition treatises, and were valued for the pedagogical benefits they offered to students of composition. Although the approach is mechanistic and the literary tone of these treatises can be light and diversionary, the underlying serious objective was to stimulate the musical imagination and transmit knowledge and skill in manipulation of musical materials.

Figure 2: melodic-rhythmic permutations (Riepel, *Grundregeln zur Tonordnung*, 1755)

4. Musical circles

A more abstract expression of the combinatorial disposition of musical space in eighteenth-century treatises is found in the circular diagrams used to depict relations of proximity and remoteness of keys for modulation within the system of 24 major and minor keys. While differing from each other in substance and presentation, such diagrams evince principles of Cartesian rationalism by identifying spatial relations in terms of relative distance from a referential point. Three "musical circles" from composition treatises of the first half of the eighteenth century are given in Figure 3. Johann David Heinichen was the first to discuss the theoretical implications of the musical circle. In his circular diagram from *Neu erfundene und gründliche Anweisung* (1711), the fifth-related major (*dur*) keys proceed in counterclockwise motion in alternation with the fifth-related minor (*moll*) keys so that each major key is followed (counterclockwise) by its relative minor [4]. Johann Mattheson, in *Kleine General-Bass Schule* (1735), presents a more elaborate pattern of interweaving of the major and minor keys in his musical circle. (The text in the interior of Mattheson's circle reads "improved musical circle, which can lead around more easily through all keys than those previously invented.") Georg Andreas Sorge's multi-layered model of concentric circles in *Vorgemach der musicalischen Composition* (1745-47) separates the fifth-related major and minor keys, enumerating the two underlying circles of twelve fifths, and in the outer circles incorporates both Heinichen's and Mattheson's pairings of relative major and minor keys [5].

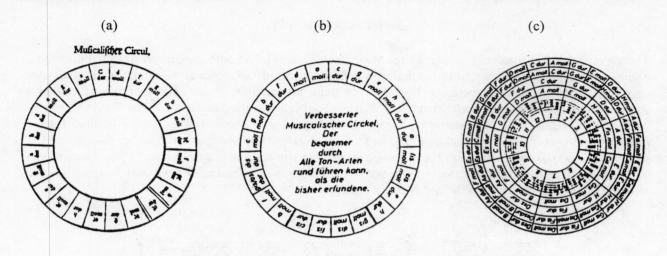

Figure 3: musical circles of (a) Heinichen (1711); (b) Mattheson (1735); (c) Sorge (1747)

5. Modular arithmetic and abstraction

The eighteenth-century musical circles represent abstract relations within a discrete system (of 24 major and minor keys) that was implicitly modular, yet were still intended as a practical aid for musical composition by rendering visible the entire range and quality of relational possibilities for modulation. The formal entry of modular arithmetic into mathematics early in the nineteenth century, in conjunction with the widespread (though not universal) acceptance of equal temperament, in theory if not practice, initiated a line of speculative thought in music theory that extends through the twentieth century and continues to occupy a central position as a component of contemporary music theory. The combined agency of modular arithmetic and equal temperament enabled the formulation of theories of pitch structures based on algebraic methods and a recovery of pure speculation in music theory, that is, theory removed from the immediate concerns of musical practice and style. The twelve pitch classes of the equal-tempered system—in which all pitches are assigned to one of twelve classes based on octave and enharmonic equivalence—are strongly affiliated with theories of atonal and serial music of the twentieth century; their manifestation in the work of earlier authors not implicated in the revolution of harmonic language in the twentieth century reveals important aspects of the generality of abstract musical space.

The earliest proponents of this new line of thought worked independently of each other, and were separated, voluntarily or involuntarily, from the mainstream of music theory and practice of their time. The explicit combinatorial mathematics that was integral to their theories did not conform to the conservatory-based norms and institutions of nineteenth- and early twentieth-century music instruction, so their work has generally not received much attention, and they remain rather obscure even among specialists in the discipline of music theory. Two of these independent thinkers will be discussed here: Anatole Loquin, writing in Bordeaux, France, and Ernst Bacon, writing in Chicago, U.S.A.

5.1 Anatole Loquin (1834-1903). Anatole Loquin, in 1871, described his objective of calculating a complete inventory of all possible combinations of notes—triads and seventh chords, as well as formations resulting from the addition of non-harmonic tones. Loquin first classified all pitches into twelve congruence or equivalence classes (commonly known since the mid-twentieth century as *pitch classes*), using the modular analogy of a circle. (Loquin's concept of the circle was intrinsically different from the eighteenth-century musical circles of key relations discussed earlier, as his represented the twelve pitch classes, devoid of triadic or functional associations.) Treating the twelve pitch classes as a discrete system, he used algebraic methods to determine the number of combinations of pitch classes in all cardinalities (the number of notes in a harmony) from 1 through 12, starting from one fixed, referential note or pitch class. Figure 4 shows Loquin's triangular table of combinations. (This table, as Loquin notes, is a reorientation of Pascal's famous triangle.) The total number of combinations of each cardinality from 1 to 12 is given along the right and bottom edges of the table. Along left edge are indicated the ordinal numbers assigned to the chords, from 1 to 2048 [6].

```
Accords.
  N°
     1.............................................  1 —    1 accord d'une seule note.
  2 à 12.  1+ 1+ 1+  1+ 1+  1+  1+  1+ 1+ 1+ 1  —   11 accords de deux notes.
 13 à 67...  10+ 9+  8+  7+  6+  5+  4+ 3+ 2+ 1  —   55 accords de trois notes.
 68 à 232.......  45+ 36+ 28+ 21+ 15+ 10+ 6+ 3+ 1  —  165 accords de quatre notes
233 à 562..........  120+ 84+ 56+ 35+ 20+10+ 4 + 1  —  330 accords de cinq notes.
563 à 1024..............  210+126+ 70+ 35+15+ 5+ 1  —  462 accords de six notes.
1025 à 1486..................  252+126+ 56+21+ 6+ 1  —  462 accords de sept notes.
1487 à 1816....................  210+ 84+28+ 7+ 1  —  330 accords de huit notes.
1817 à 1980......................  120+36+ 8+ 1  —  165 accords de neuf notes.
1981 à 2035........................  45+ 9+ 1  —   55 accords de dix notes.
2036 à 2047..........................  10+ 1  —   11 accords de onze notes.
2048..............................  1  —    1 accord de douze notes.

     1  11  55  165  330  462  462  330  165  55  11  1 {  2048 accords.
```

Column totals (read vertically, left to right): accords de douze notes. / accords de onze notes. / accords de dix notes. / accords de neuf notes. / accords de huit notes. / accords de sept notes. / accords de six notes. / accords de cinq notes. / accords de quatre notes. / accords de trois notes. / accords de deux notes. / accord d'une seule note.

Figure 4: table of combinations of pitch classes (Loquin, *Apperçu sur la possibilité d'établir une notation répresentant…les successions harmoniques*, 1871)

Loquin outlined a manual process for computing the figures in the rows and columns of the table; the referential note is assigned the number 1; the number of 2-note combinations, 11, is computed by adding successively each of the remaining notes; the number of 3-note combinations, 55, is computed by adding, to each of the 11 2-note combinations in succession, each of the remaining notes above the highest note of each 2-note combination. The numbers of combinations for the remaining cardinalities are computed in the same way, each row and column of the table comprising an arithmetic progression. By speaking of summing "notes" rather than numbers, and by not providing the algebraic equations to symbolize the calculations of the entries on his table, Loquin's terminology and methodology may seem to lack mathematical rigor, but it is likely that he attempted to simplify the information, suppressing the extent of its mathematical foundation in arithmetic progressions, in order to include mathematically untrained musicians and musical scholars in his readership.

Loquin's total of 2048 combinations is the result of a recursive algorithmic process in which combinations of each cardinality are generated by systematic accretion to the already computed combinations of the next smaller cardinality. His results, although internally consistent, reveal that he was unaware of the group-theoretic basis of the system. For example, his table omits the important cardinality of 0, the empty set (ι), that is required to balance the complementary relationships of the system; the identical totals of combinations in pairs of cardinalities in Loquin's table do not match up with complementary cardinalities within the aggregate of 12. With a little manipulation, however, Loquin's numbers of combinations within each cardinality can be shown to correlate to the correct figures, computed without recourse to a referential pitch class. His grand total of 2048 combinations is exactly half of the total number of pitch-class (pc) sets, 4096 (=2^{12}), in the universe of 12 pitch classes. Figure 5 shows the number of sets in each cardinality from 0 to 12 (the number of combinations of 12 elements taken k at a time, where k is the cardinality) as well as the number of equivalence classes under transposition within each cardinality. The number of sets of each cardinality can be calculated by summing adjacent totals in Loquin's table, counteracting the effect of the referential pitch class. That is, the total of 12 monads (single pitch classes) result from summing Loquin's figures 1 and 11; the total of 66 dyads (2-note sets) results from summing 11 and 55; the total of 220 trichords (3-note sets) results from summing 55 and 165; the 495 tetrachords (4-note sets) from summing 165 and 330; and so on.

Cardinality	0	1	2	3	4	5	6	7	8	9	10	11	12
# of pc sets	1	12	66	220	495	792	924	792	495	220	66	12	1
# of equivalence classes	1	1	6	19	43	66	80	66	43	19	6	1	1

Figure 5: numbers of pitch-class (pc) sets and transpositionally equivalent pc-set classes by cardinality

5.2 Ernst Bacon (1898-1990). Across the Atlantic and a few decades later, Ernst Bacon, a young American piano student in Chicago, similarly employed combinatorial methods within the modular system of 12 pitch classes in an unusual monograph entitled "Our Musical Idiom" (1917) [7]. Bacon bypassed the process of unraveling all possible combinations of notes or pitch classes by conceiving of sets of pitch classes in terms of the intervals separating them. He developed an elegant algorithm that efficiently amalgamates combinations of pitch classes into equivalence classes whose members are all related by transposition. (That is, they contain exactly the same succession of intervals, but begin on a different note or pitch class.) The algorithm consists of three steps: (1) the intervals between successive pitch classes are conceptually arranged as if around the perimeter of a circle marked with 12 equidistant points (the distance between adjacent points representing a semitone) in order to render any harmony in a space smaller than an octave; (2) the interval succession, which must always sum to 12 (including the complementary interval that returns to the point of origin), is recorded; (3) cyclic permutations of interval successions are eliminated, as these represent transpositions—reorderings of the same interval succession. Following these steps, each remaining, unique, permutation of the interval succession represents a harmony, a class of transpositionally equivalent pitch-class sets. Interval successions that share the same terms (but are not cyclic permutations) are grouped together as, for example, the successions <1-1-1-4-5>, <1-1-1-5-4>, <1-1-4-1-5>, and <1-1-5-1-4>, representing four five-note harmonies [8]. Bacon's notation of a representative of each of the four harmonies in this combination is shown in Figure 6. (Note that he omits the complementary interval completing the octave.)

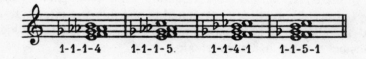

Figure 6: four five-note harmonies formed from non-cyclic permutations
of one combination of intervals (Bacon, "Our Musical Idiom," 1917)

Bacon's table showing his computations of the 43 transpositionally equivalent classes of four-note harmonies, or tetrachords, is given in Figure 7. On the left side of the table appears the tabulation of intervals (from 1 to 9 semitones, or from a minor second to a major sixth) comprising each combination.

Just to the right of the middle of the table appears the formula that applies to each combination (depending on such circumstances as whether the component intervals are unique or whether there are repetitions). Finally, the far right column gives the total number of harmonies (unique, non-cyclic permutations) within each combination, and these are shown to sum to 43. Bacon provides analogous tables for all cardinalities from 2 through 10. Although he notes some large-scale symmetries within the system, like Loquin, he does not pursue the group-theoretical implications of his data.

	1	2	3	4	5	6	7	8	9	CALCULATIONS OF HARMONIES	H
1	3								1	$H = 3!/3!$	1
2	2	1						1		$H = 3!/2!$	3
3	2		1				1			"	3
4	2			1		1				"	3
5	2				2					By trial	2
6	1	2					1			$3!/2!$	3
7	1	1	1			1				$3!$	6
8	1	1		1	1					"	6
9	1		2		1					$3!/2!$	3
10	1		1	2						"	3
11		3				1				$3!/3!$	1
12		2	1		1					$3!/2!$	3
13		2		2						By trial	2
14		1	2	1						$3!/2!$	3
15			4							By trial	1
										Total	43

Figure 7: computation of the 43 transpositionally equivalent tetrachord classes (Bacon, "Our Musical Idiom," 1917)

6. Conclusion

Both Loquin and Bacon were driven by an impulse to construct a taxonomy of all possible combinations of notes or pitch classes within a discrete system of 12 objects, even though many of the available combinations were not syntactically acceptable in the compositional practice of their time. Certainly these authors were motivated by the increasing chromaticism of later nineteenth- and early twentieth-century music, which introduced sonorities that were unexplainable in familiar theoretical terms, but their mathematical methods transcended music-stylistic boundaries and theoretical conventions. While Loquin and Bacon are unusual and intriguing for their independence from mainstream currents in music theory, their role in this study is to document the tenacity in music theory of abstract or metaphorical space articulated through combinatorial mathematics.

The atonal revolution in harmonic language in the early twentieth century and the evolution of the twelve-tone or serial method of composition, spearheaded by Arnold Schoenberg and extended by others, brings a new chapter to the study of the increasingly more significant role played by combinatorics in music theory. While the limited scope of this paper makes it impossible to address this new chapter in any depth, it can be said that the compulsion to classify all possible combinations of pitches within the discrete, modular system of twelve pitch classes proved to be much more than a curiosity. A taxonomy of equivalence classes in the universe of 4096 pitch-class sets is entrenched as a foundational component of contemporary atonal music theory, first codified by Allen Forte, and relations and transformations within equivalence classes of pitch-class sets are shown to be firmly grounded in combinatorial mathematics as well as in set theory and group theory [9]. The spatial metaphor maintains a strong presence in music theory in analogies of geometric transformations and transformations of pitch-class sets, in the writings of Robert Morris on compositional and other musical spaces [10], and in the theories of generalized musical intervals and transformational networks of David Lewin [11].

In accord with the Weyl quotation that introduced this essay, space constitutes an important point of interdisciplinary intersection, exemplified vividly in the powerful resource of combinatorics at the crossroads of music theory and mathematics.

References

[1] Hermann Weyl, *The Philosophy of Mathematics and Natural Science* (New York: Atheneum Publishers, 1949), 67. Cited in Bill Hillier and Julienne Hanson, *The Social Logic of Space* (Cambridge: Cambridge University Press, 1984), 29.

[2] Marin Mersenne, *Harmonie universelle, contenant la théorie et la pratique de la musique* [1636-37], facsimile edition (Paris: Editions du centre national de la recherche scientifique, 1965).

[3] Leonard Ratner, "*Ars Combinatoria*: Chance and Choice in Eighteenth-Century Music," In H. C. Robbins Landon and R. Chapman, eds., *Studies in Eighteenth-Century Music* (London: Allen and Unwin, 1970), 347.

[4] Joel Lester, *Between Modes and Keys: German Theory 1592-1802* (Stuyvesant, N.Y.: Pendragon Press, 1989), 108.

[5] A. Neveling, "Geometrische Modelle in der Musiktheorie: *Mos geometricus* und Quintenzirkel," in H. Schröder, ed., *Colloquium: Festschrift Martin Vogel zum 65. Geburtstag* (Bad Honnef: G. Schröder, 1988), 108.

[6] Anatole Loquin, *Aperçu sur la possibilité d'établir une notation représantant d'une manière à la fois exacte et suffisamment abréviative les succession harmoniques* (Bordeaux: Féret et fils, 1871).

[7] Ernst Bacon, "Our Musical Idiom," *The Monist* 27 (1917): 560-607.

[8] The four harmonies represented by these interval successions can be further grouped into two larger equivalence classes (<1-1-1-4-5> and <1-1-1-5-4>; <1-1-4-1-5> and <1-1-5-1-4>) based on inversional as well as transpositional equivalence.

[9] Allen Forte, *The Structure of Atonal Music* (New Haven: Yale University Press, 1973).

[10] Robert D. Morris, *Composition with Pitch-Classes: A Theory of Compositional Design* (New Haven: Yale University Press, 1987); idem, "Compositional Spaces and Other Territories," *Perspectives of New Music* 33.1 & 2 (1995): 328-58.

[11] David Lewin, *Generalized Musical Intervals and Transformations* (New Haven: Yale University Press, 1987).

The Millennium Bookball

George W. Hart, sculptor

george@georgehart.com
http://www.georgehart.com/

George W. Hart's *Millennium Bookball* is a geometric sculpture, five feet in diameter, commissioned by the Northport (New York) Public Library. The work is a spherical assemblage of sixty wooden "books," and bronze connecting elements, hanging in the library's two-story main reading room. Its structure is based on the geometry of the rhombic triacontahedron, with components rotated to generate visually interesting internal coherences. The books are made of various hard woods, with the titles and authors of "the best books of the century" carved and gold leafed. These titles were voted on by library patrons, and the sculpture was assembled at a community assembly event, like a barn raising, but for art. Dedicated on December 12, 1999, it is on permanent display, celebrating great books and geometry.

1. Introduction

As a sculptor of constructive geometric forms, my work deals with patterns and relationships derived from classical ideals of balance and symmetry. I use a variety of media, including paper, wood, plastic, metal, and assemblages of common household objects. Mathematical yet organic, I try to create abstract forms that dance with motion, inviting the viewer to partake of the geometric aesthetic. [1-5]

Figure 1. *Millennium Bookball*

In October, 1998, I wrote a proposal to the New York State Council for the Arts, (NYSCA) requesting a grant to produce the *Millennium Bookball* as a community art project. The sculpture would celebrate the best books of the millennium. The proposal also explained that I was interested in the geometry, materials, design, and other sculptural aspects of the form, but was not concerned about the exact titles of the books, so a community vote would select the titles.

A key aspect of the design is that one person alone, or even a small group of people, could not assemble the sculpture. Many people had to hold a large number of components together in their relative positions, and slide them together simultaneously. The sculpture embodies the idea that art needs a community to support it, in this case not just monetarily, but for its very existence.

The proposal was funded by NYSCA with an individual artist's award, administered through the Huntington Arts Council. I approached the Northport Public Library for co-funding and a permanent site. Their management and board of directors enthusiastically supported the idea and it was agreed that the library would organize a community fund-raiser to support a large-scale permanent sculpture that was more substantial than my original proposal. The library created voting booths, (allowing three write-in votes per patron) arranged publicity, contracted for a hook in the ceiling, and took care of all other matters, allowing me to focus on the fun part, sculpture.

2. Three Designs

In my sculpture, I seek to create forms that are enriched by an underlying mathematical depth. While viewers sometimes describe my works as spheres, I see them not *as* spheres, but *on* spheres. One would not say that oil painters traditionally paint rectangles, but they paint on rectangles. So, I see the sphere as a canvas to sculpt upon. From the design perspective, a sphere presents a significant challenge, because there is no convenient edge to the canvas. The form must meet with itself "around the back" in a coherent manner—and every side is the back.

I mainly design by visualization, but also use paper and computer models [5-6]. Like most of my work, this piece went through many design iterations. For illustration, three significant stages in the evolution of the *Millennium Bookball* are presented here.

2.1 Initial design

In my initial design, there are sixty directly interlocking books. The books are each identical, with five slots cut in each. Figure 2 shows the paper model that I presented with the proposal. The figure does not make clear its subtle coloring pattern: there are ten books in each of six different colors, with each book's five neighbors being of the other five colors. Figure 3 is the template for each book. If you make sixty photocopies of the template onto colored card stock, cut them out, slit the slits, and slide them together, you will duplicate my activities on the night before the proposal was due.

Figure 2. *Initial design*

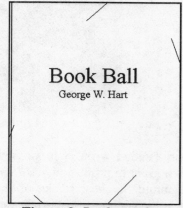

Figure 3. *Book template*

Part of what is visually interesting about the form in Figure 2 is that it is not immediately clear to the casual observer that the books are identical. Because of their many different orientations, this strong relationship between the components is only noticed on longer viewing. Some viewers of the paper prototype expressed a feeling that it is impossible to make a "jigsaw puzzle" like that where all the pieces are identical.

To understand the structure behind the form in Figure 2, it helps to be familiar with the polyhedron known as the *pentagonal hexecontahedron*, Figure 4. The most familiar variant of this polyhedron is the dual to the Archimedean snub dodecahedron. The snub dodecahedron, Figure 5, is composed of 12 regular pentagons and 80 equilateral triangles. There are sixty identical vertices, each incident to one pentagon and four triangles. This form was presumably known to Archimedes, though his text concerning it is lost, and was rediscovered around 1600 by the astronomer and mathematician Johannes Kepler [7]. The French mathematician Eugene Catalan discovered its dual, the pentagonal hexecontahedron of Figure 4, in 1865 [8].

 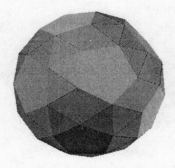 

Fig. 4 *Pentagonal hexecontahedron*   **Fig. 5.** *Snub dodecahedron*   **Fig. 6.** Mobius triangles

This hexecontahedron can be derived by imagining the snub dodecahedron inscribed in a sphere. Where each vertex touches the sphere, a tangent plane can be constructed, and the region interior to all sixty planes is the hexecontahedron. Because the sixty vertices of the snub dodecahedron are identical, the sixty faces of the hexecontahedron are identical. Because the vertices of the snub dodecahedron each connect to five neighbors, the faces of the hexecontahedron each have five sides. Because the snub dodecahedron has edges of a single length, this hexecontahedron has all its dihedral angles equal.

To go from the polyhedron of Fig 4 to the template of Fig. 3, it is just necessary to make a drawing of one face and overlay it with a rectangle. Imagine sixty books, one on each face, extending slightly beyond each face. Where they cross each other, the books have slits.

It is not generally appreciated that there is an infinite number (a two parameter family) of geometrically distinct polyhedra like Figure 4, with the same connectivity and symmetry, and each having sixty identical pentagonal faces. They can be derived starting from the sphere divided into 120 Mobius triangles, Figure 6. Each triangle corresponds to one tenth of a dodecahedron face or, equivalently, one sixth of an icosahedron face. Black and white indicates left and right-hand triangles. Pick any point inside one of the sixty white triangles, and then mark the corresponding point in the other fifty-nine white triangles. The region interior to the sixty tangent planes at these points will be a pentagonal hexecontahedron. Figure 4 corresponds to a special choice of initial point—the snub dodecahedron vertex—giving a "canonical polyhedron."

[9] By adjusting the initial point within this triangle, many other pentagon shapes are possible, giving the artist a range of choices. The design in Figure 2 is based upon a different point, chosen by experimentation with an interactive software tool of my own construction.

2.2 Intermediate design

The initial design, above, was conceived with a pedestal mounting. Although very strong structurally, due to the tight interlocking, it is very closed visually. After learning that I could make a hanging sculpture, I decided that I wanted a more open design, which would have the effect of floating.

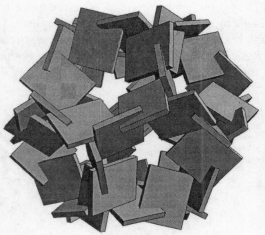

Figure 7. *Second design*

Figure 7 is an illustration of an intermediate design, produced by special-purpose software which I wrote using the *Mathematica* software package. This form keeps the notion of sixty books, and the overall icosahedral symmetry. In effect, each book of the initial design is simply rotated in place, allowing visual access to the interior. The rotations are chosen so each book has edge-contact with two other books, resulting in a rigid structural integrity. This can be fabricated by pre-assembling twenty groups of three books, with one slit per book. The twenty groups can then slide into each other simultaneously and be fastened.

This form strongly suggests a dodecahedron. Although opening access to the interior was a step in the right direction, I found it somewhat chunky and overly simple. I quickly saw what I liked about it, and kept going in the same direction.

2.3 Final design

My final design involves hollow wooden books connected via bronze hubs. The thirty-two hubs are the "donuts" visible in Figure 1. Casting bronze components was a possibility that arose because of the financial support of the library fundraiser. It makes sculptural sense because the hubs provide a framework in which each book stands on its own, inviolate, un-slit. Structurally, the strength of metal hubs allows for many design options. Although it has a very different visual impact, this structure is actually very similar to the two earlier designs, with each book simply rotated in place into a different orientation from the above.

From hub-to-hub is a structural steel rod passing through the inside of each book, which carries the load. The toroidal hubs were drilled very precisely for the rods and tapped for setscrews, which lock everything together. Twelve of the hubs are 5-fold connectors and twenty are 3-fold.

Although I did not begin with this thought, the final design is based upon the rhombic triacontahedron (RT), Figure 8. This polyhedron consists of thirty equal rhombi and sixty edges. Each edge corresponds to the steel rod through one of the books. In Fig 8, I have rendered the RT with open faces, a style of illustration invented by Leonardo da Vinci. Each face has enough body to give a sense of its planarity, easily distinguishing front surface from back, (avoiding the "Necker cube" effect) yet the openness allows the whole structure to be grasped at a glance.

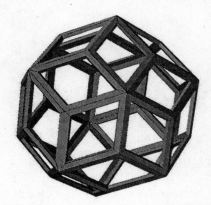

Figure 8. *Rhombic triacontahedron*

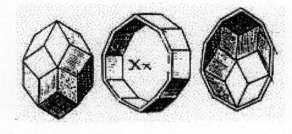

Figure 9. *Kepler's illustration*

The RT was first described by the astronomer and mathematician Johannes Kepler. [7] (However, it was forgotten and independently rediscovered by Catalan 250 years later. [8]) Figure 9 shows the illustration from Kepler's 1619 book, *Harmonice Mundi*. While not in accurate perspective, Kepler's illustration highlights a key property of the RT: It can be assembled as an "equator" and two "caps," with ten rhombi in each component. This property is key to understanding the assembly process described below.

Mathematically, the RT is a member of the class of polyhedra called *zonohedra*, first recognized by the Russian crystallographer, Evgraf Federov in the 1890's [10]. These polyhedra have a number of interesting properties. [11] A key characteristic of zonohedra is that their edges are grouped into sets of parallel edges called *zones*. The RT has six zones of ten edges (or ten faces) each. In each zone, there are ten parallel edges. Kepler's illustration points out one zone. In the sculpture this translates into six sets of ten books; in each set the books' ten spines are parallel.

If we look along the direction of an edge, we see the projection of the RT illustrated in Fig. 10. The ten dots represent a zone of ten parallel edges, seen on end. The lines are the edges of one cap of ten rhombi. The rectangles indicate the orientation of ten books. To add visual coherence to the sculpture, each book is rotated about its spine to be coplanar to another book on the opposite side, e.g., books A and C are aligned.

Figure 10 shows how the appropriate angle is determined. Because the ten rods are spaced 36 degrees apart about the center, O, of the circle, angle BOC is 72 degrees. BAC is half that because the angle seen by a point on the circumference is half the angle seen at the center. So each book is rotated 36 degrees from a radially outward, alternately left and right.

To fix this 36 degree rotation angle, there is a steel bracket inside each book, screwed to the wooden interior. The bracket was brazed to its rod in a fixture that accurately set its position relative to two flats milled near the ends of the rods. The flats are locked by the setscrews in the bronze connectors.

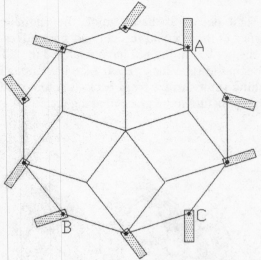

Figure 10. *End view of ten books.*   **Figure 11.** *View along 5-fold axis.*

There are several natural ways that the books could have been colored with a six-color or a five-color pattern. To choose among them, I made several paper models before selecting the final six-color pattern. Each set of ten books is constructed from a different hard wood. The six woods are maple, cherry, sapele, bubinga, purpleheart, and walnut. The effect of this arrangement is that at each 5-fold meeting point, a combination of five different woods meet to form a "propeller." At each 3-fold meeting point, three different woods make a "propeller" which spins in the other direction. Each book connects a 5-way and a 3-way connector, and appears to spin clockwise in one yet counterclockwise in the other.

3. Assembly

This sculpture can not be assembled by a single person. One might begin easily enough. But at the time for placing the final piece, there would be no way to insert it, as *both* ends of its rod would have to slide into connectors. However, with a large group of people, this problem can be circumvented.

Figure 12. *Books at start of Assembly*   **Figure 13.** *Initial Connections*

The geometry of the zonohedron suggests the method for its assembly. It is easiest to understand the assembly in reverse sequence. Imagine taking the sculpture apart, referring to Kepler's Fig. 9. Because the cap of ten rhombi fits on to ten parallel edges, the final step in the assembly is to join the cap of ten rhombi onto the cap-plus-equator structure. Ten connections—ten rods into ten donuts—must be made simultaneously. This required ten people to guide the junction, ensuring that the penetration was uniform and nothing jammed at an angle. Previously to this, the cap had to be assembled, by making a connection of five simultaneous joints analogously. The subparts of the caps were similarly built up by simultaneous parallel connections.

The Northport Public Library arranged for a community sculpture-raising event at which this occurred. See Figures 12-15. Dozens of volunteers came and worked together for the two hours that it took. Although I was somewhat nervous that someone might drop, scrape, scratch, or otherwise damage components, everything went very smoothly.

Figure 14. *Half-way done*   **Figure 15.** *Final connection of ten struts*

Two months previously, the library had arranged a title-announcement party, at which the winning books were unveiled in a countdown from number sixty up to number one. The title voted number one is *The Catcher in the Rye*; the complete list can be found in [1]. In the intervening months, I had carved all the titles and authors into the covers with a computer-controlled router, and gold-leafed them. For the assembly, we only had to connect everything in the proper positions and tighten the 120 setscrews. As a construction guide, I brought along a small Zometool model of the rhombic triacontahedron, [12] with colored paper indicating the positions of the books.

The *Millennium Bookball* is now suspended by chain from one of the 5-fold connectors. Weighing 160 pounds, raising it was straightforward with a block and tackle. It is set near a balcony so it can be seen from its own height or from below. All the titles are visible from somewhere, but if one seeks a particular title, it may not be obvious. In order that viewers might easily find their favorite books, I positioned the books in numerical sequence in groups of ten. The books voted 1-10 are placed in order in the walnut equator, the darkest wood. Books 11-20 are in order in the purpleheart equator, a slightly lighter wood. And so on up to 51-60, in maple, which is the lightest. So looking up any title in the brochure which the library provides, one can determine its number and the wood of the equator it in which to find it. Then examining whatever books one sees in that wood, finding their numbers on the list, one can count around the sculpture to the desired book.

Conclusion

The *Millennium Bookball* is a sculptural celebration of great books and geometry. It is a community artwork in three ways: the "best books of the century" titles were selected by vote, a fundraiser supported it financially, and it was physically assembled at a "barn raising." I feel the geometric basis of its form gives it a depth, so viewers discover more and more of its properties over time, while it also serves to instill an appreciation for mathematics.

For more images of the sculpture and its assembly, a list of the sixty titles, press clippings, acknowledgments of many who helped, and other examples of my geometric sculpture, see [1].

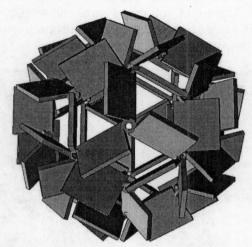

Figure 16. *View along 3-fold axis (compare to Fig.1)*

References

[1] G.W. Hart, *Geometric Sculpture*, http://www.georgehart.com/

[2] G.W. Hart, "Icosahedral Constructions," in Proceedings of *Bridges: Mathematical Connections in Art, Music and Science*, Southwestern College, Winfield, Kansas, July 28-30, 1998, pp. 195-202.

[3] G.W. Hart, "Zonish Polyhedra," Proceedings of *Mathematics and Design '98*, San Sebastian, Spain, June 1-4, 1998.

[4] G.W. Hart, "Loopy," to appear in *Humanistic Mathematics*.

[5] G.W. Hart, "Sculpture based on Propellorized Polyhedra," Proceedings of *MOSAIC 2000*.

[6] G.W. Hart, "Reticulated Geodesic Constructions," *Computers and Graphics*, to appear 2000.

[7] Johannes Kepler, *The Harmony of the World,* 1625, transl. E.J. Aiton, A.M. Duncan, and J.V. Field, American Philos. Society, 1997.

[8] M. E. Catalan, "Memoire sur la Theorie des Polyedres," *Journal de L'ecole Imperiale Polytechnique,* Vol. 24, book 41, pp. 1-71 plus plates, 1865.

[9] G.W. Hart, "Calculating Canonical Polyhedra," *Mathematica in Research and Education,* Vol. 6 No. 3, Summer, 1997, pp. 5-10.

[10] E. S. Federov, *Symmetry of Crystals,* transl. David and Katherine Harker, American Crystallographic Assoc., reprint 1971.

[11] G.W. Hart, "Zonohedrification," *The Mathematica Journal,* vol. 7 no. 3, 1999.

[12] G.W. Hart and Henri Picciotto, *Zome Geometry: Hands-on Learning with Zome Models,* Key Curriculum Press, to appear Spring, 2000.

A Topology for Figural Ambiguity

Thaddeus M. Cowan
Kansas State University

Abstract

A topology for figural ambiguity (braid theory) is proposed which describes figure-figure, figure-ground, and figure-ground-figure ambiguities with the defining relations of the topology. The power of the topology is demonstrated by the classification of ambiguities and the revelation of new ambiguous forms.

Introduction

Ambiguous figures, configurations that have more than one perceptual interpretation (Figure 1), have wide appeal; they are often regarded with surprise, amazement, curiosity, bewilderment, and even amusement.

Figure 1. *Agule's ambiguity*[1]. *Is this man a truth teller or a liar. The answer is written all over his face (rotate 1/3 counterclockwise).*

They have been around since antiquity, but only recently have there been attempts at theoretical explanations of these enigmatic patterns. Any theoretical account of them should, at a minimum, incorporate the following in its narrative:

1. *Stimulus Constancy:* The stimulus pattern does not change.
2. *Stimulus Segmentation:* The stimulus pattern is either partitioned (e.g., figure-ground) or it is not (e.g., figure-figure).
3. *Response Multiplicity:* There is more than one interpretation of the stimulus pattern.
4. *Response Saliency:* One and only one interpretation of the stimulus pattern can be attended to at any given moment.

The logical result of 3 and 4 is *Response Abruptness:* The change in salience is sudden and complete–all or none.

Topology, a mathematics that concerns itself with invariance under conditions of continuous distortion or change (*homotopic* equivalence), addresses both stimulus constancy and response multiplicity.

To use an example that has almost become a cliché, a torus can be smoothly and continuously transformed into a cup shape, but, topologically, the cup remains, fundamentally, a torus; thus a topologist cannot tell the difference between the doughnut he eats and the cup from which he drinks his coffee. The ambiguous stimulus does not change, but the multiple perceptual interpretations of it are as different as the coffee cup and the doughnut.

Algebraic Topology and the Theory of Braids

The topology of braids [2] is an algebraic topology particularly appropriate for the purposes at hand, since it appears helpful in describing "impossible" figures, seen in the artistry of Oscar Reutersvaard and M. C. Escher [3,4], and perceptual impossibilities and ambiguities are likely related [5].

Consider a finite number of strings stretched between two frames g_1 and g_2 (Figure 2a). If there are

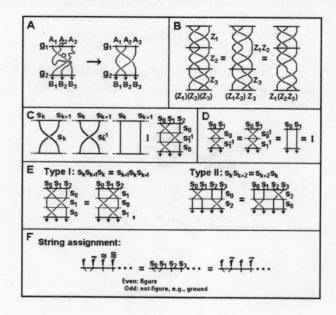

Figure 2. *The topology of braids. (a) An unacceptable and acceptable braid.   (b) The associativity of briads. (c) Right. A pigtail braid partitioned into single nodes.   The other figures represent the three possible forms of a single node, s_n, s_n^{-1}, I . (d) Inverse nodes. A tug on the frames creates an identity braid. (e) Type I and Type II relations. These are topological equivalences (homotopies) that help define the algebra of the topology. (e) String assignment for the ambiguity problem (see text for description).*

n strings, then g_1 and g_2 are divided into segments by points $A_1 A_2 A_3 ..... A_n$ and $B_1 B_2 B_3 ..... B_n$. The points are ordered from left to right. The strings can cross, but no string crosses the frame, and the courses of the strings are such that any line parallel to g_1 and g_2 crosses each string exactly once. Thus the loops and whorls in any of the lines shown in Figure 2a, left, are not allowed; neither is the frame crossing of A_2. A crossing of more than two strings at a single node (Figure 2a, left) is also unacceptable. The strings are moved to produce a distribution of crossings such that any line parallel to g_1 and g_2 never passes through more than one crossing. The order of the crossings is read from g_1 to g_2 (Figure 2a, right).

If two braids Z_i and Z_j have an equal number of strings, they can be combined by placing the upper frame of one against the lower frame of the other then suppressing the two overlapped frames. Such a binary operation is called a concatenation (product) of Z_i and Z_j and is expressed by $Z_i Z_j = Z_k$. It is obvious that this operation is associative. That is, $(Z_1 Z_2)Z_3 = Z_1(Z_2 Z_3)$ (Figure 2b). The commutative law does not hold in general: $Z_i Z_j \neq Z_j Z_i$.

Since no two crossings lie on any one line parallel to g_1 and g_2, any braid, no matter how complex, can be divided into a series of n connected braids (Figure 2c, right), each with one of the following properties: (1) The kth string from the left crosses over the k + 1st string. This operation is called s_k and is shown in Figure 2c, left. (2) The kth string from the left passes under the k + lst string. This operation is labeled s_k^{-1} and is shown in Fig. 2c, second from the left. (3) There are no crossings (Fig. 2c, second from the right). This braid is denoted by I. It is a unit braid or identity braid since ZI = IZ = Z; the joining of I to any braid does not alter its structure. It is clear that s_k and s_k^{-1} are inverse operations since $s_k s_k^{-1} = s_k^{-1} s_k = I$ (Figure 2d). (Just as s_k has an inverse, so does the more complex braid, Z. The inverse of Z, Z^{-1}, is a mirror image of Z.) This notation allows the braids to be treated symbolically. For example, the common pigtail braid shown in Figures 2a and 2c, right, is given by $s_0 s_1^{-1} s_0$. The two braids in Figure 2e describe two relations, $s_0 s_2 = s_2 s_0$ and $s_0 s_1 s_0 = s_1 s_0 s_1$, that help define the algebra, and they also play a central role in the description of figure-ground ambiguity.

Braid Rank and File Assignments for Figure Ambiguity. The vertically ordered rows within the braid are given position numbers, and the horizontal string positions (Figure 2f) are labeled $f \tilde{f} \widetilde{\tilde{f}} \cdots$ so that a count is imposed on the strings by the number of negations (~). This count defines the string indices. The sequence of events $f \tilde{f} \widetilde{\tilde{f}} \cdots$, of course, is equivalent to $f f f \cdots$. The only assignments imposed on the braid are the status of a string at the node (s_k, s_k^{-1}), the vertically arranged horizontal sections created by the separation of nodes into one node per section, and two mutually exclusive entities, s_{even} and s_{odd} or f and $\tilde{f}$.

Neither the labeling scheme nor the topology makes any assumptions about perception, but the assigned properties can serve as a template onto which perceptual primitives can be mapped. For example, f or s_{even} is read as figure, and $\tilde{f}$ or s_{odd} as non-figure, commonly regarded as ground in figure-ground problems $(s_0 s_1 s_2 \cdots \equiv f g f \cdots)$. Figure and ground, of course, are mutually exclusive. In the case of figure-figure ambiguities, f and $\tilde{f}$ represent two mutually exclusive interpretations. Left string dominance, imposed by a positive exponent (s_k), represents perceptual salience (e.g., f) and its inverse (s_k^{-1}) perceptual non-salience (e.g., $\sim f$). Finally, the vertical braid segments with their internal frames are logical assignments of left to right figure segments with their partitions.

The assumptions concerning perception are clearly minimal, viz., the existence of mutually exclusive perceptual interpretations (e.g., figure-ground separation or horizontal string assignments), salience (nodes), and partitioned segments (node separation). The complexities of ambiguity are derived from the topology. These offer a unique look at perceptual interpretation and salience interplay found in figural ambiguity.

Simple Ambiguities

Three segment figures. The face-vase problem appears in Figure 3a, left. Reading from left to right the observer encounters a face-ground-face or ground-vase-ground perceptual organization. During the perceptual switch left *face* passes to left *ground*, center *ground* to center *vase*, and right *face* to right *ground*.

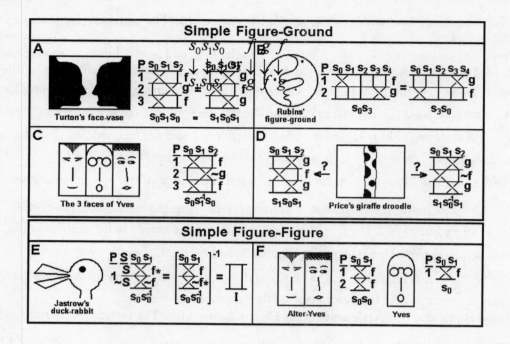

Figure 3. Simple ambiguities. (a) Three segment figure-ground ambiguity. (b) Two segment figure-ground ambiguity. (c) Stable figure-figure configuration. (d) Roger Price's giraffe droodle.   (f) Figure-figure ambiguity with its inverse representation. (e) Representations of stable figure-figure configurations.

Everything that passes is *contra*; that is, there is consistency in the type of transition for all segments, which is something one would expect in a *global* transformation where everything changes at the same time.

$$
\begin{array}{ccc}
s_0 s_1 s_0 & & f\ g\ f \\
\downarrow \downarrow \downarrow & \text{or} & \downarrow \downarrow \downarrow \\
s_1 s_0 s_1 & & g\ f\ g\ .
\end{array}
$$

The Type I relation seems to represent this ambiguity:

$$
s_k s_{k+1} s_k = s_{k+1} s_k s_{k+1} : \ s_0 s_1 s_0 = s_1 s_0 s_1 \ \text{or}\ fgf = gfg. \tag{1}
$$

The switch is reflected in the homotopic shift between the two twisted forms (*multiplicity*), and this topological equivalence implies they are taken from an unchanging stimulus (*constancy*). Two internal frames are fixed implying two fixed partitions within the figure (*segmentation*). Topological equivalence is not assumed but is revealed directly by the homotopic shift of the relation strings. The string flexion effecting this string movement is smooth and continuous, but shifting node location is not; it is sudden and discrete like the perceptual shift observed in figural ambiguity (*abruptness*). Finally, all nodes are overpasses (*salience*).

Two segment figures. A figure-ground ambiguity appears in Figure 3b. Reading from left to right, a face is encountered, then its background. Suddenly one sees the face on the right with its background on the left. It is difficult to see both faces at the same time. The left *face* passes to left *ground*, and right *ground* passes to right *face*

$$
\begin{array}{ccc}
s_{even} s_{odd} & & f\ g \\
\downarrow \ \downarrow & \text{or} & \downarrow \ \downarrow \\
s_{odd} s_{even} & & g\ f\ .
\end{array}
$$

Again, everything is *contra* supporting a global shift in perception. A Type II relation with nodes $s_0 s_3$, their even (f) and odd (g) indices appropriately separated for positive global saliency, correctly describes alternating, salient perceptual states:

$$s_k s_{k+i+2} = s_{k+i+2} s_k : \quad s_0 s_3 = s_3 s_0 \text{ or } fg = gf. \tag{1}$$

The extended separation of the nodes is necessary since $s_0 s_2$ is figure-figure and $s_0 s_2^{-1}$ is figure-not-figure ($f \sim f$ not $f \, \tilde{f}$) which disallows global salience. That is, figure-not-figure is a local description and figure-ground is global.

Again, topological equivalence is not merely assumed; it is a consequence of node shift, which implies an unchanging stimulus (*constancy*) in the presence of changing perceptual interpretation (*multiplicity*). Again, the internal fame is fixed indicating a permanent partition in the figure-ground pattern (*segmentation*). As before, all nodes are overpasses (*salience*).

Cognition: Braids say little about cognition, "the ghost in the machine," and this is part of the appeal of the representation. While braids say nothing about what the cognitive process is, beyond salience and figure-ground separation, they suggest where cognition lies.

First, in the case of *local* salience, piecemeal perception, the appropriate braids are simple pig-tail braids (Figure 3c) with immutable patterns, and they possess fixed internal frames implying fixed partitions ($s_1 s_0^{-1} s_1$). Now consider Figure 3d, one of Roger Price's "droodles" [6], a giraffe passing by a second story window, a seemingly stable figure if there ever was one. However, the stable pigtail braid $s_1 s_0^{-1} s_1$ ($g \sim fg$) does not describe the figure, but the flexible relation, $s_1 s_0 s_1$ or gfg, does. Yet, the droodle does not appear ambiguous, and the flexible string is perplexing.

However, the topology does not demand that the string be flexed–that the homotopy be manifested; it only provides the possibility, and the assumption that the middle string in $s_1 s_0 s_1$ can show resistance from full flex to rigidity seems viable. The process responsible for this resistance (or lack of it) is cognition.

For a more direct demonstrable link between cognitive influence and relation shift, an attempt will be made to influence the reader's perceptual predisposition: Price's giraffe is not a giraffe at all but a view from a basement apartment window, through parted curtains, of a leopard giving chase. This transforms the droodle into an ambiguity represented by the Type I relation braid with an adjustable middle string that formerly resisted flexing.

One segment figures: Figure-figure ambiguities, like their figure-ground cousins, reveal two interpretations that are independent and mutually exclusive; when one is seen the other is not. We begin by concatenating salience and non-salience forming two sections of a single braid, which represent overt and covert percepts. The frame between sections ($|$) divides overt and covert perceptual responses and replaces the segment partition seen in figure-ground patterns. The braid forms $[s_0 | s_0^{-1}]$ or $[f | \sim f]$ appear next to Jastrow's ambiguity in Figure 3e.

Rabbit has been assigned to f. If Rabbit is salient, Duck is non-salient and assigned to $\sim f$. Rabbit is distinguished from Duck by $*$; when $*$ is encountered, read "Rabbit." How is Duck revealed? There are three braid functions available for the ambiguity problem: Type I and Type II relations and the inverse. Since the two relations have already been spoken for, we are forced to choose the inverse. Braid inversion is effected by appropriately distributing the negative exponent then inverting the order of the terms. This moves the hidden duck and its perceptual non-salience to the top of the braid where the sign for non-salience is lost from the duck, and gained by the rabbit. Here figure passes to figure, i.e.,

$$\begin{array}{cc} s_{0*}s_0^{-1} & f \sim f \\ \downarrow\downarrow & \text{or} \quad \downarrow\downarrow \\ s_0 s_{0*}^{-1} & f \sim f \end{array}.$$

Everything that passes is *ipsi*. The fact that braid forms $s_{0*}s_0^{-1}$ and $s_0 s_{0*}^{-1}$ are topologically equivalent, stimulus *constancy* under conditions of a changing perceptual interpretation is observed (*multiplicity*).

As for "Alter Yves," $s_0 s_0$ (ff) captures the essence of two equally salient faces. Furthermore, there is one internal frame which matches the internal partition separating the faces. Yves, a figure without partition, is represented by a single node with no internal frame. If internal frames represent internal partitions what is to be done with the internal frame of $s_{0*}s_0^{-1}$ representing Jastrow's figure, a figure without partition? The braid group provides an elegant solution to the *partition* problem. The two braids for the two interpretations each reduces to an identity braid in the manner of all inverse pairs,

$$[s_{0*}s_0^{-1}]^{-1} = s_0 s_{0*}^{-1} = \mathbf{I}, \tag{3}$$

and identities have no internal frames, hence no partitions are present in the figure..

Complex Ambiguities

A complex ambiguity is composed of simple ambiguities. Figure 4a is a figure-figure ambiguity, two profiles or two halves of a full face. When one observes a vase in Turton's figure (Figure 3a), ground is seen either as two flanking regions or a single background passing behind the vase by the Gestalt principle of good continuation; here there is no such choice.

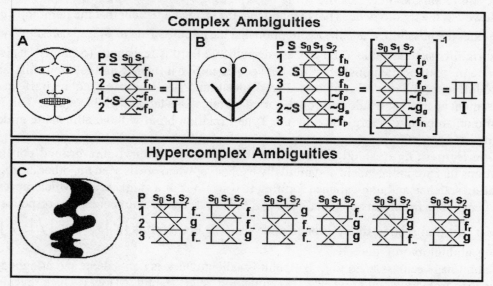

Figure 4. Complex and hypercomplex ambiguities. (a) Complex figure-figure ambiguity. (b) Complex figure-ground-figure ambiguity with figure-figure components. (c) Hypercomplex figure-ground-figure, figure-figure-ground, and figure-ground-ground ambiguity with their homotopic braid forms. This figure possesses fifteen ambiguities.

There are three interpretations of Figure 4b. Taking things segment by segment: profile, grins +

separation, profile; half face, full smile, half face, and ground, psi, ground. The number of ambiguities equals the number of interpretations taken two at a time giving three ambiguities. These include two profiles and two halves of a full face each perceptually switching (homotopically) with the Greek letter psi. That is, from left to right, *figure* passes to *ground*, *ground* passes to *figure*, and *figure* passes to *ground* again,

$$\begin{matrix} f & g & f \\ \downarrow & \downarrow & \downarrow \\ g & f & g \end{matrix} \ .$$

Everything that passes is *contra*. Each ambiguity in relation form,

$$s_0 s_1 s_0 = s_1 s_0 s_1 : \mathit{fgf} = \mathit{gfg} \ . \tag{9}$$

The third ambiguity is figure-figure consisting of two *half faces* and a full *smile* passing to two *profiles* and a pair of *grins* as seen in Figure 5a in part. That is,

$$\begin{matrix} f & g & f \\ \downarrow & \downarrow & \downarrow \\ f & g & f \end{matrix} \ \text{or} \ \begin{matrix} h & s & h \\ \downarrow & \downarrow & \downarrow \\ p & g & p \end{matrix} \ . \tag{10}$$

Everything that passes is *ipsi* (same). In terms of braids,

$$s_{0h} s_{1s} s_{0h} | s_{0p}^{-1} s_{1g}^{-1} s_{0p}^{-1} = s_{0p} s_{1g} s_{0p} | s_{0h}^{-1} s_{1s}^{-1} s_{0h}^{-1} \ . \tag{11}$$

The expression for the complete set of ambiguities would be

$$s_{0h} s_{1s} s_{0h} | s_{0p}^{-1} s_{1g}^{-1} s_{0p}^{-1} | s_{1g} s_{0\psi} \, s_{1g} \ .$$

(Changing to the homotopy of both facial forms, $s_{1g} s_{0\psi} s_{1g} | s_{1g}^{-1} s_{0\psi}^{-1} s_{1g}^{-1}$, produces a vacuous figure-figure ambiguity at best.)

Hypercomplex Ambiguities

A crucial test of a model is whether or not it can create something new–here a different ambiguous form. Consider the ambiguity in Figure 4c and its interpretation of two faces looking right with ground at the right (ffg). The braid $s_0 s_0 s_1$ cannot represent this since it is inflexible, and we need at least two homotopic forms to represent two faces looking left (gff) and two faces looking at each other (fgf).

A variant of the Type II relation is $s_k s_{k+1} s_k^{-1} = s_{k+1}^{-1} s_k s_{k+1}$, a braid whose middle string passes under the other two strings rather than between them. That is,

$$s_1^{-1} s_0 s_1 = s_0 s_1 s_0^{-1} \ \text{or} \ {\sim}\mathit{gfg} = \mathit{fg}{\sim}\mathit{f}$$

an odd mix of figure-ground and salience and non-salience. If we interpret this as

$$s_1^{-1} s_0 s_1 = s_0 s_1 s_0^{-1} \ \text{or} \mathit{ffg} = \mathit{fgg}$$

we find two faces looking left are homotopic to one face looking left toward two grounds. Thus a new ambiguity is revealed: The middle portion can be seen as ground (shadow?) to the left-most face or a figure, a face looking left with its ground to the right . This is a weak ambiguity, more interpretive than perceptual, but, nonetheless, it is an ambiguity.

Thus far there are two ambiguities which serve as two figure-figure composites. That is,

$$s_1^{-1} s_0 s_1 | s_1 s_0 s_1^{-1}$$

or ffg and its homotopy fgg in a figure-figure relationship with gff and its homotopy ggf.
Add to this two faces looking at each other (fgf) with its paired interpretation, a wire loop with a black

ribbon attached across the loop in the middle (gfg) or $s_1 s_0 s_1 = s_0 s_1 s_0$. This gives the final expression,

$$s_1^{-1} s_0 s_1 | s_1 s_0 s_1^{-1} | s_0 s_1 s_0.$$

There are six interpretations here (see Figure 4c), and this seemingly simple looking figure boasts 15 different ambiguities.

Conclusions

Topology is proposed as a logical mathematical approach to figural ambiguity. It describes the perceptual *experience* (phenomenology) of figural ambiguity rather than cognitive influence. The topology of braids as a model for ambiguity meets all the criteria set forth at the beginning of this paper: It reflects stimulus constancy through topological equivalence. It satisfies stimulus segmentation through fixed internal frames and, in the case of partitionless figure-figure ambiguities, a reduction to $\mathbf{I}$. Response multiplicity is revealed by the relations and the inverse operation. Saliency is represented by string dominance at the nodes. Finally, response abruptness follows from the braid rule that no two nodes occupy the same segment during a homotopic shift.

Braid topology reveals characteristics of ambiguities that have been ignored: It suggests complex and hypercomplex ambiguities and relates these to simple ambiguities thereby creating a classification that, heretofore, has not been seen. Finally, it points the way to new ambiguous forms.

References

[1] P. Agule in J. R. Block and H. E. Yuker *Can You Believe Your Eyes?* Gardner Press, New York, 1989.

[2] E. Artin, Theory of braids. *Annals of Math.*, **48**, 101 (1941).

[3] T. M. Cowan, The theory of braids and the analysis of impossible figures, *Journal of Mathematical Psychology*, **11**, 190, 1974.

[4] T. M. Cowan, Turning a Penrose triangle inside out. *Journal of Mathematical Psychology*, **26(3)**, 252 (1982).
Reprinted in *Knots and Applications*, L. Kauffman, Ed. (World Scientific Press, Singapore, 1995), pp. 465-475.

[5] E. H. Gombrich, *Art and Illusion: A Study in the Psychology of Pictorial Representation* (Princeton University Press, Princeton, NJ, 1969).

[6] R. Price, *Droodles* (Simon and Schuster, New York, 1953).

Synetic Structure

F. Flowerday
Rodney, Michigan
flowerday@tucker-usa.com

Abstract

This paper presents a new method for the resolution of structure into discrete patterns of tension and compression. Synetic designs are scaleable from atomic through molecular levels, finding many applications in domes and spherical structures. The system also provides a force interaction model that is applicable to a broad range of real and theoretical problems. Synetic principles operate from micro- to macro-scales and in a wide variety of appropriate materials. As energy is increasingly invested along Synetic paths, a structural continuum, a bridge, is established from virtual pattern to strong, resilient construction.

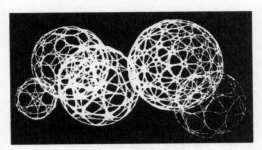

Synetic spheres.

Introduction

Synetics is a flexible yet triangulated structural system, consistently applicable to polyhedral, lattice and manifold geometries. Discrete patterns of tension and of compression act in energetic balance to produce models or structures of unlimited complexity or size. In this system, compressive paths are curvilinear, following the tendency of material to curve under compression, as does a bow. Tensile paths are straight, according to the behavior of material to straighten under tension, as does a bowstring.

In Synetic terms, curvature is the norm and linear is the exception, requiring energy to maintain. In this system, vectors of tension conform to lines of minimum distance, and trajectories of compression follow curves of minimum energy. Reaction and deformation as well as load are efficiently shared throughout Synetic structure. Tension and compression balance in dynamic equilibrium. Forces are resolved through closed circuits of compressive material braced by closed circuits of tensile material. Spheres, domes, tubes and toroids behave as tough pneumatic membranes, bouncy and resilient even at large diameters and capable of rebounding strongly from extreme distortion.

The strung bow represents the first structural differentiation of material into discrete balanced elements of tension and compression. Bows are connected and continued to complete structural circuits in foam-like and bubble-like structure.

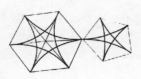

Building with bows.

Designs

Synetic design utilizes compressive material to the fullest advantage. Tensile elements may be apparent and plainly involved or found to be employed less visibly, as they act to support compressive arcs. Attachment of compressive arcs is entirely by tangency. Operating in simple thrust, tangent articulation provides a structurally integral and uniform connection as the basis for a widely applicable modularity.

Arcs and chords are variously overlapped and systematically joined to form lattices and open arrays or to form closed polyhedral systems and their familiar combinations, distortions and globally symmetric development.

Patterns of tangent circles provide particularly important circuits of compression. Digitally manipulated as circles or built as strong hoops, tangent assemblies react to stress in a predictable hierarchy of transformation. Radial patterns of tangent hoops inscribed in central angles of polyhedra are 'dual' to patterns inscribed in polyhedral faces.

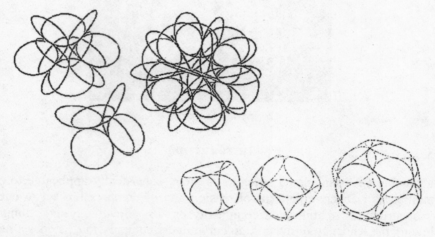

Radial patterns of circles form a tetrahedron, octahedron and icosahedron.
Circumferential patterns produce the structurally sound dual figures of
tetrahedron, cube and pentagonal dodecahedron. Elements of these patterns are
used to develop lattice, radial, or manifold geometries.

Circular duals in combination.

Models

Synetic models demonstrate the dynamic behavior of a particular system as well as its spatial behavior. Tensile paths define linear structures with vertexial or nodal connections. Tension conforms to triangular meshes, to the edges, axes and chords of conventional polyhedra. A lattice node is treated as a polyhedron, through and around which curved compressive paths are taken without particular reference to a central point.

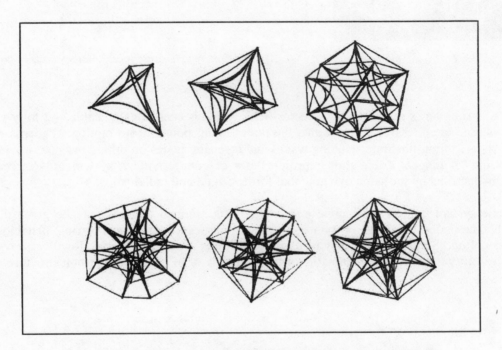

Curved compressive paths in polyhedra defined by linear tensile paths.

Behaviors of whole systems derive directly from topology and from the periodicity of modular units which self-adjust in angle and distance into structures of increasing complexity.

A stellated icosadocecahedron.

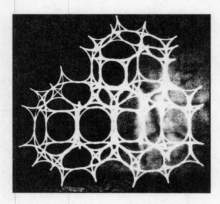

A bubble lattice made of unit tetrahedra.

An indefinitely extensible vector matrix made of bows.

A matrix of strung bows might model a vector-field, which is energetically stable, yet fully vibrational in both tensile and compressive components, the nodes being points of no motion. Patterns of tangent circles on a sphere might illustrate standing waves and resonant modes on orbital orientation. In modular sub-units, points of tangent articulation represent sites of connectivity, a system of vectored valence symmetrically disposed by the same dynamic that forms orbital and radial paths.

The Synetic system models only angle and energy. Its minimal patterns may be given the various energetic characteristics of fields or bonds or the properties of mass and spring behavior used in molecular modeling. Similarly, they may be assigned features of strong material. Regular arrays have the controlled flexibility to model the pneumatic forms and behavior of nano-scale carbon structure.

Octahedra form a cubic lattice.

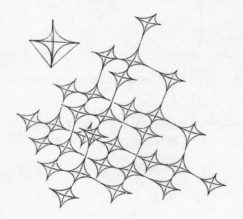

A Synetic diamond lattice.

Icosahedron in a dodecahedron:
two ways to arrange the same twelve circles on a sphere.

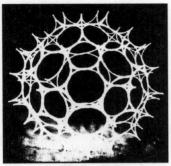

Synetic tetrahedra
model carbon-60.

Twenty tetrahedra
form a dodecahedron.

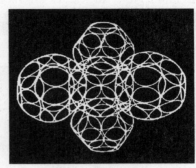

Tetrahedra model a clathrate
structure.

The system conforms to the interstitial architecture of packings and to the irregular cellular assemblages of biological structure. It self-adjusts to the fundamental angular characteristics of bubbles and foam, modeling surface-tension phenomena and the relations between membranes. At the macro-scale, tangency appears as the universal cohesive principle in the relation between fiber and membrane, and between fiber and fiber, occurring in minutely accretive structure as well as in branching, tree-like growth.

Woven Domes

A Synetic dome framework is an airy and lace-like basketry of thin arcs patterned in curvilinear triangulation. Extremely light in weight, such frames display strength and resilience with minimal material, bounce without much mass, surface tension without much surface.

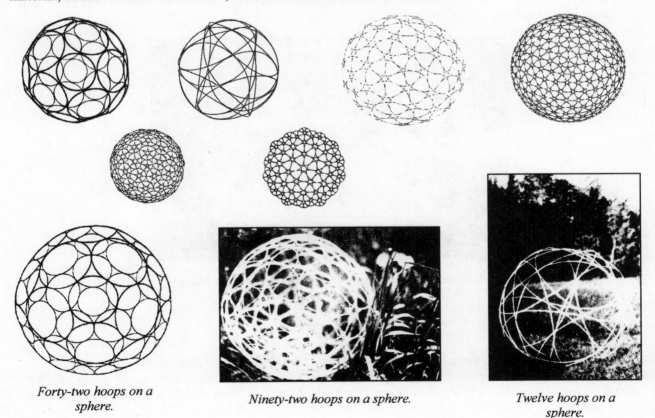

Forty-two hoops on a
sphere.

Ninety-two hoops on a sphere.

Twelve hoops on a
sphere.

Linking and interweaving of compressive paths provides minimal attachment between hoops. The strength and rigidity of frames are increased by further tensile involvement, by lashing and tying of lines and nets, or by attachment of fabric or membrane. Frames stand without the structural aid of a covering, allowing the use of thin fabric or membrane, cut to simple templates or merely wrapped.

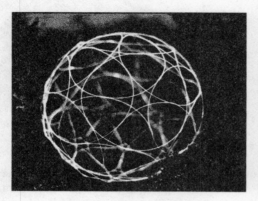

Thirty-two hoops on a sphere.   *Forty-two hoops on a sphere.*

Curvilinear patterning gives structural advantages to compressive material similar to those conferred on tensile material by fine division, bundling and networking. The compressive function is taken from the single column or spar and spread through a manifold or lattice along continuous networks.

Synetic design is singularly effective in low-tech applications, being tolerant of distortion, inaccuracy and inconsistency of material. An important feature of the system is its potential for strengthening by the incremental addition of compressive as well as tensile materials. Other advantages appear in the modularity and simplicity of the system, and in its low cost.

Hemispherical dome of 3/8" diameter fiberglass rod.

Summary

Synetics offers a new approach to minimal construction with potential application in many fields. It combines rigid triangular connectivity with flexible circular integrity into a strong and useful building system. Synetics provides a bridge of dynamic modeling between virtual design and verifiable structure.

BRIDGES
Mathematical Connections
in Art, Music, and Science

From the Circle to the Icosahedron

Eva Knoll
evaknoll@netscape.net

Abstract

The following exercise is based on experiments conducted in circular Origami. This type of paper folding allows for a completely different geometry than the square type since it lends itself very easily to the creation of shapes based on 30-60-90 degree angles. This allows for experimentation with shapes made up of equilateral triangles such as deltahedra. The results of this research were used in Annenberg sponsored activities conducted in a progressive middle school in Houston TX, as well as a workshop presented at the 1999 Bridges Conference in Winfield, KS. Not including preparatory and follow up work by the teacher, the activities in Houston were composed of two main parts, the collaborative construction of a three-yard-across, eighty-faced regular deltahedron (the Endo-Pentakis Icosi-dodecahedron) and the following exercise. The barn-raising was presented last year in Winfield, and the paper folding is the topic of this paper.

1.Introduction

Circular Origami introduces a whole different set of possibilities from the traditional square sort. The geometry of the circle allows for the creation of a convenient regular triangular grid using a few simple steps repeated around the center. These steps are all based on the properties of the circle and of the grid itself. In the first part of the exercise, the lines of the grid will be folded. Later, the grid will be used to move from the 6-triangles-per-vertex flat plane geometry to the 5-triangles-per-vertex geometry of the regular icosahedron. Using this method to create an icosahedron gives insight into its properties by showing how it can grow out of the flat plane in a simple progression. Later, it would be possible to expand this method to other polyhedra made of a single kind of faces (the cube, the tetrahedron, the octahedron and other deltahedra).

2. Preparing the folds.

In the first part of the construction, the disk is folded into a triangular grid by making use of the specific geometry of the circle. All folds are on the same side.

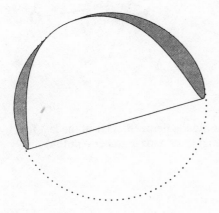

Figure 1: Fold the circle in half by bringing together the opposite edges

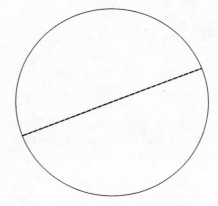

Figure 2: Unfold the paper

The first fold, because it can be made anywhere along the circle, shows the infinite rotational symmetry of the circle. Because of this property, the circle can of course be subdivided into any number of sections, including 2, 3, 4, 6, and 12 which are the ones needed for this exercise. A further advantage of the circle will be exploited here: the fact that the radius of a circle subdivides it into exactly 6 arcs comes into play in steps 5 and 6 (in trigonometric terms, the fact that *sin*(/6) = ½ allows for the subdivision of the circle into 6 without the use of measuring instruments).

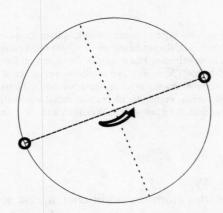

Figure 3: Fold the perpendicular by bringing together the ends of the first fold—unfold

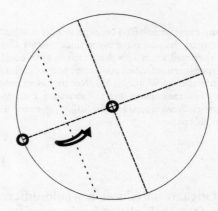

Figure 4: Fold one extremity of the first fold towards the middle as shown—unfold

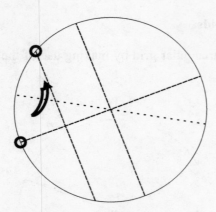

Figure 5: Fold one end of the last fold towards the nearest end of the perpendicular fold—unfold

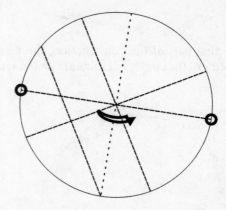

Figure 6: Fold the perpendicular of the last fold by bringing its ends together—unfold

The fold of step 4 not only subdivides the perpendicular radius in half, but also subdivides the quarter arcs into 2/3 and 1/3 on either side. This introduces the factor 3 into the subdivisions of the entire circle. In step 5, the new points are used to begin transferring this 30° angle all around. When the exercise was performed with the children, the teacher pointed out angles of 90°, 60° and 30° as they appeared, as well as equilateral and 30-60-

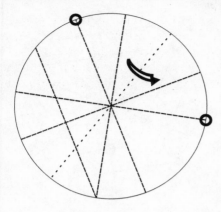

Figure 7: Fold the bisector as shown—unfold

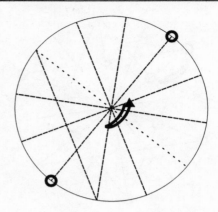

Figure 8: Fold the perpendicular of the last fold—unfold

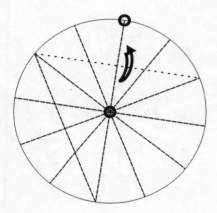

Figure 9: Fold in the end of the fold shown —unfold

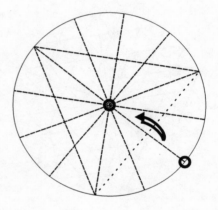

Figure 10: Fold in the last side of the inscribed triangle as shown—unfold

In step 9, we build the grid itself. This begins with an inscribed triangle started in step 4 (where it served to subdivide the circle) and ending in step 10. Note here not only the presence of several 30-60-90 triangles but also the reflectional and 3-fold rotational symmetries. Then a star of David is folded using the existing inscribed triangle. Note the new 6-fold rotational symmetries. This is a good time to stop and open a discussion with the children about visible symmetries. Finally, in steps 12 and 13, the smaller grid is folded, giving us enough triangles to construct the icosahedron.

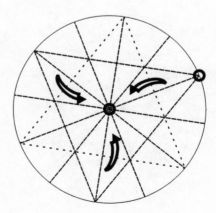

Figure 11: Fold all three points of the inscribed triangle towards the center—unfold

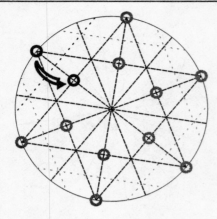

Figure 12: Fold all six vertices of the two inscribed triangles to the middle of the closest edge of the other triangle as shown—unfold

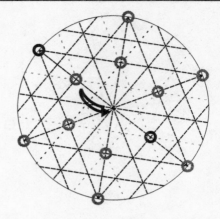

Figure 13: Fold all six vertices of the two inscribed triangles to the middle of the opposite edge of the same triangle as shown—unfold

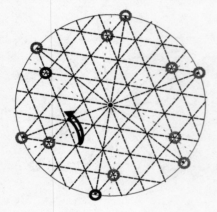

Figure 14: Fold six partial folds as shown—unfold

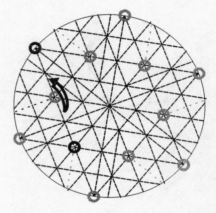

Figure 15: Fold six partial folds as shown —unfold

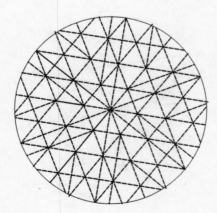

Figure 16: At this point your paper should look like this

An advantages of this method is that children can create their own regular triangular grid without measuring instruments (including a compass), and without difficulty (the most challenging fold is in step 4!) Once the grid is ready, any deltahedron can be constructed, provided there are enough triangles in the circle. See polyhedron being built up from the plane

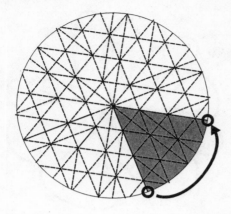

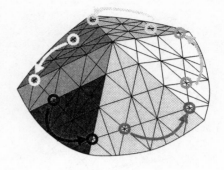

Figure 17: Join up two ends of radii as shown tuck the extra material under

Figure 18: Join five pairs of points as shown tucking the material under

3. Cutting out the net

After step18, you would normally repeat that same procedure again for the third row of 5-vertices, but the whole process becomes complicated by the fact that there begins to be too much material tucked in. At this stage, it becomes necessary to remove some of the excess material that would otherwise get tucked under. To this end, it is best to flatten out the circle again and make a few strategic cuts. Although this is counter to traditional Origami practice, it is necessary for the successful closing of the shape. Furthermore, it will help emphasize some of the properties of the icosahedron and its relationship with the net. As we know, an icosahedron is composed of exclusively 5-vertices. So the first cut needs to reduce a 6-vertex to a 5-vertex. In an ideal world, starting in the middle, we would therefore cut a 60° wedge into the circle (represented by the shaded are of figure 19). This way, we are left with only 5 triangles at the first vertex. For practical purposes, however, we will only cut out half of that wedge so as to keep some material as a tab. Repeating the exercise for the next 5 vertices (the first row around the center), we get 6 new cuts as shown in figure 20 (there are 6 cuts because 2 of them will in fact overlap when the shape is assembled).

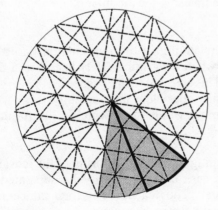

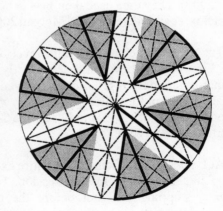

Figure 19: 60° wedge and cutting line

Figure 20: the second row of wedges and cuts

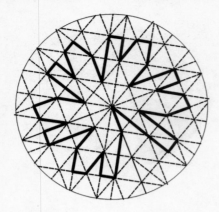

Figure 21: The cutout shape

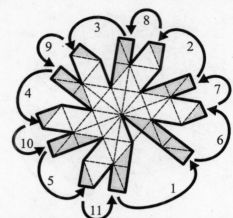

Figure 22: Assembling the icosahedron using the tabs.

Continuing the same process will yield the cutout of figure 21 (some additional cust have to be made due to overlapping areas). It is interesting to note that the cutting exercise itself emphasizes the shape of the icosahedron: the cuts are performed in sets of 1, then 5 (6 with the overlap), then 5, just as the vertices on the icosahedron are, starting with a vertex, in a sequence of 1, then 5, then 5 and then 1(the last vertex doesn't have a corresponding cut because it is the point of joining of the icosahedron).

Experience has shown that the best way to assemble the shape is using 'blue tack' to stick the tabs to the backs of the triangles. This is particularly effective if the children are going to disassemble and reassemble the net into the deltahedron for further exercises and to experience the process.

4. Conclusion

This exercise is great for demonstrating the symmetry of the icosahedron based on the structure of the net. In the flat state, all the vertices except the last are visible, clearly showing their position relative to the first one. In fact, the net clearly shows how the polyhedron emerges as well as its internal structure. This particular net has such a close relationship to the closed shape that it was used in the Houston Middle School event, after the children had done the paper folding activity, to plan how to color an icosahedron regularly with 5 colors [Morgan, 2000]. They then used their nets to build giant icosahedra using the modular elements made for the endo-pentakis icosi-dodecahedron [Morgan 2000].

The children achieved a high level of intuitive understanding of the icosahedron based on this exercise. First of all, they could see exactly what happens when a 6-vertex (flat) becomes a 5-vertex (pentagonal-pyramidal) by removing a 60° wedge (1 triangle). This step can then be repeated at every vertex, creating an algorithm for building the shape, which coincidentally does close up! Secondly, the fact that the vertices occur in sets of 1, 5, 5,and 1 show the 5-fold symmetry of the whole shape which holds true no matter which vertex you start with.

At Lanier Middle School, the entire exercise (not including the regular coloring of the icosahedron) was completed in one 90 minute session. However, the process can very easily be separated into 2 components, stopping the first before the cutting of the net. The teacher can then concentrate on 2-dimensional geometry in the first session and 3-dimnesional geometry in the second.

Results attained so far in schools justify large scale implementation of the use of these materials and activities and research into their effectiveness both in terms of geometry teaching and of motivating under-achievers to perform better in mathematics. For school classrooms, we recommend having a central project for each area which has a set of triangles and one or two people to go out to schools to use them with children. For museum displays, we recommend having shapes you can get inside and organized shape building activities for visiting school groups. My partner, Simon Morgan, and I have been developing and implementing lesson plans and exercises both using the circular Origami exercises and the giant triangles. Simon is presenting some results in his paper in the same procedings.

References

[Morgan, 2000] Simon Morgan, *Polyhedra, Learning by Building: Designing a Math-Ed Tool*. Bridges 2000 proceedings.

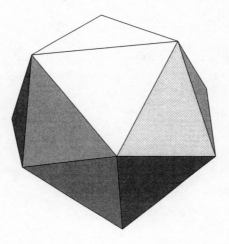

Figure 23: The assembled icosahedron

Uniform Polychora

Jonathan Bowers
11448 Lori Ln
Tyler, TX 75709
E-mail: HedronDude@aol.com

Abstract

Like polyhedra, polychora are beautiful aesthetic structures - with one difference - polychora are four dimensional. Although they are beyond human comprehension to visualize, one can look at various projections or cross sections which are three dimensional and usually very intricate, these make outstanding pieces of art both in model form or in computer graphics. Polygons and polyhedra have been known since ancient times, but little study has gone into the next dimension - until recently.

Definitions

A polychoron is basically a four dimensional "polyhedron" in the same since that a polyhedron is a three dimensional "polygon". To be more precise - a **polychoron** is a 4-dimensional "solid" bounded by cells with the following criteria: 1) each cell is adjacent to only one other cell for each face, 2) no subset of cells fits criteria 1, 3) no two adjacent cells are corealmic. If criteria 1 fails, then the figure is **degenerate**. The word "polychoron" was invented by George Olshevsky with the following construction: poly = many and choron = rooms or cells. A polytope (polyhedron, polychoron, etc.) is **uniform** if it is vertex transitive and it's facets are uniform (a uniform polygon is a regular polygon). Degenerate figures can also be uniform under the same conditions. A **vertex figure** is the figure representing the shape and "solid" angle of the vertices, ex: the vertex figure of a cube is a triangle with edge length of the square root of 2. A **regiment** is a set of polytopes that have the same vertex and edge arrangements. Polytopes in the same regiment will have vertex figures with the same corners.

Shorter Names

As anyone who studies polyhedra would know - polyhedron names can get very long - writing down "great quasitruncated icosidodecahedron" fifty times can get wearisome. I would usually write an abbreviation, so instead of "great quasitruncated icosidodecahedron", I would write GQTID, and eventually I started to pronounce the abbreviations. Now I call the GQTID - gaquatid (pronounced GAH qua tid). Shorter polyhedron names - like dodecahedron - were simply shortened - doe. Following are the short and long names of the 9 regular polyhedra as well as a few other uniforms:

Tet - tetrahedron, Cube - cube, Oct - octahedron, Doe - dodecahedron, Ike - icosahedron, Gad - great dodecahedron, Sissid - small stellated dodecahedron, Gissid - great stellated dodecahedron, Gike - great icosahedron, Co - cuboctahedron, Oho - octahemioctahedron, Cho - cubihemioctahedron, Did - dodecadodecahedron, Quit Sissid - quasitruncated small stellated dodecahedron, Sirsid - small inverted retrosnub icosicosidodecahedron.

Just for fun - here is a sampling of polychoron short names: Gogishi, quit gashi, gun hidy, gixhinxhi, ican phix, frogfix, gnappoth, gaquapac, thaquitpath, sidhiquit paddy, irp xanady, madtapachi hiccup, gotpodady, quicdady, sirc, gixhi fohixhi, paquerfix, gaqrigagishi, wavhiddix, dod honho, and spiddit.

Discovery of the 8186 Uniform Polychora

More like "known" polychora - there may still be undiscovered ones. As been mentioned above, there are 75 uniform polyhedra as well as the infinite series of prisms and antiprisms, these polyhedra are pictured in Magnus J. Wenninger's book Polyhedron Models [1]. Many of these polyhedra are very interesting expecially the monster sirsid (#118) which has 3060 pieces. Now the question is - what happens in four dimensions? - what are the uniform polychora like? In the early 1990's I wondered the same thing - also how many were there? So I started to search every book or article related to the subject, and found very little except for the list of regular polychora and the grand antiprism [2] as well as nice color illustrations of the convex regulars [3]. Earlier discoveries were made by Alicia Boole Stott using generative processes she discovered to visualize new polytopes from known ones, as well as John Conway who completed the discovery of convex uniform polychora, and Norman Johnson who found 4-D antiprisms. As it turns out, before 1990, only 130 or so were known, these included the 16 regular polychora, the 41 "Archemedian" polychora, the 74 polyhedral prisms, and a few extra concave ones - mainly antiprisms and those with ditrigonal polyhedra as vertex figures. So I started the search and several years later the number of uniform polychora grew to 8186 - where eight of the latest ten were discovered by George Olshevsky this past year, these latest finds are the swirlprisms. George and I have been corresponding quite regularly recently, George is the author of the primary polychoron web site [4]. Some of the methods that I used to find polychora included Coxeter's graphics notation [5] as well as "digging" into the vertex figure for more polychora - basically this involves finding all polychora within the regiment and then digging into the cells to find cells in deeper regiments.

Many of the vertex figures look bizarre, some have holes, atleast two instances have vertex figures with a complete tunneling system inside, some look like encased spindles, another appears to have a small tetrahedron dangling by corners inside a strange encasing, one looks like a football trophy, another looks like a geometric bird skull (I nick named it "Phoenix"), and several appear to have faces - not quite the nice simple bowtie, and trapezoid shapes of the polyhedron vertex figures!

Uniform Polychoron Categories

Following are the categories of the uniform polychora:

1. The 16 regulars and 3 miscellaneous facetings (19)
2. The truncates and quasitruncates (21)
3. The triangular rectates (7 regiments of 3) - vertex figures are triangular prisms and facetings.
4. The ico regiment, 14 in regiment, ico is the regular 24-cell and has been counted (13)
5. The pentagonal rectates (4 regiments of 22) - vertex figures are pentagonal prisms and facetings.
6. The sphenoverts which include small rhombates (24 regiments of 7, three already counted in category 3) - vertex figures are wedges and facetings.
7. The bitruncates (11) - vertex figures are tetrahedra with 2 pairs of isosceles triangles, two have tripled up vertices.
8. The great rhombates and related (23) - vertex figures are tetrahedra with bilateral symmetry.
9. The maximized polychora (includes great prismates) (22) - v.f's are irregular tetrahedra.
10. Prismatorhombates and related (30 regiments of 3) - v.f's are trapezoid pyramids and facetings.
11. Triangular small prismates and related (1 regiment of 5 and 5 regiments of 7) - v.f's are triangular antiprisms and facetings.
12. Frustrumverts (4 regiments of 7 and 2 more from a regiment in category 3) - v.f's are triangular frustra and facetings.

13. Spic and giddic regiments (2 regiments of 22) - v.f's are square antiprisms and facetings.

14. Skewverts (4 regiments of 15) - v.f's are skewed wedges and facetings.

15. The afdec regiment (52) - afdec has a v.f. with 6 faces, 2 parallel rectangles in orthogonal positions seperated by 4 alternating trapezoids. Cells are 48 coes and 48 goccoes (great cubicuboctahedra).

16. The affixthi regiment (96) - affixthi's v.f is similar to afdec's, except one rectangle is a square and the other long. Cells are 600 octs, 120 dids (dodecadodecahedra), 120 gaddids (great dodekicosidodecahedra), and 120 gidditdids (great ditrigonal dodekicosidodecahedra).

17. The sishi regiment - sishi is the small stellated 120 cell and has a dodecahedron v.f. (34) - there are actually 44 in this regiment, 2 are regulars and been counted, the remaining 8 are swirlprisms.

18. The ditetrahedrals (3 regiments of 71) - v.f's are truncated tetrahedral for 2 regiments and "cuboctahedral" in the other (this "cuboctahedron" has rectangles replacing the squares) plus the facetings, many are very intricate.

19. Prisms (74) - prisms of uniform polyhedra, cube prism has been counted as the regular tesseract.

20. The 6 antiprisms, 6 normal snubs, and the 10 swirlprisms (22). Swirlprisms have a strange swirly symmetry.

21. The padohi super-regiment (354) - padohi's v.f is a pentagonal antiprism with a larger base. Cells are 120 sissids, 120 ikes, 720 5/2Ps, and 1200 3Ps.

22. The gidipthi super-regiment (354) - gidipthi's v.f is a pentagonal frustrum, its cells are 120 sissids, 120 ikes, and 120 gaddids.

23. The rissidtixhi super-regiment (316) - its v.f is a "ditrigon" prism - a ditrigon is a semiregular hexagon. Its cells are 120 gids (great icosidodecahedra), 600 octs, and 120 sidtids (small ditrigonal icosidodecahedra).

24. The stut phiddix super-regiment (238) - it's v.f. has a triangle top and a ditrigon base, it's cells are 120 sidtids, 600 tets, 720 5/2Ps, and 600 coes.

25. The getit xethi super-regiment (238) - it's v.f. is like stut phiddix's but squattier, it's cells are 120 sidtids, 120 gaddids, 120 quit gissids, and 600 tets.

26. The blends (16) - this regiment is formed by the fact that 3 regiments have the same convex hull and can blend together, the v.f's resemble a "3-D pentagon" and facetings.

27. Baby monster snubs (2 regiments of 17) - the "sidtaps" are formed by compounding 5 polychoron A's and 5 polychoron B's where A and B are in the rox regiment (the rectified 600-cell which has a 5P v.f), both need octs for cells so this conglomerate compound can "blend" together to form a true polychoron - if A and B are different - the result is a non-Wythoffian chiral snub. One of the sidtaps (formed by blending 10 roxes together) is hollow inside - it's v.f looks like 2 pentagonal prisms attached to a removed square which both have in common. The "gidtaps" are formed the same way but with the raggix regiment (rectified grand 600-cell).

28. Idcossids (2749) - one of the ultimate monster snub regiments formed by the blending compound of 5 padohi A's and 5 padohi B's. Nearly all of these are chiral.

29. Dircospids (2749) - the other ultimate monster snub regiment, formed by the blending compound of 5 gidipthi A's and 5 gidipthi B's. Nearly all of these are chiral.

There are also two infinite categories not counted in the main count - these are the duoprisms and the antiprism prisms (also called antiduoprisms).

Figure 1 shows various vertex figures of the polychora. Even the vertex figures make great models to build, in which I have constructed over 160 of them.

Just for the record, here are the short names of the 16 regular polychora - I use Coxeter's notation for identification: Pen {3,3,3}, tes {4,3,3}, hex {3,3,4}, ico {3,4,3}, hi {5,3,3}, ex {3,3,5}, fix {3,5,5/2}, gohi {5,5/2,5}, gahi {5,3,5/2}, sishi {5/2,5,3}, gaghi {5,5/2,3}, gishi {5/2,3,5}, gashi {5/2,5,5/2}, gofix

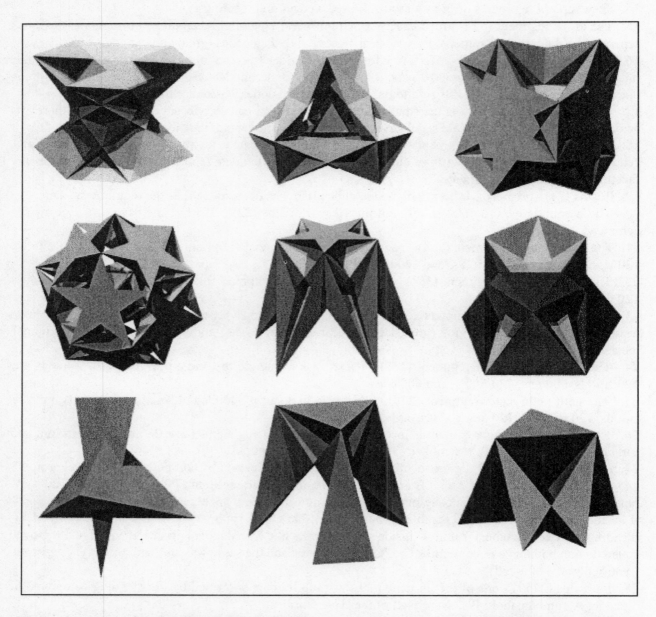

Figure 1: *Vertex figures of various polychora, the top left one belongs to the rissidtixhi regiment, the next two are ditetrahedrals, left middle is in the sishi regiment and contains many holes, the central one belongs to the padohi regiment, next is a gidipthi regiment member, bottom left is the vertex figure of one of the blends, the next two belong to the stut phiddix and getit xethi regiments respectively. As you can see, even the vertex figures alone are very interesting.*

{3,5/2,5}, gax {3,3,5/2}, and gogishi {5/2,3,3}. This notation works as follows: the cells of {a,b,c} are {a,b}'s and its vertex figure is a {b,c}, the faces of {a,b} are a-gons, and the v.f is a b-gon. Ex. {4,3} is a cube, and {5,5/2} is a gad.

Rapsady

The most interesting normal snub is "rapsady" = retroantiprismatosnub dishecatonicosachoron. Rapsady is the only known polychoron other than the prisms that contain snub polyhedra as cells - and guess what they are, it has 120 "sesides" (the small snub icosicosidodecahedron) and 120 sirsids! as well as 1440 pentagonal antiprisms as snub cells. The sirsids are on the surcell and connect to each other by ten vertices, the vertex figure has triangular symmetry - containing three sirsid vertex figures on the sides. Rapsady probably has over a million pieces. Rapsady, like sirsid is not chiral. Figure 2 shows rapsady's vertex figure as well as one of it's cells - sirsid.

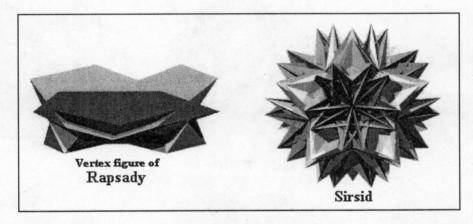

Vertex figure of
Rapsady

Sirsid

Figure 2: *The vertex figure of rapsady and it's most intricate type of cell - sirsid (also known as Yog-Sothoth by George O.)*

Idcossids and Dircospids

The ultimate polychora are the extremely intricate great monster snubs - these are the idcossids and dircospids (these names are shortened forms of the short names for the first idcossid and dircospid named). There is a uniform compound of 5 padohis as well as 10 padohis, same for gidipthi - the compound of 5 is chiral. The compound of 10 has vertices that double up, and this is what forms the idcossids (from padohi compound) and the dircospids (from gidipthi compound). The vertex figure of an idcossid looks like a padohi member's vertex figure, blended with another member which is flipped 180 degrees - they blend about the vertex figure of co, oho, or cho which are cells that each component needs in order to be a true polychoron. Same thing with dircospids, except using the gidipthi members. Each idcossid and dircospid has 7200 vertices. They also have some of the most horrendous vertex figures known, making these the ultimate monstrosities in the fourth dimension. There are 2749 of each, and 2732 of each are chiral! So there are 5538 snubs where 5492 are chiral - that means over 2/3 of the uniform polychora are monster non-Wythoffian chiral snubs! Just to show how bad these snubs get, they can have not just one - but as many as five different kinds of snub cells - where one kind shows up in two sets of orientations. These snubs could have tens of millions if not hundreds of millions of pieces!

Idcossids

Dircospids

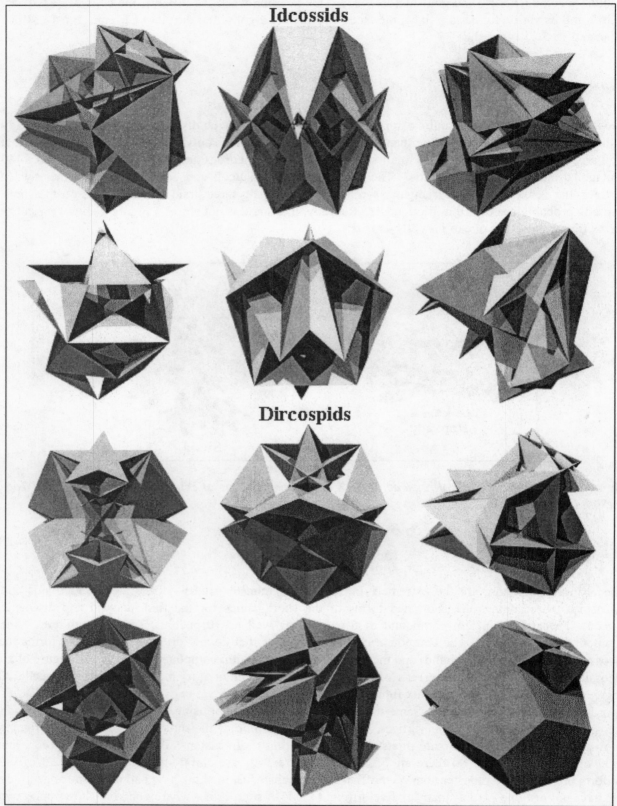

Figure 3: *Vertex figures of various idcossids and dircospids. Phoenix is on bottom left.*

Figure 4: *Cross sections of nine various polychora, the numbers in parenthesis represent the slicing realm (0 at top, 1 at bottom, .5 halfway). Sirdtaxady is a ditetrahedral, ragaghi is a triangular rectate, girpith goes with category 10, taggix is a truncate, gidtixhi is in the sishi regiment, sirc is in the spic regiment, icannixady is in category 5, affic is in the afdec regiment, and finally is gax – a regular polychoron.*

Certainly the idcossids and dircospids are the grand-daddies of all regiments! I discovered these non-Wythoffian snubs in 1997. Figure 3 shows vertex figures of various idcossids and dircospids.

Cross Sections

The best way to visualize polychora are by using cross sections - however even this would be quite puny compared to actually seeing a polychoron in full 4-D splendor. Presently I have wrote a program to section polychora and regularly add new cells and polychoron code, so far over 140 polychora are sectioned including all 98 with tesseractic symmetry, all 16 regulars and many others. They also have been animated by John Cranmer. Figure 4 shows a cross section of several different polychora.

Further Searches

There may yet be more uniform polychora - however I do believe that all of the "normal" ones are found, any more will most likely be bizarre cases - the search continues!

Meanwhile in the fifth dimension, I began the search for the uniform polytetra. There are only 3 regulars here and we lose the larger symmetries, however the regiments get fairly large - one of the hixic regiments (hix = hexatetron the 5-D symplex) has 42 polytetra - surprisingly large for a "small" regiment. There are 164 known polytetra with hixic symmetry - I have yet to give a count on the penteractic and hemi-penteractic ones. There are also 8111 known prisms and 75 infinite waves of duoprisms which are the cross-product of the uniform polyhedra and uniform polygons. Bizarre things start happening in six dimensions. There is a symmetry which is not at all related to the simplex and the hexeract symmetries - it is the sporadic "yaz" symmetry (yaz = heptacontidiicosiheptapenton (72 + 27 "pentons" - I use y and z to represent 72 and 27 respectively). Yaz was discovered by H.S.M. Coxeter [6] as 2$_{21}$. It has 27 vertices and 27 "tacs" and 72 hixes as its pentons (5-D version of cell), tac = tricontiditetron and is the 5-D version of oct. Yaz resembles some sort of 6-D "pentagon". Seven and eight dimensions also have sporadics, but they have even number of vertices. Beyond eight dimensions - the sporadics are gone, however with that much space - bizarre symmetries bound to show up elsewhere, but thats .. another dimension.

References

Acknowledgments: Thanks to George Hart for invitation to the Bridges Conference and some helpful information, and to George Olshevsky for condensing the computer images to GIF format, and to Norman Johnson , George O., and Robert McClure for saving the day when my computer crashed.

[1] Magnus J. Wenninger, *Polyhedron Models,* Cambridge University Press, 1971
[2] H.S.M. Coxeter, *Regular Complex Polytopes,* Cambridge University Press, 1974, p.167
[3] Thomas Banchoff, *Beyond the Third Dimension: Geometry, Computer Graphics, and Higher Dimensions,* Scientific American Library, 1990
[4] http://members.aol.com/Polycell/uniform.html
[5] H.S.M. Coxeter, *Regular Complex Polytopes,* pp. 14-18
[6] H.S.M. Coxeter, *Regular and Semi-Regular Polytopes. III,* Mathematische Zeitschrift **200**, p. 22, 1988.

The Square, the Circle and the Golden Proportion
A New Class of Geometrical Constructions

Janusz Kapusta
1060 Ocean Avenue, Apt. D5
Brooklyn, NY 11226, U.S.A.
E-mail: kapusta@earthlink.net

Introduction

The reason behind taking another look at the number Phi is its overwhelming appearance in art, nature and mathematics [1,2,3]. I feel that such a power must have a deep basis. As a result of this investigation I have discovered a new world of geometrical relationships residing within the square and the circle. This picture essay can be read as an example of how complexity emerges inexorably from simplicity.

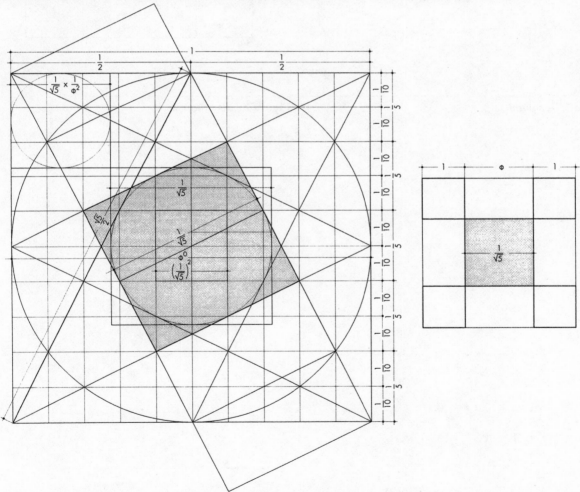

Figure 1a and 1b: *The square, the circle and 8 ($\sqrt{5}/2$) diagonals forming eight pointed star. Notice how the 10 x 10 grid appears naturally from it. The shaded square in the middle has side ($1/\sqrt{5}$). When rotated so as to be vertical, the golden proportion appears as shown to the right and above. Many properties of this star have been investigated by T. Brunes and J. Kappraff [4,5].*

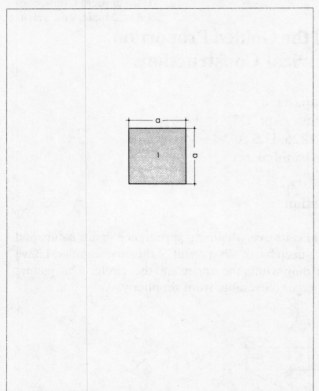

Figure 2: *A square.*

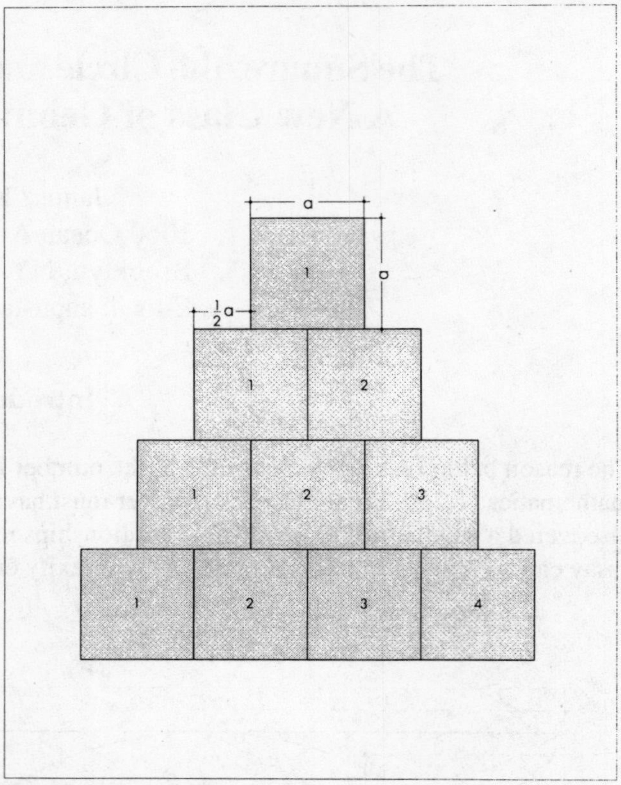

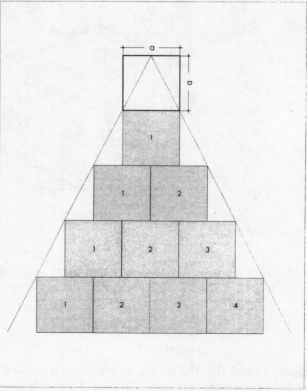

Figure 3: *10 squares.*

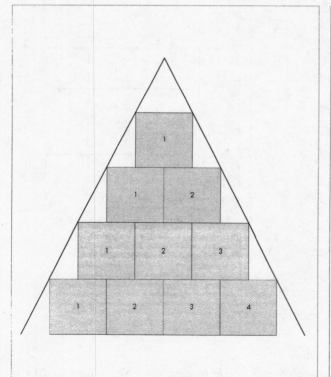

Figure 4: *2 tangent lines to 10 squares.*

Figure 5 : *Appearance of a new square with upward pointed triangle.*

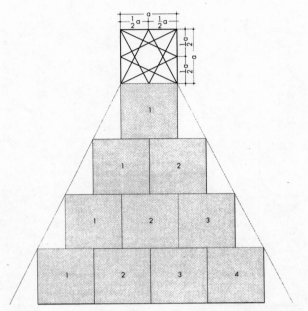

Figure 6 : *Upward, downward and sidewards triangles form an 8-pointed star .*

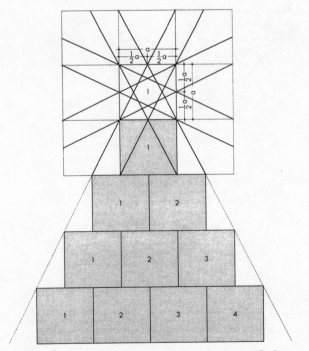

Figure 7 : *The 8-pointed star is expanded to a nine-square grid.*

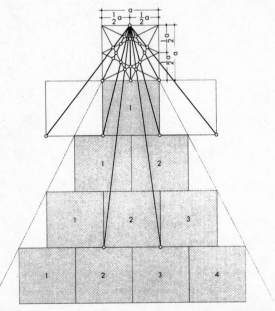

Figure 8 : *Notice how the 8-pointed star is related to the original sequence of squares.*

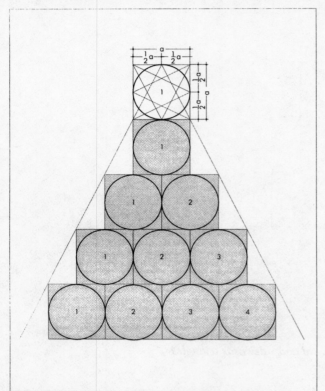

Figure 9 : *Circles are placed within the squares.*

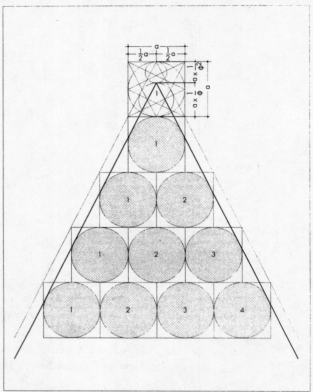

Figure 10 : *A triangle is formed tangent to the circles from which a pair of circles are defined with diameters in the golden proportion.*

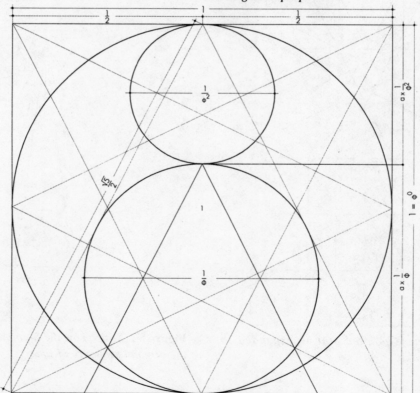

Figure 11 : *The upper square is seen in exploded view.*

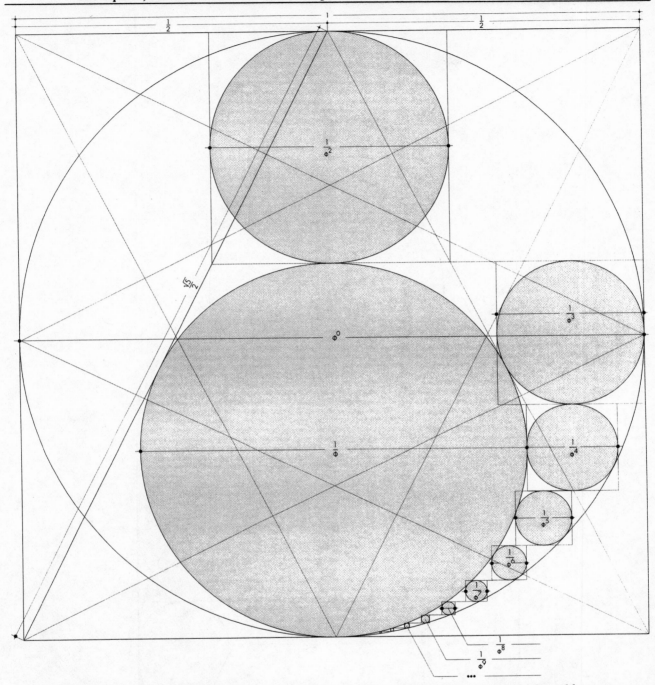

Figure 12 : *A sequence of "kissing" (tangent) circles are created with the negative powers of the golden mean.*

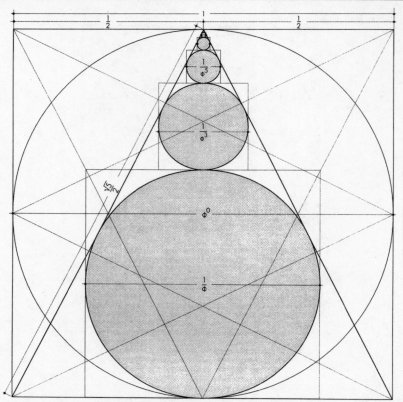

Figure 13 : *A visual proof that the odd negative powers of the golden mean sum to unity [5].*

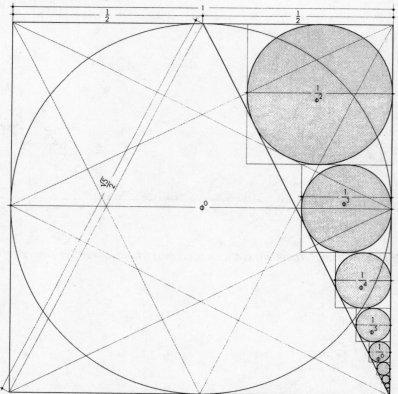

Figure 14 : *All the negative integer powers of the golden mean with the exception of 1/ϕ sum to unity.*

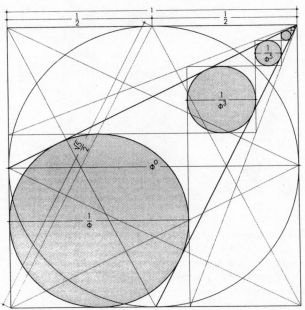

Figure 15 : *Another way to view the odd negative powers of the golden mean as a sequence of circles.*

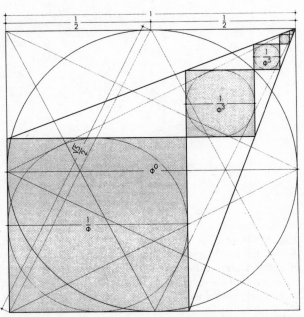

Figure 16 : *They can also be seen as a sequence of squares.*

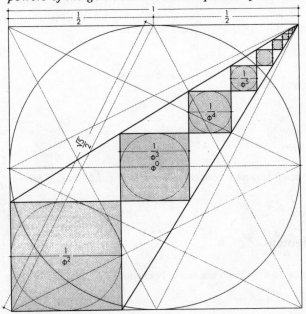

Figure 17 : *The other infinite sequence is seen as geometric series of squares and circles of decreasing size.*

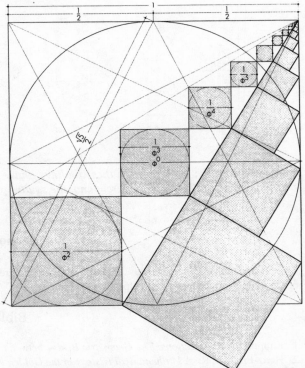

Figure 18 : *The Pythagorean theorem is expressed by this sequence of squares. Notice how a sequence of vertices of the squares upon the hypotenuse lie against the right edge of the framing square.*

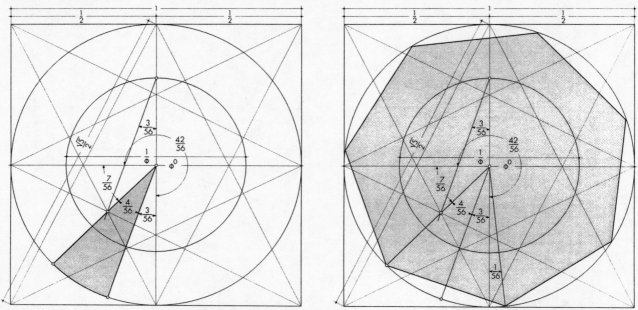

Figure 19 and 20: *An approximate compass and straight-edge construction of the angle of 3/56 x 360 degrees can be related to the golden mean. I found the error to be 0.12 %. Since 7/56 x 360 degrees equals 45 deg. this means that the circle can be subdivided by compass and straight-edge into 56 equal angles to close approximation. It should be noted that the Aubrey circle at Stonehenge has 56 equally placed stones [6]. The 56 subdivisions enable a heptagon to be constructed with compass and straight-edge to within 0.73 % error.*

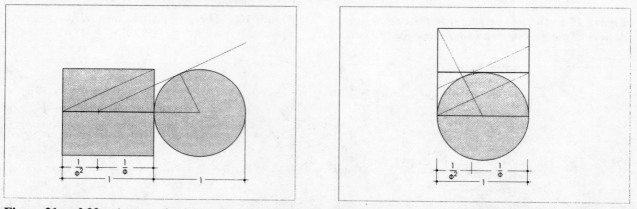

Figure 21 and 22 : *As a result of these findings I have come upon two new constructions of the golden mean based on the relationship between the circle and the square.*

Acknowledgment

I would like to thank Professor Jay Kappraff who showed great interest in my search and not only encouraged me to present part of the findings in this paper but helped in its editing.

Bibliography

1. Kappraff, J. *Connections: The Geometric Bridge between Art and Science.* New York: McGraw-Hill (1991)
2. Hertz-Fischler, R. *A Mathematical History of the Golden Number.* Mineola, New York: Dover Publications (1998)
3. Huntley, H.E. *The Divine Proportion: A Study in A Mathematical Beauty.* New York: Dover Publications (1970)
4. Brunes, T. *The Secrets of Ancient Geometry and its Use.* Copenhagen: Rhodos (1967)
5. Kappraff, J. *Beyond Measure: A Guided Tour through Nature, Myth, and Mathematics.* (In press)
6. Hawkins, B. *Stonehenge Decoded.* New York: Dell (1966)

A Fresh Look at Number

Jay Kappraff
Department of Mathematics
New Jersey Institute of Technology
Newark, NJ 07102
Kappraff@aol.com

Gary W. Adamson
P.O. Box 124571
San Diego, CA 92112 - 4571

Abstract

A hierarchy of rational numbers is derived from the integers and shown to be related to naturally occurring resonances. The integers are also related to the Towers of Hanoi puzzle. Gray code is introduced as a tool to aid in understanding Towers of Hanoi and also used to predict the symbolic dynamics of the logistic equation of dynamical systems theory. The Towers of Hanoi and Gray code are both generalized to number systems base n and used to derive a probability density function for the divisibility of integers. The number system based 4 expressed in generalized Gray code is shown to be a natural framework for the representation of the 64 codons of DNA.

1. Introduction

What do the divisibility properties of the positive integers have to do with the Towers of Hanoi puzzle, Gray code, dynamical systems theory, and the structure of DNA? This paper explores these relationships.

2. A Natural Hierarchy of Numbers

It is not commonly known that when one counts the positive integers one also counts a hierarchy of rational numbers in lowest terms in which the "most important" rational numbers appear higher on the list. Our meaning of "most important" will be defined below, but first we describe a procedure for determining location of a number in the hierarchy by giving an example. What is the 19th rational number in this hierarchy? To answer this, first write 19 in binary, i.e., $19 = (10011)_2$. Next duplicate the last digit and separate the contiguous 1's and 0's as follows:

$$1\ 00\ 111 \text{ corresponds to } [1,2,3]$$

where the numbers in brackets are the number of 1's and 0's in each contiguous group, i.e., 1 one, followed by 2 zeros, followed by 3 ones.

The numbers in brackets are the indices of the continued fraction expansion [1],[2] of 19th rational number in the hierarchy, i.e.,

$$19 \text{ corresponds to } [1,2,3] = \cfrac{1}{1+\cfrac{1}{2+\cfrac{1}{3}}} = \cfrac{1}{1+} \cfrac{1}{2+} \cfrac{1}{3} = 7/10$$

Note that the boldface indices appear as elements of the continued fraction. This leads to the following algorithm for determining the rational number corresponding to any integer of the hierarchy.

Algorithm 1:

a) Write the number in binary.

b) Duplicate the last digit and write the numbers of 0's and 1's in each contiguous group, referred to as the indices, beginning from left to right.

c) These are the indices of the continued fraction expansion of the rational number in lowest terms, i.e., $p/q = [a_1, a_2, ..., a_n] = \dfrac{1}{a_1} + \dfrac{1}{a_2} + ... + \dfrac{1}{a_n}$

Note that $[a_1, a_2, ..., a_n] = [a_1, a_2, ..., a_n - 1, 1]$. By duplicating the last digit of the binary notation we have chosen the first rather than the second representation. This procedure can also be carried out in reverse to determine the hierarchy number of a given rational number. Using this algorithm $1/2 = [2]$ corresponds to the integer 1 after eliminating the duplicated last digit. Therefore 1/2 sits atop the hierarchy. Table 1 lists the first 15 numbers in the hierarchy. In this Table, the 2^{n-1} integers with n digits are grouped in blocks and their corresponding rational numbers have continued fraction representations with indices that sum to n+1 ,e.g, there are 8 integers with 4 digits whose corresponding rationals have indices that sum to 5.

Table 1. A Hierarchy of Rational Numbers and their Representations as Binary, Gray Code, and Tower of Hanoi Positions

N	Binary	Gray 3210	Moduli	Indices	Fraction	Pegs A	B	C	TOH
0	0	0		[0]	0/1	(Start)			
1	1	1 0		[2]	1/2	1			1
2	10	11 1		[1,2]	2/3	1	2		2
3	11	10 0	3	[3]	1/3		1/2		1
4	100	110 2		[1,3]	3/4	3	1/2		3
5	101	111 0	3	[1,1,2]	3/5	1	3	2	1
6	110	101 1	5	[2,2]	2/5	1	2/3		2
7	111	100 0	3	[4]	1/4	1/2/3			1
8	1000	1100 3		[1,4]	4/5	1/2/3	4		4
9	1001	1101 0	3	[1,2,2]	5/7	2/3	1/4		1
10	1010	1111 1	5	[1,1,1,2]	5/8	2	3	1/4	2
11	1011	1110 0	3	[1,1,3]	4/7	1/2	3	4	1
12	1100	1010 2	7	[2,3]	3/7	1/2	3/4		3
13	1101	1011 0	3	[2,1,2]	3/8	2	1	3/4	1
14	1110	1001 1	5	[3,2]	2/7	1	2/3/4		2
15	1111	1000 0	3	[5]	1/5		1/2/3/4		1

It should be noted that, strictly speaking, it is the blocks in Table 1 that are ordered in the hierarchy. Within each block there is no strict ordering of "importance," e.g., in block 4, 3/4 is no more "important" than 1/4, 3/5, or 2/5. The numbering of rational numbers within each block of Table 1 follows the well-known Farey sequence shown in Table 2 [2], [3],[4].

Table 2. Farey Sequence

	0/1					1/2						1/0
Row 0							1/2					
Row 1				1/3					2/3			
Row 2		1/4			2/5			3/5			3/4	
Row 3	1/5	2/7	3/8	3/7		4/7		5/8	5/7		4/5	

...

In this table each rational number is generated from the two that brace it from above by adding numerators and adding denominators of this pair, e.g., 5/8 is braced by 3/5 and 2/3 so that 5/8 = (2 +3)/(5+3) and 5/7 is braced by 2/3 and 3/4. Beginning in row 0 and counting right to left, we find that 5/8 is the 10th rational in the hierarchy. Applying the above procedure, 10 corresponds to 1 0 1 00 (with last digit duplicated) which corresponds to the continued fraction [1,1,1,2] = 5/8. Continuing to the next row, you can check that 7/10 is indeed the 19th fraction in the hierarchy. It should also be noted that any term x in a row of the Farey sequence gives rise to two terms 1/(1+x) and x/(1+x) in the next row, e.g., x= 2/5 in row 2 gives rise to 5/7 and 2/7 in row 3.

Figure 1 illustrates the so-called devil's or satanic staircase generic to almost all dynamic systems [3]. This figure graphs the winding number ϖ vs a

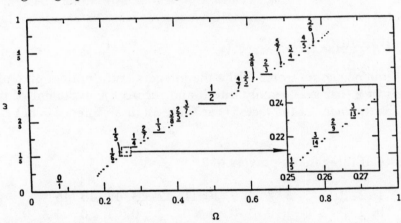

Figure1. Devil's staircase with plateaus at every rational number. From Fractals, Chaos and Power Laws by M. Schroeder. By permission of W.H. Freeman and Company.

frequency ratio Ω that represents the ratio of a driving force frequency and the resonance frequency of an oscillator for a system known as the circle map [3]. In this map a sequence of points z_0, z_1, z_2 ... are generated by the application of the function,

$$z_{j+1} = R(z_j) \text{ where } R(z) = z + \Omega - k/2\pi \sin 2\pi z.$$

The j indices can be thought of as time intervals. The winding number ϖ of the map is defined as the limit of $(z_n - z_0)/n$ as $n \rightarrow \infty$. Map R depends on a parameter k related to the energy of the system. As k$\rightarrow$1, a critical value, the system approaches a chaotic state in which every winding number from the unit interval [0,1] is obtained depending on the value of Ω with rational values of ω *phase locked* to finite intervals of Ω. By phase locking we mean that the same winding numbers are manifested for a finite interval of Ω values. The phase locked intervals for the irrationals have zero width and so the irrationals form a kind of "dust" between the rationals. We see that the higher a rational number is in the hierarchy the wider is the phase locked interval corresponding to it. Winding numbers represented by larger intervals correspond to resonances of dynamical systems with greater stability justifying our reference to the rationals corresponding to these intervals as being "more important." For example the three widest plateaus occur for 1/2, 2/3, and 1/3 corresponding to the first three rationals of the hierarchy. The relative widths of the plateaus are ordered according to the terms of the Farey sequence with elements within each row having approximately the same width.

3. Gray Code, Divisibility, and the Towers of Hanoi

Notice that the strings of 0's and 1's (bit strings) in the third column of Table 1 are labeled as Gray code. Gray code is a system of representing integers as bit strings such that from integer to integer only a single bit changes in its representation unlike binary in which more than one bit can change from one integer to the next, e.g., 7 = 111 whereas 8= 1000. To go from one integer to the next in Gray code the value of the bit in the least significant digit changes (0 to 1 or 1 to 0) to give a bit string not already listed. The digits of the bit strings have been labeled 0,1,2,... from right to left and the digit of the Gray code that changes is listed in column 4; it is the sequence,

$$0102010301020104\ldots \qquad (1)$$

Where does this sequence come from? Take the integers and divide a given integer by the by the highest powers of 2 that goes evenly into it and record the exponent of the highest power which we refer to as the *index* of the factor (see Table 3). If an integer is not divisible by 2 then its index is 0.

Table 3. Divisibility of integers by powers of 2

Integer N:	1	2	3	4	5	6	7	8	9	10	11	12	13	14	15	16 ...
Index :	0	1	0	2	0	1	0	3	0	1	0	2	0	1	0	4 ...

You will notice that sequence 1 has made its appearance again. Now if you remove the 0's, or alternatively, add 1 to each term, you get another familiar sequence,

$$121312141213121\ldots \qquad (2)$$

This corresponds to the sequence of moves required to solve the Towers of Hanoi puzzle described below. Next reduce each integer in this series by 1 to get 10201030102010... a replication of Series 1. Of course this transformation (remove the zeros and reduce by one each of the remaining numbers in Sequence 1) can be repeated ad infinitum so that we have found a self-similar pattern within the number system related to division by 2.

Note that if the modulus of any pair of adjacent numbers a/b and c/d in the Farey Sequence of Table 2 is defined to be |ad-bc|, then the Towers of Hanoi sequence is asymptotically developed for Row n as n->∞. For example, the sequence of moduli for Row 3 of Table 2 is

3537353 which corresponds to 1213121 with the following replacements: 3->1, 2->5, and 7->3. The moduli of the first 15 entries to the Farey sequence are shown in column 5 of Table 1.

The final idea in this cycle of ideas requires us to describe the Towers of Hanoi puzzle (abbreviated TOH). The Towers of Hanoi puzzle, an invention of the French Mathematician Edouard Lucas in 1883, is rich in number theoretic relations [2], [5], [6]. N disks of increasing

Figure 2. The Towers of Hanoi Puzzle with posts arranged in a circular fashion

sizes, numbered from 1,2,3,...,N from smallest largest are placed on one of three posts (see Figure 2). The object of the puzzle is to move all the disks to another post in such a manner that a large sized disk never lies atop a smaller sized disk during the transfer. If the posts are labeled A,B,C in a clockwise manner and the first move is in a clockwise direction, then the shortest path to the final configuration is unique. The sequence of moves is shown in column 8 of Table 1. Here the configuration

$$d_1/d_2/d_3/...d_m \text{ signifies that disk } d_1 \text{ lies atop disk } d_2 \text{ atop } d_3 \text{ etc., e.g., 1/2/3 means that}$$
$$\text{disk 1 lies atop 2 which lies atop 3.}$$

An optimal transfer satisfies the following rules:

1) A smaller disk must always lie atop a larger disk;
2) The parity of two adjacent disks must also be different, i.e., adjacent disks must be even – odd or odd-even numbered, e.g., in the example, 1 lies on 2 and 2 lies on 3 in the above example.
3) During any transfer, an odd numbered disk moves clockwise (CW) while an even numbered disk moves counterclockwise (CCW).

Observe that the TOH positions are correlated with the integers. In fact the following algorithm enables one to convert from the binary representation of the integer to a TOH position:

Algorithm 2:

a) For a given integer, express its binary representation in terms of its number of repeated digits. For example, 22 = 1 0 11 0 is written as {1121} since the initial 1 and 0 and the final 0 are singletons whereas the middle 11 is a pair.

b)   Starting at the right, this set of indices describes how to uniquely place n disks onto the three TOH posts. Starting on the right, the quantities of disks corresponding to the first index are placed on a post in the order 1,2,3,… The number of disks corresponding to the second index then go on another post.

c)   The number of disks corresponding to the third index are then placed on a different post than the previous move. If a choice arises between an occupied post or a vacant post, always choose the occupied post. At all times the three rules stated above must be satisfied.

Following these rules for {1121} the TOH position is :

$$5 \quad 2/3 \quad ¼$$

or, starting at the right, the first disk (#1) must go onto a post. Then the next two disks (#2 and #3) must go onto a different post. Then disk #4 must go onto a different post from the preceding one. However, according to rule c, we now have a choice:

Vacant post or occupied post

Always choose the occupied post as long as the parity rule is not violated. Therefore, #4 goes beneath #1. Then #5 must go onto a different post since it cannot go beneath disk #3 because of parity (no two odds together), so it must go onto the vacant post.

In this manner, each decimal number n is uniquely associated with a TOH position. Let us now make a short digression to convert a binary number directly to its equivalent Gray code representation. Consider 19, represented in binary as 10011. Multiplying by 2 and adding the next digit we get the sequence 1, 2, 4, 9, 19. These are the decimal representations, B_{19}, of the family of binary numbers: 1, 10, 100, 1001, 10011.   Next take the differences of adjacent terms of the B_{19} sequence which we refer to as the Gray code sort,

```
            n: 5  4  3  2  1
       Binary: 1  0  0  1  1
         B₁₉: 1  2  4  9  19
Gray code sort: 1  1  2  5  10
```

Taking each number of the Gray code sort mod 2 (i.e., even is 0 and odd is 1) yields the Gray code equivalent, 11010.   However, perhaps more significantly, the Gray code sort indicates the number of times the nth disk moved, up to the Nth move in the transfer to the final TOH position, e.g., up to the 19th move: disk 1 moves 10 times, while disks 2,3,4, and 5 move 5,2,1, and 1 times respectively.

Binary numbers can also be converted directly to Gray code by Algorithm 3.

Algorithm 3:

a) Record the leftmost digit of the binary representation.

b) If the k+1-th digit of binary is equal to or larger than the k-th digit then the k+1-th digit of the Gray code representation is the difference between these digits.

c) If the k+1-th digit is less than the k-th digit, then add 2 (base number of binary) to the k+1-th digit, and then subtract the k-th digit from it to obtain the k+1-th digit of Gray code.

4. The Generalized Towers of Hanoi Problem

We refer to the Towers of Hanoi puzzle as TOH:2 because, as we have seen, it is based on the binary numbers. We have been able to generalize this puzzle to the transfer of disks between n+1 posts which we refer to as TOH:n. Once again the n posts are arranged in a circular formation and, once again, we find that in the optimum transfer odd number disks move clockwise while even numbered disks move counterclockwise. Also the sequence of moves are related to the exponent of the greatest power of n, referred to as the index, that divides evenly into a given integer. Table 4 illustrates the divisibility sequence TOH:3.

Table 4. Divisibility by powers of 3

Integer N :	1 2 3 4 5 6 7 8 9 10 11 12 13 14 15 16 17 18 ...
Index :	0 0 1 0 0 1 0 0 2 0 0 1 0 0 1 0 0 2 ...

Eliminating the zeros results in the TOH:3 sequence 112112... It should be noted that in this sequence of moves, a disk is permitted to move either clockwise or counterclockwise from one post to an adjacent post. Thus if we wish to move a disk from one post to another clockwise from it by two post, we must transfer it in two steps.

We can also determine the Gray code sort corresponding to the n-ary, or base n, representation of the number of moves N and use this to predict the number of moves each sized disk makes during the optimal transfer. To do this express the TOH:n position in the number system base n. For example, step number 137 is represented in base 3 as $(12002)_3$. The sequence B_{137} in base 3 is then obtained by multiplying each digit of the binary representation by 3 and adding the next digit to get: 1,5,15,45,137 which gives the decimal values of the base 3 sequence: 1,12,120,1200, 12002. Next take difference to yield the Gray code sort:

n:	5	4	3	2	1
3-ary:	1	2	0	0	2
B_{137}:	1	5	15	45	137
Gray code sort:	1	4	10	30	92

Thus, disk 1 moves 92 times and disks 2,3,4,5 move 30, 10, 4, 1 time respectively. If the Gray code sort numbers are expressed mod 3 we get the generalized Gray code representation of 137 or 11100.

We can get the generalized Gray code of integer N directly from the base n representation by slightly modifying Algorithm 3 so that in step c the base number n is added to the k+1-th digit if it is less than the k-th digit. Just as for Gray code, a single digit changes between successive integers represented by generalized Gray code and the place value of the change numbered 0,1,2,...from the right represents the highest power of n that divides evenly into N. For example, expressing the following numbers in the base 4, $31 = (133)_4$ and $32 = (200)_4$ with Gray code representations, 31 = (120) and 32 = (220). Notice that a single digit changes in the third place (i.e., place value 2). Therefore 4^2 is highest power of 4 that divides evenly into 32. Also note that 31 mod 4 is gotten by adding the digits of the Gray code representation mod 4, e.g., 31 mod 4 = (1+2+0) mod 4 = 3. This holds for any integer N written in any base n. Table 5 lists the generalized Gray code for the first 64 numbers of the 4-ary system.

Table 5. 4-ary Gray Code and its Relationship to the DNA Codons

0	000	CCC	16	130	AUC	32	220	GGC	48	310	UAC
1	001	CCA	17	131	AUA	33	221	GGA	49	311	UAA
2	002	CCG	18	132	AUG	34	222	GGG	50	312	UAG
3	003	CCU	19	133	AUU	35	223	GGU	51	313	UAG
4	013	CAU	20	103	ACU	36	233	GUU	52	323	UGU
5	010	CAC	21	100	ACC	37	230	GUC	53	320	UGC
6	011	CAA	22	101	ACA	38	231	GUA	54	321	UGA
7	012	CAG	23	102	ACG	39	232	GUG	55	322	UGG
8	022	CGG	24	112	AAG	40	202	GCG	56	332	UUG
9	023	CGU	25	113	AAU	41	203	GCU	57	333	UUU
10	020	CGC	26	110	AAC	42	200	GCC	58	330	UUC
11	021	CGA	27	111	AAA	43	201	GCA	59	331	UUA
12	031	CUA	28	121	AGA	44	211	GAA	50	301	UCA
13	032	CUG	29	122	AGG	45	212	GAG	61	302	UCG
14	033	CUU	30	123	AGU	46	213	GAU	62	303	UCU
15	030	CUC	31	120	AGC	47	210	GAC	63	300	UCC

5. A Probability density function for divisibility of integers

We have previously shown that if each integer from the TOH:2 series , 12131214…, is reduced by 1 unit, the indices of Table 2 : 01020103… result. If the zeros are removed from this series, the TOH:2 series is replicated. This process can be repeated ad infinitum and also applied to the generalized TOH:n series to reveal the self-similarity of the TOH:n series. The following theorem is the result of this self-similarity process.

Theorem: The probability P that a randomly chosen integer is divisible by n^d but no higher power of n is given by,

$$p(n^d) = (1-x)x^d \qquad (3)$$

where $x = 1/n$.

Proof:

   Since every n-th integer is divisible by n, the probability x that an integer is divisible by n is x = 1/n , and the probability that an integer is not divisible by n is 1-x. Therefore, the probability that an integer has index 0 (not divisible by n) in the TOH:n table corresponding to Table 4 is 1-x while the probability that it has a non-zero index (divisible by n) is x. By the self-similarity of the TOH:n series, the probability that an integer has index 1 (divisible by n^1 but no higher power of n) is (1-x)x, while the probability that the index is 2 or higher (divisible by all powers of n greater than 1) is x^2. That the probability of an integer having index d is $(1-x)x^d$ follows by induction.

Corollary: $1 = \sum\limits_{d=0}^{\infty} p(n^d)$

Proof: Since $1/(1-x) = \sum\limits_{d=0}^{\infty} x^d$,

$$1 = (1-x) \sum\limits_{d=0}^{\infty} x^d \ .$$

The corollary follows from $p(n^d) = (1-x)x^d$.

Example 1: The probability that an integer is divisible by 3^2 but no higher power of 3 is $(1-1/3)(1/3)^2 = 2/27$.   Therefore the number of integers smaller than 137 and divisible by 3^2 but no higher power of 3 is approximately: $(2/27)(137) = 10.148 \cong 10$ as derived above. The numbers are: 9,18,36,45,63,72,90,99,117,126. This result is approximate since the probability distribution applies asympotically as the integer approaches infinity.

The probability distribution function given by Equation 1 is :

$$\overset{0 \qquad\quad 1 \qquad\qquad 2 \qquad\qquad 3}{1 = (2/3) + (2/3)(1/3) + (2/3)(1/9) + (2/3)(1/27) + \ldots}$$

where the numbers above the terms in this equation refer to the probability distribution of $1,3,3^2,3^3,\ldots$e.g., the total number of integers less than 137 divisible by any power of 3 is the sum of terms 1,2,3, and 4 of this probability distribution multiplied by 137 or $(137)(2/3)(1/3+1/9+1/27+1/81) = 45.102 \approx 45$.

Example 2: The relative frequency of clockwise (CW) movements of the TOH:2 disks is gotten by adding the even terms of Equation 1, to get $1/(x+1)$, while the relative frequencies of the counterclockwise (CCW) terms are gotten from the odd terms of Equation 1, or $x/(1+x)$. For example, for TOH:2 $x = 1/2$, and two-thirds of the moves are CW while one-third are CCW, a ratio of CW to CCW movements of 2:1. Note that $1/(1+x)$ and $x/(1+x)$ are the pair of terms of the Farey series resulting from the fraction $x = 1/n$.

6. Relationship between 2-ary Gray Code and the Logistic Equation of Dynamical Systems Theory

The logistic equation of dynamical systems theory [2] is given by,

$$x_{n+1} = a\, x_n(1-x_n) \qquad\qquad (4)$$

Given a sufficiently small value of 'a' and beginning with x_0 on the interval [0,1], the successive iterates (considered to be time intervals) of Equation 4 constitute a *trajectory* on [0,1]. It is well known that as values of 'a' are increased to the critical value of 3.5699..., the trajectories approach periodic orbits of periods 2^n for n = 1,2,3,... For values of 'a' exceeding the critical value, the system enters the realm of 'chaos' in which small changes in the initial conditions result in wildly different trajectories given a sufficient number of iterates (time). If we label values of x_n by 0 if they lie either at the maximum point or to the left of the maximum point of the logistic function $y = ax(1-x)$, and 1 if they lie to the right of the maximum, the resulting string of 0's and 1's is known as the *symbolic dynamics* of the logistic equation.

The symbolic dynamics of the logistic equation can be generated from the binary sequence of integers in two steps:

1. Add the digits of the binary numbers mod 2 to get the so-called Morse-Thue sequence [Schroeder 1991] , e.g,
    0, 1, 10, 11, 100, 101, 110, 111, 1000, 1001, 1010, ... → 0 1 1 0 1 0 0 1 1 0 0 ...
2. The symbolic dynamics is gotten by transforming the Morse-Thue sequence to Gray Code by Algorithm 3. This is equivalent to adding adjacent binary digits mod 2 to get what we refer to as the *Complementary Morse-Thue sequence* (CMT).
    1 0 1 1 1 0 1 0 1 0 ...

Notice that the symbolic dynamics of the CMT sequence is equivalent to the parity of the Towers of Hanoi sequence 121312141213121... where even digits are assigned 0 and odd digits 1 (i.e., CW = 1 and CCW = 0). Therefore, the 1's and 0's of CMT are governed by the probability density function of Equation 3, i.e., the ratio of 1's to 0's is 2:1.

7. The Relationship between 4-ary Gray Code and DNA

It is well known that DNA is composed of 64 codons. Each codon is a three letter "word" made up of one of four bases C (Cytosine), A (Adenine), G (Guanine), U/T (Uracil/Thymine). There is a natural way to relate the DNA codons to Gray Code. Use the following assignments:

C= 0, G = 1, A = 0, U/T = 1 .
0   1   1   0

For example CUG = 011 and GAC = 100. GAC is called the anti-codon of CUG since
                001   110
1 of the codon is replaced by 0 and 0 by 1 of the codon to get the anti-codon. Notice that the upper and lower bit strings of both the codon and anti-codon differ in a single bit, e.g., they have a Hamming distance of 1. Subtracting the Hamming distance from 9 yields the number of hydrogen bonds per codon/anti-codon pair [7]. For example, both CUG and GAC have 9 – 1 = 8 hydrogen bonds.

In Table 6 the codons are arranged in an 8x8 square pattern along with their number of hydrogen bonds. In this square both the row and column numbers are labeled 0 to 7 in the standard Gray Code, e.g., 000, 001, 011, 010, 110, 111, 101, 100, and each element of the table is listed by a 6-bit representation. This is equivalent to a Karnaugh map for a Boolean system with six variables. The Karnaugh map is a commonly used tool to simplify compound statements in Boolean (2-valued) logic [8]. In this table, as in all Karnaugh maps, adjacent elements to the left, right, up, down, or wrap-around of any element differs from that element in a single bit. Also to find the location of an anti-codon given the position (row and column) of a codon, or vice versa, use the following algorithm: row (column) 0 matches row (column) 5, 1 matches 4, 2 matches 7, 3 matches 6. For example, GGU is located in row R_4 and column C_5 which becomes R_1 and C_0 for the anti-codon CCA and each have 8 hydrogen bonds.

Table 6. Gray Code DNA Matrix

```
        0      1      2      3      4      5      6      7
       000    001    011    010    110    111    101    100
       000    000    000    000    000    000    000    000
 0 000
       CCC 9  CCU 8  CUU 7  CUC 8  UUC 7  UUU 6  UCU 7  UCC 8

       000    001    011    010    110    110    101    100
 1 001 001    001    001    001    001    001    001    001
       CCA 8  CCG 9  CUG 8  CUA 7  UUA 6  UUG 7  UCG 8  UCA 7

       000    001    011    010    110    111    101    100
 2 011 011    011    011    011    011    011    011    011
       CAA 7  CAG 8  CGG 9  CGA 8  UGA 7  UGG 8  UAG 7  UAA 6

       000    001    011    010    110    111    101    100
 3 010 010    010    010    010    010    010    010    010
       CAC 8  CAU 7  CGU 8  CGU 9  UGC 8  UGU 7  UAU 6  UAC 7

       000    001    011    010    110    111    101    100
 4 110 110    110    110    110    110    110    110    110
       AAC 7  AAU 6  A GU 7 AGC 8  GGC 9  GGU 8  GAU 7  GAC 8

       000    001    011    010    110    111    101    100
 5 111 111    111    111    111    111    111    111    111
      AAA 6  AAG 7  AGG 8  AGA 7  GGA 8  GGG 9  GAG 8  GAA 7

       000    001    011    010    110    111    101    100
 6 101 101    101    101    101    101    101    101    101
       ACA 7  ACG 8  AUG 7  AUA 6  GUA 7  GUG 8  GCG 9  GCA 8

       000    001    011    010    110    111    101    100
 7 100 100    100    100    100    100    100    100    100
       ACC 8  ACU 7  AUU 6  AUC 7  GUC 8  GUU 8  GCU 8  GCC 9
```

It has been found that the amino acids are formed from contiguous groups of codons, e.g., proline: CCC, CCU, CCA, CCG; glutamine: CAA, CAG; leucine: CUU, CUC, CUG, CUA, UUA, UUG; etc. [7]. Apparently Gray code arises in genetics as a means of minimizing the "cliffs" or mismatches between the digits encoding adjacent bases and therefore the degree of mutation or differences between nearby chromosome segments. The requirement in an encoding scheme is that changing one bit in the segment of the chromosome should cause that segment to map to an element which is adjacent to the premutated element.

There is a natural relationship between 4-ary Gray Code and the DNA
codons which can be understood by making the following correspondences: $0 \rightarrow C$,

$1 \rightarrow A$, $2 \rightarrow G$, $3 \rightarrow U/T$ in Table 5 and $C \rightarrow 0$, $A \rightarrow 0$, $G \rightarrow 1$, $U/T = \rightarrow 1$ in Table 6.
       0   1   1   0

This enables Table 6 to be rewritten as Table 7 using the 4-ary Gray code in Table 5. The corresponding decimal values are listed in Table 8. For example, the bit string in Row 3, Column 7 of Table 6 is 101 which corresponds to UAG or Gray Code 312 according
                          011
to Table 7 or decimal 50 according to Table 8. Notice that Table 7 inherits the property that each codon differs from an adjacent codon: up, down, right, left, wrap-around, in a single bit. Table 7 reveals the 4-ary number system as the natural system with which to characterize the DNA codons. The integers from 0 to 63 are divided into four quadrants. Each quadrant is subdivided into four compartments of four codons.

Table 7. 4-ary Gray Code Matrix

000	003	033	030	330	333	303	300
001	002	032	031	331	332	302	301
011	012	022	021	321	322	312	311
010	013	023	020	320	323	313	310
110	113	123	120	220	223	213	210
111	112	122	121	221	222	212	211
101	102	132	131	231	232	202	201
100	103	133	130	230	233	203	200

Table 8. Decimal Values for 4-ary Gray Code Matrix

0	3	14	15	58	57	62	63
1	2	13	12	59	56	61	60
6	7	8	11	54	55	50	49
5	4	9	10	53	52	51	48
26	25	30	31	32	35	46	47
27	24	29	28	33	34	45	44
22	23	18	17	38	39	40	43
21	20	19	16	37	36	41	42

8. Conclusion

Galileo stated that, "Nature's great book is written in mathematical symbols." We have shown that we can uncover the secrets of number by holding it up to the light in the proper way. Information about naturally occurring resonances, the nature of dynamical systems, and the structure of DNA are already built into the number system through a natural hierarchy of rational numbers. The Farey sequence and continued fractions were introduced as tools to facilitate an understanding of this hierarchy. We have also shown that the structure of the integers is isomorphic to the Towers of Hanoi puzzle and its generalizations. Gray code and its generalizations was introduced as an aid to understand this puzzle and also as the natural framework for systematizing the 64 codons of DNA.

Bibliography

[1] I.A. Khinchin, *Continued Fractions*, Chicago: Univ. of Chicago Press. 1964

[2] J. Kappraff, *Beyond Measure: A Guided Tour through Nature, Myth, and Number*, In Press.

[3] M. Schroeder, M.Fractals, *Chaos, and Power Laws*, New York: W.H. Freeman. 1991

[4] A. Beck, M.N. Bleicher, and D.W. Crowe, D.W., *Excursions in Mathematics*, New York: Worth.1969.

[5] M. Gardner, *Towers of Hanoi: in the Icosian Game*, Sci. Am. May 1957.

[6] J. Kappraff, and G.W. Adamson, Unpublished Manuscript. 1999.

[7] K. Walter, *Tao of Chaos: Merging East and West*, Austin: Kairos Center. 1994

[8] K. Rosen, *Discrete Mathematics and its Applications*, McGraw-Hill (1999).

On Growth and Form in Nature and Art:
The Projective Geometry of Plant Buds and Greek Vases

Stephen Eberhart
Department of Mathematics, California State University, Northridge
Northridge CA 91330-8313

Abstract

D'Arcy Thompson's pioneering book *On Growth and Form* showed how a square grid could be laid over the profile of one species of animal and then made to fit that of another related animal by a suitable deformation of space, thus allowing e.g. the shapes of missing bones to be estimated when reconstructing the fossil skeleton of an unknown species. Where Thompson had resorted to quite arbitrary spatial distortions for his examples, George Adams realized that the kind of conformal maps first discussed by Felix Klein (collineations of plane and space with invariant "path curves") would fit at least certain parts of plants and animals by conservative means, being in effect linear transformations in a non-Euclidean setting. This was then extended by Lawrence Edwards to quadratic models, showing how certain pairs of parts of a given plant or animal can be formally related in a species-true manner. I have applied these approaches to Greek amphoræ, showing how their body and base and/or neck shapes are similarly related.

Varying the Canon: Dürer's Manuals Applied by Thompson

In last year's contribution to these proceedings [1, p. 130], I showed how ancient Egyptian artists used a square grid laid over a sketch both to determine the ideal proportions of the various figures in the scene and to facilitate enlargement of the sketch to full size on a similarly gridded wall. Their name for this procedure, ⬤◠ = kh·r·t (possibly vocalized *kharet*), has come down to us via Greek *khartē* and Latin *charta* for the "card" of papyrus on which such a "chart" was drawn, augmented in Italian to *cartone* to become the "cartoons" drawn by Renaissance artists such as Leonardo DaVinci and Michelangelo Buonarroti as preliminary sketches for large paintings.

That both uses — determination of proportions and enlargement of sketches — were still practiced in 16th century Italy is evinced by the technique manuals published by German artist Albrecht Dürer upon his return from study there in the first decade of that century. Prior to his publications, only guild members had been allowed to be taught the technical methods of such a trade. For example, at a meeting of architects and builders at Regensburg in 1459 it was resolved that "no worker, master, polisher, craftsman; no one, no matter what he is called, unless he belongs to our trade organization, shall be taught how to build or erect structures from a ground plan" [2, p. 7]. When no less than the builder of Regensburg cathedral published a little treatise on the design of ornamental towers in 1486 he was severely criticized by his guild for this breach of professional confidence, but a precedent had been set.

Of the two traditional uses of the grid, the determination of proportions had become much more complicated by Dürer's era, incorporating numerous subdivisions of each part of the body. Nevertheless, inspection e.g. of the ideal man in Book III of his *Four Books of Human Proportions* [3, folio R verso] lets us verify that, if the distance from ground to hairline is taken as 18 units, then 6 units is still *ob dem knye* (on the knee), 9 is still the wrist (lined but not labeled), 12 still *in der weichen* (in the waist), ca. 14.4 still the armpit (unlabeled), and while 16 is labeled *kin* (chin) rather than shoulder (both of these being in relaxed lowered attitude compared with the Egyptians' stiffer uprightness) the same line actually runs across the top of the shoulder at level of the collar bone. All five key points of the Egyptian canon are therefore recognizably still observed.

But when Dürer attempted to describe the general form of a bird's egg in Book I of his *Painter's Manual* [2, p. 78], he employed the same *ad hoc* means he would apply at length in Book III for shaping Roman letters, going back to the ultimate Roman authority on such matters, Vitruvius. That is to say, he pieced the egg form together quite arbitrarily out of arcs of various circles, in the same manner as a calligrapher pieces together the several parts of a capital letter [2, p. 262].

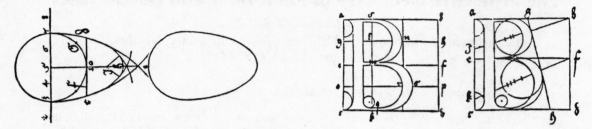

While such results may be æsthetically pleasing to some extent, their technique remains highly arbitrary, useful in its adaptability much the way cubic splines and wire-framing have become in modern computer graphics, but without any power to convince that the description thereby given of the egg is anything more than superficial — there is no appeal to biology, no connection with the egg's conception in the mother bird's ovary and propulsion, growing all the while, along her oviduct muscle to its eventual laying from her uterus, no "natural philosophy," no science.

When Thompson tried to describe "the shapes of eggs and certain other hollow structures" [4, p. xiii], there was scarcely more science involved. The descent along the oviduct was described merely as by means of "peristaltic waves" (formless in themselves), having the supposed result of blunting the foremost end, since eggs were known to be laid blunt-end first. Even as Thompson wrote this in 1917, there was mounting evidence that later proved conclusively in 1951 that eggs in fact travel pointed-end first along the oviduct and are only turned around afterwards in the uterus before laying [4, p. xiv]. The section on egg forms was accordingly one of those deemed better deleted from the 1961 abridged edition as no longer scientifically supportable, thereby begging the question as to the true, natural, form of the egg, and the role of the oviduct in shaping it.

Studying Invariance: Klein's Lectures Realized by Adams

As it happened, the necessary mathematics to address this question was already at hand, but hidden in the part of Felix Klein's lecture legacy which the fickle fates of two world wars conspired to prevent from being translated. His three volumes on elementary mathematics for training of *Gymnasium* (high school) teachers have long been available in English, but his two further volumes on non-Euclidean geometry and higher geometry for *Hochschul* (university) research have remained relatively inaccessible in German.

Two German-speaking mathematicians (one Swiss, the other Anglo-German) were aware of these latter lectures by Klein. One had even had the advantage of asking Rudolf Steiner (a graduate of Vienna's technical university and editor of natural scientific portion of Kürschner's complete edition of Goethe's works, among many other things) for his advice in attempting to model the egg form and been told (ca. 1924, in a conversation) that he would need to study Lobachevsky space. But this was the pure mathematician, Louis Locher-Ernst, whose career kept him busy with professorial duties at the universities of Winterthur and Zürich, and he did not pursue either the natural science questions or the connection with Lobachevsky space. He did write a study of projective geometry [5] including Klein's approach to classifying distance measurement according to whether the cross-ratio involved was with respect to a pair of fixed points which were real and distinct (multiplicative Lobachevsky metric), real and coincident (additive Euclidean metric), or complex conjugate (angular Riemannian metric). This was the heart of Klein's Erlangen Program: *studying transformations by determining the elements which remain invariant under them.* But when he later attempted a study of egg forms in profile [6, pp. 99-105], he restricted himself to eggs which were even more *ad hoc* constructs than Dürer's — arbitrary free-hand ovals.

The other man, George Adams, had done interpreting for Steiner in England but the subject of eggs apparently never came up. Adams' academic background was in physical chemistry (Cambridge), but left him unsatisfied. Like Locher-Ernst, he had grown up loving to hike in the mountains; unlike Locher-Ernst, he wanted to pursue his mathematics in them, not get away from it. Unaware of Locher-Ernst's work, he wrote his own version of a didactic study of projective geometry, using many illustrations from art history to show how human perception of space had gradually changed over historical time. In it [7, pp. 198 ff. & 434], he mentions Klein's *W-Kurven* but does not yet pursue them further. These were intended by Klein to be *Wurf-Kurven*, referring to von Staudt's theory of harmonic "throws," but Adams creatively re-interpreted them as *Weg-Kurven* from their occurrence as *continuous limit orbits of iterated projective transformations*, differentially applied in ever smaller steps as studied by Sophus Lie, and they have remained known as "path curves" in the English literature of those continuing Adams' work. It took a day in Regent's Park, breathing in the spring air of peace after W.W.II, for Adams to suddenly become aware that the buds on the bushes all around him were 3-dimensional cases of just such path-curve forms, living in nature, to which he later added path-curve descriptions of egg shapes as similarly symmetrical, venturing as far as describing musculature of the left ventrical of the heart as asymmetrical variation of the same basic bud or egg form, and taught these things to Lawrence Edwards, who had had his university studies cut short by W.W.II the way Adams' had been curtailed by W.W.I. The distance metric involved in these profile models was multiplicative like that illustrated last year in our study of the Egyptian canon [1, pp. 126-130] — i.e. it was Lobachevskian. Steiner had been right, but it was the man without benefit of this tip who found it, seen as "observation" (the original sense of Greek *theōrēma*) out of doors, in nature.

The Three Projective Scales

The three kinds of scales at which we looked last year were characterized by their terms forming arithmetic, geometric, and harmonic sequences, respectively, as sensed to be equispaced by human touch, hearing, and sight [1, p. 128]. If a,b,c is a subsequence, they satisfy $b = \frac{1}{2} \cdot (a+c)$ additively, $b = (a \cdot c)^{\frac{1}{2}} = \sqrt{ac}$ multiplicatively, and $1/b = \frac{1}{2} \cdot (1/a + 1/c)$ inverse-additively, respectively.

This year's trio is related but different, arrived at by study of transformations and what they leave invariant, following Klein's Erlangen Program. At first, we will find the geometric or multiplicative case now at one extreme, while the arithmetic and harmonic become conflated as central watershed case, and an angularly rotational case appears as new other extreme. But from a deeper view we will come to recognize that all three are multiplicative and all three are rotational when their appropriate arithmetics and geometries are recognized, and all three are related to human sight.

The basic phenomenon of projective geometry as study of perspective vision is that moving objects do not retain the same apparent size; approaching they seem to loom larger, and retreating they dwindle. Furthermore, rotating objects do not retain the same apparent proportions of sizes; the short end of a rectangle seems longer when turned toward us, while the longer sides taper in foreshortening. *It is only when one goes to a third level of comparison, measuring proportions of the apparent proportions* (as measured in turn in proportion to units on some standard foot or meter stick laid onto the pictorial image from without) *that invariance is attained.* If M,A,B,N are four points along a line, then [(B-M)/(A-M)]/[(B-N)/(A-N)] is constant, no matter how one views the line in perspective, or projects it onto other lines; the transformed points will have the same ratio of ratios, known as the cross-ratio of A and B with respect to M and N, abbreviated {A,B;M,N}.

1. The Hyperbolic Case (Lobachevsky)

In the sequence of canonical Egyptian measurements studied last year $0, ..., 6, 9, 12, 14\frac{2}{5}$, $16, ..., 18$ (then associated respectively with ground level and heights to knee, waist, elbow, armpit, shoulder, and hairline), we may take ground and hairline as M = 0 and N = 18, and investigate the cross-ratios of successive pairs of other measurements with respect to them. Taking A = 6 and B = 9 yields {6,9; 0,18} = [(9-0)/(6-0)]/[(9-18)/(6-18)] = (9/6)/(-9/-12) = (3/2)/(3/4) = (1/2)/(1/4) = 2. Similarly, taking A = 9 and B = 12 yields {9,12; 0,18} = (12/9)/(-6/-9) = (4/3)/(2/3) = 4/2 = 2, and the same for $\{12, 14\frac{2}{5}; 0,18\}$ and $\{14\frac{2}{5}, 16; 0,18\}$; they all yield 2 as cross-ratio. When parallel projected as shown in [1, p. 127] so that M remains finitely placed but N is sent infinitely far away (in Euclidean view), then the intermediate points are found to be spaced as powers of 2, measured from new M as 0, with arbitrary choice of which step is the unit as 0th power.

Algebraically, there is a <u>linear fractional transformation</u> or l.f.t. $f(x) = (ax+b)/(cx+d)$ which sends successive scale points to one another.   In the case of the Egyptian sequence $6, 9, 12, 14\frac{2}{5},$ 16 it is $f(x) = 36x/(x+18)$.   Since such a ratio is unique up to proportion of coefficients, it is clear that any one of the four coefficients a,b,c,d (provided non-0) may be taken as unit and three equations in three unknowns solved to determine the other three.   *Any three points of a line may thus be sent to any three other points of that line*, of which this is the special case with $6{\to}9{\to}12{\to}14\frac{2}{5}$ as our choice of scale steps; once these three moves have been made, our three degrees of freedom are used up but it can be verified that $14\frac{2}{5}{\to}16$ also works, i.e. that $f(14\frac{2}{5}) = 16$.   Setting $f(x) = x$ evidently results in a quadratic equation in x, so there will necessarily be *two fixed points* or *invariants*, namely the roots of this 2nd degree equation.   In this case, setting $36x/(x+18) = x$ yields $36x = x^2 + 18x$, $x^2 - 18x = x(x{-}18) = 0$, whence $x = 0$ and 18 are the *real and distinct* invariants.   Applying the transformation repeatedly yields $16{\to}16.94{\to}17.45{\to}17.72{\to}17.86{\to}\cdots$ (to 2 place accuracy) approaching 18 as limit but never reaching it; values $x > 18$ are also drawn toward 18 by f so that 18 is said to be an "attractor" or a sink of f, while values $\pm x \approx 0$ move away from 0 as "repeller" or source.   Applying the inverse transformation $f^{-1}(x) = 18x/(36{-}x)$ reverses these roles, sending $6{\to}3.6{\to}2{\to}1.06{\to}0.54{\to}\cdots$ closer to 0 as sink and away from 18 as source.

Analytically, as we saw last year [1, p. 130], if we follow this $f(x)$ by a further $g(x) = (x{-}9)/9$, then the sequence $0, ..., 6, 9, 12, 14\frac{2}{5}, 16, ..., 18$ is sent to $-1, ..., -\frac{1}{3}, 0, \frac{1}{3}, \frac{3}{5}, \frac{7}{9}, ..., 1$ which are the values of $\tanh ku$ for $k = -\infty, ..., -1, 0, 1, 2, 3, ..., \infty$ and $u = \frac{1}{2}\ln 2$. The sequence is thus not only projectively equivalent to powers of 2, terms of a *geometric sequence* generated by *repeated multiplication* by 2 as predicted by $\{6,9; 0,18\} = 2$; it is also projectively equivalent to *hyperbolic rotation* through angle $u = \frac{1}{2}\ln 2$.   This is what led Klein to define the distance from A to B, with respect to M and N as real and distinct inaccessible points, as $\frac{1}{2}\ln\{A,B;M,N\}$; it is the distance metric belonging to 1-dimensional Lobachevsky space [8, p. 164].

2.   The Parabolic Case (Euclid)

Suppose instead that the scale consisted of $..., 6, 9, 12, 15, ...,$ terms of an *arithmetic sequence* generated by *repeated addition* of 3, so its l.f.t. is $f(x) = x+3$.   Setting $f(x) = x$ yields $x = \pm\infty$, identified projectively as *real and coincident* point $M = N$, so the cross-ratio of such a sequence (and any to which it is projectively equivalent, including harmonic sequences obtained by inversion) must be 1, since numerator and denominator of $[(B{-}M)/(A{-}M)]/[(B{-}M)/(A{-}M)]$ are now identical.   But $\frac{1}{2}\ln 1 = 0$, so the corresponding notion of *parabolic rotation* must be through a vanishingly small angle about that infinitely remote point, successive "radii" appearing as parallel lines. This is the distance metric belonging to 1-dimensional Euclidean space, whose unit size is arbitrary (thus the necessarily free choice of conventions for inches, centimeters, etc.).   There are also almost-but-not-quite trivial parabolic versions of trigonometric functions given by $\sin p\, a = \tan p\, a = a$ and $\cos p\, a = 1$ which satisfy the analog of DeMoivre's theorem etc., argument $2a = y =$ double area of right triangle with corners $(0,0)$, $(1,0)$, $(1,y)$, analogous to viewing $2u$ as double area of hyperbolic segment with corners $(0,0)$, $(x,0)$, (x,y), with $x^2 = 1$ playing the role of unit parabola analogous to $x^2 \pm y^2 = 1$ as unit circle or hyperbola.   There is even a short-and-sweet (2-term) Fourier expansion analog available by taking $e^{2\phi a} = \cos p\, 2a + \phi\sin p\, 2a = 1 + 2a\phi$ if ϕ is a nilpotent $\phi^2 = 0$ causing all higher power terms to vanish, analogous to taking $e^{2\varepsilon u} = \cosh 2u + \varepsilon\sinh 2u = 1 + 2u\varepsilon + \frac{1}{2}(2u)^2 + \frac{1}{6}(2u)^3\varepsilon + \cdots$ with $\varepsilon^2 = 1$.   If ε is the real unit 1, then we have Klein's $\frac{1}{2}\ln e^{2u} = u$; if ε is something else (one of the Pauli matrices, like i as one of the quaternions) then we must modify the definition to be $u = (\frac{1}{2}\ln e^{2\varepsilon u})/\varepsilon$, and similarly in the parabolic case $a = (\frac{1}{2}\ln e^{2\phi a})/\phi$, both just like the formula $\theta = (\frac{1}{2}\ln e^{2i\theta})/i$ for circular rotation first discovered by Laguerre [9, p. 158].

3.   The Circular Case (Riemann)

Suppose lastly that the scale consisted of $..., 6, 9, 12, 15\frac{3}{4}, ...$ The l.f.t. granting these three movement wishes is $f(x) = (18x+162)/(-x+36)$, or $18(x+9)/(-x+36)$.   Setting this $f(x) = x$

yields a quadratic equation $x^2 - 18x + 162 = 0$ with *complex conjugate* roots $x = 9 \pm 9i$. Plotting them in the complex plane, we recognize them as the vantage points from which the terms of the sequence subtend *equal angles* of *circular rotation*. What is the angle? Evaluating the cross-ratio $\{6,9; 9+9i, 9-9i\}$ is laborious, but careful work shows it to be $\frac{4}{5} + \frac{3}{5}i = \cos 2\theta + i \sin 2\theta = e^{2i\theta}$ for $2\theta = \tan^{-1}\frac{3}{4}$, whence $\theta = \tan^{-1}\frac{1}{3} \approx 18.4349...°$, as seen in the diagram below showing the fixed point $9+9i$ at center of counterclockwise motion (which would be mirrored by clockwise motion about other fixed point $9-9i$), confirming Klein's modified definition that $\theta = (\frac{1}{2}\ln\{A,B;M,N\})/i$.

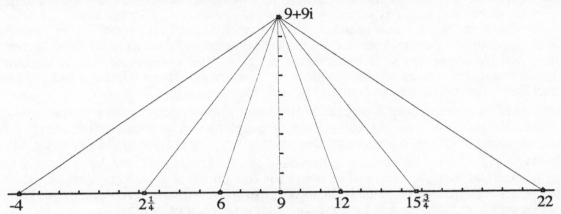

It is even possible to find the analog of the earlier construction which revealed powers of 2: As before, the idea is to send one fixed point to a new 0, the other to ∞, and one of the intermediate scale points (arbitrary which) to the new 1 (as 0th power). Instead of using parallels, however, we use the trick familiar to students of complex analysis whereby any three points are easily sent to $0, 1, \infty$ by a <u>conformal map</u> (another name for l.f.t.). Here, we send $x = 9+9i, 9, 9-9i$ to $0, 1, \infty$ via $g(x) = (x - 9 - 9i)/(9 - 9i - x)$. The sequence $..., 6, 9, 12, 15\frac{3}{4}, 22, ...$ is then found to be mapped to $(\frac{4}{5}+\frac{3}{5}i)^k$ for $k = ..., -1, 0, 1, 2, 3, ...$ as result of *repeated multiplication* by the base predicted by the cross–ratio $\{A,B;M,N\}$, appearing now as rotation through $2\theta = \tan^{-1}\frac{3}{4}$ around unit circle about 0.

It had been the suggestion of Christian von Staudt (in his pictureless 1847 *Geometrie der Lage*) that *one could in effect picture otherwise invisible imaginary or complex elements by taking them to be the implied invariants of a circling motion among the visible real elements*, thus anticipating (and perhaps even suggesting) Klein's Erlangen approach to studying transformations of all kinds. The general form of a linear fractional transformation that leaves complex conjugate elements $M,N = m \pm ni$ invariant and moves any real element $A = x$ to $B = f(x)$ in such a way as to subtend angles $\theta = \pm 180°/n$ from M,N when viewed in complex plane is given by the ratio

$$\frac{(-m + n\cot\theta)x + m^2 + n^2}{-x + m + n\cot\theta}.$$

Since any n-cycle would serve the same purpose of locating a given set of complex conjugate invariants, in practice it is common to use the n = 2-cycles or <u>involutions</u> satisfying $f^2 = $ id. For since $\theta = 90°$ in that case and $\cot 90° = 0$, these have the simpler form $f(x) = (m^2 + n^2 - mx)/(m-x)$.

Taking $u = \frac{1}{2}\ln k$, $m = \frac{1}{2}(M+N)$, and $n = \frac{1}{2}(N-M)$, it is also possible to give an analogous formula for the general l.f.t. that leaves real and distinct elements $M,N = m \mp n$ invariant and moves any other real element $A = x$ to $B = f(x)$ so as to yield $\{A,B;M,N\} = k$ as base of power sequence:

$$\frac{(m + n\coth u)x - m^2 + n^2}{x - m + n\coth u}.$$

[These details, while fiddly, are given here as Bridge between the usual elementary and advanced expositions which leave them out (as too advanced or too elementary). Anyone who wishes may skip them, but it is hoped that anyone attempting to work in the field will find them useful.]

Path Curves as Invariants of Collineations in 2 and 3 Dimensions

Non-guild-members of 15th century Regensburg were not allowed to "be taught how to build or erect structures from a ground plan," but we are about to learn how to do just that! The ground plans and elevations of the buds and eggs we seek to erect will be formed in the simplest possible way. First, we will draw the equivalent of a square grid on each of two charts, then declare each bud or egg's outlines to be the straight lines traversing those grids diagonally, like bishops' moves on a chessboard. The geometry will appear to be curvilinear, to make the bud or egg properly oval; but the algebra will remain linear, restricted to 1st degree occurrences of each variable. The catch — what makes this magic possible — is the word "equivalent"; for instead of the usual Cartesian grid with x and y axes that bear Euclidean or arithmetically-measured scales, we shall be using homogeneous x,y,z coordinates for the projective plane and their axes will bear one or other of the two non-Euclidean scales: Lobachevskian or geometrically-measured for the elevation and Riemannian or angularly-measured for the plan. The bud with its petal windings and the egg with its oviduct musculature will then arise in 3-D by their interweaving.

Instead of 3, we now have 4 degrees of freedom: *Any 4 points of the plane may be moved to any other 4 points*, and a unique <u>collineation</u> of the plane can be found which carries this out, moving every other point one-to-one to another point, every straight line to another straight line.

Instead of 2, there will now be 3 invariant points, joined pairwise by 3 invariant lines, forming an *invariant triangle*. Each of its corners is fixed in place by the collineation, while lines through them (except for the two side lines apiece) will move; likewise each of its sides is fixed in place, while points on them (except for the two corner points apiece) will move.

Rather than moving 4 points arbitrarily and then hunting for the 3 fixed points, it is easier to use 3 our our 4 degrees of freedom by specifying the fixed points X,Y,Z to begin with, and then moving one further point (not collinear with any two of X,Y,Z) from W to W'. This fourth point casts a moving shadow on each side of the invariant triangle, when projected from the opposite corner. The two corners on each side can be taken in turn as M & N, and projections of the two positions of the moved point on that side as A & B, thereby determing a unique l.f.t. on each side.

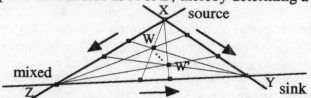

The three parts of the invariant triangle, however, are not all alike! The induced motions along two sides (here XY and XZ) flow away from their common corner (X), making it a source; those along two other sides (XY and ZY) flow toward their common corner (Y), making it a sink; while those on the remaining pair of sides (XZ and ZY) flow contrarily, making their common corner (Z) of mixed nature. The induced scales along the two sides through the mixed corner are used to create a fan of lines apiece from the source and sink corners. They form the non-Euclidean grid or chessboard, and the further (and prior) positions of W,W',W",... are traced along along it diagonally (ultimately with differential refinement to become smooth), forming stages of a *path curve* that departs tangentially from the source and moves tangentially toward the sink. Depending on nature of source and sink, the resulting trajectories look like one of the following [8, p. 105]:

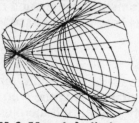
X & Y real & distinct

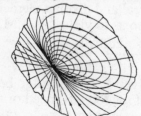
X & Y real & coincident

X & Y complex conjugate

Klein's illustrations above [loc. cit., repeated on p. 205], however, show one fan of lines from mixed corner Z (which remains real in every case).   The interweaving of the two fans from source X and sink Y is shown by Edwards [10, p. 206, or 12, p. 37], below left.   He also shows what happens if Z is placed centrally and X & Y are sent to ∞ orthogonally, below right [12, p. 40].

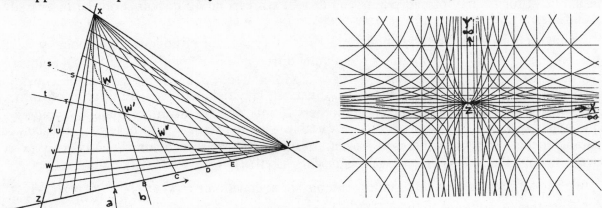

As special cases, the curves can be ellipses when Z is remote and ZX//ZY, or hyperbolæ when Z is central and ZX⊥ZY; in both cases, to be conic sections, the growth measure (the constant k at base of repeated multiplication) must be the same along both scales through Z. *In general, when different growth measures occur along these two sides, the two ends of the ovals will be different, one more or less pointed and the other more or less blunted* — both more or both less so, not independently choosable.   Unlike Dürer's method, the entire oval is determined in one piece.   It is also not unique but a member of a family of similar forms, affinely stretched or compressed [12, p. 42].

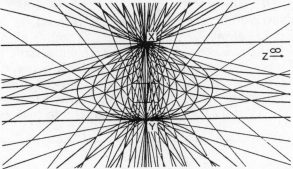

For the plan view, we must also generalize Klein's special case.   Instead of ellipses or circles (when X,Y become I,J, the "circling points at ∞") with k = 1 preserving projective radii, general spirals that become logarithmic spirals arise from radial growth measures > 1 [12, pp. 43-44].

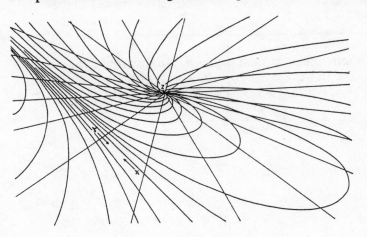
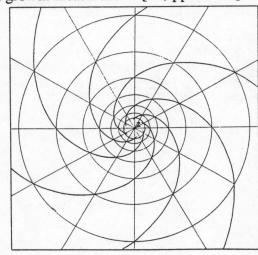

To make good on the claim that all of these path curves may be considered "straight" lines only made to look curved by their non-Euclidean settings, we must verify that they are all given by algebraically linear equations in their coordinates. These must be <u>homogeneous coordinates,</u> i.e. triples (x,y,z) or (x_1,x_2,x_3) only determined up to proportionality, with (x,y,1) corresponding to Cartesian (x,y) for finitely-placed points and (x,y,0) corresponding to points on the inaccessible line at infinity.

Conic sections with Cartesian equations such as xy = C for orthogonal hyperbolæ, y = Cx^2 for parabolæ, and x^2+y^2 = C for circles seem quite different, but Klein [9, pp. 168-169] points out that all three are of homogeneous form $x^a y^b z^c$ = C with a+b+c = 0, hence all have linear equations alnx + blny + clnz = 0 in the logs of those coordinates. The trick is to view all three as of form xy = Cz^2, pictured as above drawn tangent to corners X and Y of the invariant triangle, avoiding Z. For the hyperbolæ, it is the homogeneous z variable which is fixed = 1 to become Cartesian; for the parabolæ, it is either x or y that is so frozen; for the circles, we first factor x^2+y^2 as (x+iy)(x–iy) = r^2, then rename these as $x_1 x_2$ = $C(x_3)^2$, satisfying $(x_1)^1(x_2)^1(x_3)^{-2}$ = C with 1+1–2 = 0.

We can also factor the Cartesian logarithmic spiral equation r = $(x^2+y^2)^{\frac{1}{2}}$ = $(x+iy)^{\frac{1}{2}}(x–iy)^{\frac{1}{2}}$ = $Ce^{k\theta}$. But x+iy = $re^{i\theta}$ and x–iy = $re^{-i\theta}$, so (x+iy)/(x–iy) = $e^{2i\theta}$ and $[(x+iy)/(x–iy)]^{-\frac{1}{2}ki}$ = $e^{k\theta}$. Substitution into the previous equation gives $(x+iy)^{\frac{1}{2}}(x–iy)^{\frac{1}{2}}$ = $C[(x+iy)/(x–iy)]^{-\frac{1}{2}k}$, whence homogeneously $(x+iy)^{\frac{1}{2}(1+ki)}(x–iy)^{\frac{1}{2}(1–ki)}t^{-1}$ = C, of same form $(x_1)^a(x_2)^b(x_3)^c$ = C with a+b+c = 0.

Bud, Egg, and Vortex Shapes

The equation a+b+c = 0 above is equivalent to $e^a e^b e^c$ = 1 which we may rename αβγ = 1 with a = lnα, b = lnβ, c = lnγ. In the all-real case for ovals, this means [cf. 12, pp. 271-272, where they are inconsistently labeled] that if α is the multiplier or growth constant along XZ and β is that long ZY, then the multiplier along YX must be γ = 1/αβ to satisfy αβγ = 1. This is true if X,Y,Z take turns being the source along each side in cyclic fashion, as stated. When re-stated so that X and Y are respective sources along XZ and YZ but Y is source along YX, then it is γ = αβ. For example, suppose α = 3 and β = 2, so that we may take successive powers of those bases as scale steps along XZ and YZ when these axes are parallel and YX is perpendicular to them. Then if the distance from Y to X is taken e.g. as 5 of the same units we can confirm that points S = 0.94, T = 2.90, U = 4.46 along YX (corresponding to heights of three successive steps along a typical path curve oval) satisfy {S,T;Y,X} and {T,U;Y,X} both equal 6 to two decimal places.

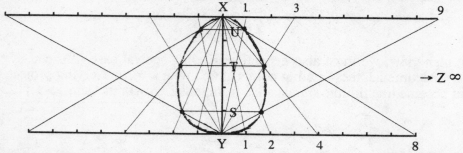

To express the relative pointed- and bluntedness of such an oval, Adams and Edwards use parameter λ = a/b = lnα/lnβ *[loc. cit.]; here,* λ = ln3/ln2 = 1.585. *If* λ = 1, *both ends are alike as ellipse;* λ → ∞ *or 0 makes ovals tend to triangles, while* 0 > λ > -1 *makes them vortical [12, p. 53].*

λ=2.0 λ=1.0 λ=0.5 λ=0.1 λ=0 λ=-0.5

The other parameter used to describe the steepness of pitch of the spiral winding about the o-void form in 3-D is $0 \leq \varepsilon \leq \infty$, yielding the transition from the analogs of horizontal latitude circles, to loxodromes spiralling from one pole to other, to vertical longitude ovals between poles, defining 2ε to be the growth from Y to X during one radian of spiral winding. To compute it [cf. 12, pp. 316-317, where cross-ratio at bottom of p. 316 is upside-down], we first find the multiplier μ of upward growth between any two points along the vertical axis (Edwards uses two points lying at about the middle 3/4 of the bud height) by evaluating a cross-ratio such as $\{T,U;Y,X\}$ in the illustration above, then take $m = \ln\mu$. Next, we measure the distances of the two spiral points from their projections onto the vertical axis, setting them each in ratio to the radius at its level and adding their inverse sines to determine the total angle turned from one to the other, expressed in radians as ω; then $\underline{\varepsilon = \frac{1}{2}m/\omega}$. With these two parameters, we have complete control over the curves.

A wide variety of plant forms such as flower buds, inflorescences, leaf buds, and seed cones were intuited by Adams to be path curve forms. After Adams' death in 1963, Edwards set about testing this hypothesis, both in his Scottish home and abroad in Australia; others have done some field work in England, Canada, and the U.S., but it is Edwards (now equipped with camera and computer link) who continues to make daily rounds, inspecting e.g. leaf buds on the same dozen or so trees from their formation in autumn to opening in spring, year after year, documenting not only their forms throughout that time but also the slight but statistically significant regular pulses in the λ values exressing those forms (a fortnightly "Is it spring yet?" incipient opening gesture, tied in to shifting lunar rhythms — cf. [12, Ch. 15], supplemented by several volumes of subsequent data analyses) akin to animal heart-beats (taking left ventrical as asymmetrical path form — [cf. 12, Ch. 8]) that lead too far to attempt to describe further here.

Suffice it to exhibit the seed cones of a Scots pine (with $\lambda = 3.03$, $\varepsilon = 0.22$, and Mean Radius Deviation = 1.3%) at left and a larch ($\lambda = 1.7$, $\varepsilon = 0.23$, MRD = 0.9%) at right [12, pp. 66-67].

Scots
Pine

Larch

One morning, Edwards observed a slight trace of a spiral imprint on the membrane between the shell and hardened albumen of his boiled egg. Visiting the Poultry Institute in Glasgow, he was shown its source in a dissected chicken: The mother hen's oviduct muscle is spirally wound.

A tracing from
photograph of
partially-formed
egg, part-way
down oviduct.

A comparison with
the nearest path curve
which could be found
[12, pp. 178-179].

The pitch is less at first (smaller ε), getting gradually more (larger ε) as the egg descends, with the effect that *the widening egg distends the oviduct muscle wall in such a way that the resulting pitch remains virtually constant throughout the descent.* The egg is spirally *wrung* from ovary to uterus (not just peristaltically pushed), the mother's musculature completing the flower bud analogy.

The geometry of 3-D path surfaces is a natural extension of what we have seen so far: There are 5 degrees of freedom, of which we may use 4 to specify the corners of an invariant tetrahedron, allowing a moving further point to project upon opposite faces as invariant triangles of induced 2-D motions, in turn projected upon edges as 1-D motions, determing the basic scales. Ignoring coincident corners, there are three principal cases: all-real tetrahedron, semi-real semi-complex, and all complex (with corners disappearing pairwise as conjugates); in all three cases, the curves pass through two corners and avoid the other two. The first case has no known natural application; the second yields the convex ovoid forms of buds and eggs (with two real tangent planes meeting at horizon and two complex conjugate planes meeting in central axis) and concave vortices; the third yields generalized helical forms, such as the oviduct muscle.

One of Edwards' most profound original discoveries is the *pivot transformation* whereby the planes tangent to a family of vortex spirals coaxal with a family of ovoid spirals are allowed to pick out the points of the latter family to which they are tangent, yielding a family of path surfaces of second degree, one member of which for a rose bud yields a rose hip, and one member of which for a St. John's wort bud yields a St. John's wort gynæcium, species-true, that also lead too far to describe further here [cf. 12, Ch's 9 & 10]. Suffice it to note that the kind of vortices required for this purpose are those with λ < -1 with ideal tips at infinite distance, which Adams intuited matched water vortex shapes and Edwards has experimentally verified by photographing controlled tank flows.

The dotted outline is traced from a photograph of an actual water vortex; the smooth line is path curve match with λ = −1.74 [12, p. 166].

The rose bud as member of ovoid family of curves, and intersection lines of planes tangent to member of vortex family, resulting in rose hip as pivot transform [12, p. 150].

Greek Amphoræ

It was one of Edwards' Australian collaborators, John Blackwood, who suggested at Easter of 1979 that one look into Grecian urns as candidates for path surface shapes. This seemed plausible to me on at least two counts — the Greeks were known to have been sensitive to mathematical æsthetics in their art and architecture, and birds have surely arrived at path shapes for their eggs out of design merit through evolution, making such shapes recommendable to be imitated (whether consciously or not) by humans for their own container designs — so I procured a copy of Arias' *A 1000 Years of Greek Vase Painting* (Abrams, N.Y., 1961) and began investigating.

The first difficulty I ran into was one also encountered by Edwards: Exactly where are the ends of the ovoid? On a leaf or flower bud as well as a seed cone, one end is attached to a stem which obscures its exact location, and the other is likely to be in the process of slightly opening; only eggs have well-defined natural ends. The vase ovoids have stands set into their bottoms and pouring necks set into their tops, interrupting the main form, so like Edwards I had to exercise judgment on selection of ideal end points (extending the given profile by dotted lines, next page).

The simplest way to arrive at a λ value for an oval profile is by means of a nomogram such as that reproduced on next page. The height of the oval is divided into 8 equal parts labeled top to bottom as X,A,B,C,T,D,E,F,Y, and each relative diameter at A,B,C,D,E,F is measured against that at T as unit and found on vertical scale of nomogram; the horizontal coordinate is then the matching λ value which would place a path curve oval through ends X and Y as well as through the profile points P', T' and their mirrors P",T", when P is one of points A through F [13, p. 3]. The seven ideal radii in this case for points A to F are 3.73, 3.83, 3.62, 3.23, 2.70, 2.03*, and 1.23.

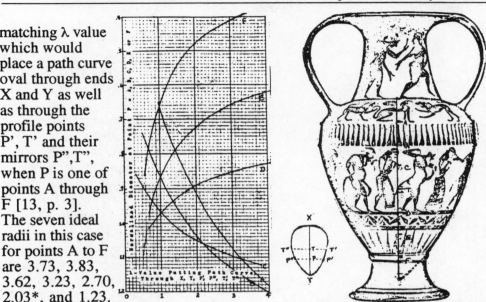

The measured radii are 3.78, 3.875, 3.625, 3.23, 2.71, 2.08*, and 1.23. Dividing by 3.23 at T gives normalized radii 1.170, 1.200, 1.122, 1.0, 0.839, 0.644, and 0.381, leading to λ values 3.86, 3.92, 3.79, —, 3.50, 3.28*, and circa 3.7. Assigning weights 4, 2,1,-,1,2,4 to these values, they average 3.71, with mean deviation .21 or 6%.

The sharp eye can detect (at least after-the-fact) that it is level E (*) which shows the greatest deviation in measure from mean λ value, pointing to a slight excess in width at that level, noticeable particularly on the left side. It is because the two sides of such objects do not always agree (glaringly so in the case of bumpy pine cones but still subtly so for vases despite their being turned smoothly on a potter's wheel) that we first measure diameters and then halve them to find radii.

The next example [13, p. 5] is painted to show a *Potnia Therōn* or Queen of the Wild Beasts. Besides being another excellent path surface shape of ovoid type (whose analysis is shown in fine print: weighted mean λ = 2.02 with mean deviation .08 or 4%), it is noticeable in this case that both stand and neck shapes are also suggestive of path surfaces of vortex types. The previous example can then also be recog-

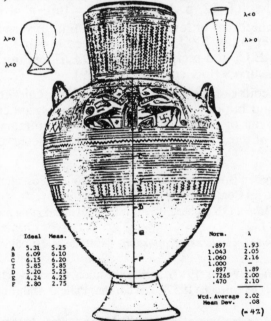

	Ideal	Meas.		Norm.	λ
A	5.31	5.25		.897	1.93
B	6.09	6.10		1.043	2.05
C	6.15	6.20		1.060	2.16
T	5.85	5.85		1.000	—
D	5.20	5.25		.897	1.89
E	4.24	4.25		.7265	2.00
F	2.80	2.75		.470	2.10

Wtd. Average 2.02
Mean Dev. .08
(= 4%)

nized after-the-fact as having at least its base of similar vortical shape though its neck narrowed rather than widened to the top. Scanning through the book's collection, it became evident that many of the vases had the vortical base shapes as standard design feature, while neck designs varied, some narrowing, others widening. A wide-bottomed base is, of course, common sense; *but when inverted we see Edwards' pivot!*

Two other examples are shown on the next page, one with a widening neck and the other narrowing, one with a base suggestive of ovoid cap and the other of an arbitrarily tiered design, to give some indication of the variety of neck and base forms extant. The bodies, however, remain very good path surface shapes, that on the left having weighted mean λ = 3.20 with mean deviation .06 or 2%, and that on the right having weighted mean λ = 2.77 with mean deviation .19 or 7%. That on the left is thus a virtually perfect path ovoid. That on the right shows its main deviation from ideal at point A, which may be due to its having handles attached differently at that level and thus confusing my measurements there. The handles are attached below A on the vase at left, above A on the vase at right. But as a little extra fillip, the vase at right has an extant lid with what appears to be a near-perfect path ovoid (inverted, in usual bud orientation) as top-notch.

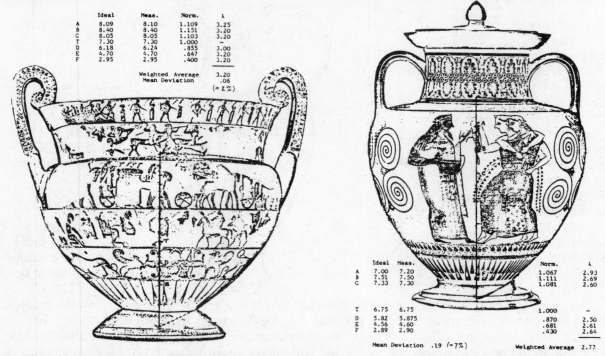

	Ideal	Meas.	Norm.	λ
A	8.09	8.10	1.109	3.25
B	8.40	8.40	1.151	3.20
C	8.05	8.05	1.103	3.20
T	7.30	7.30	1.000	–
D	6.18	6.24	.855	3.00
E	4.70	4.70	.647	3.20
F	2.95	2.95	.400	3.20

Weighted Average 3.20
Mean Deviation .06
(= 2%)

	Ideal	Meas.		Norm.	λ
A	7.00	7.20		1.067	2.93
B	7.51	7.50		1.111	2.69
C	7.33	7.30		1.081	2.60
T	6.75	6.75		1.000	–
D	5.82	5.875		.870	2.50
E	4.56	4.60		.681	2.61
F	2.89	2.90		.430	2.64

Mean Deviation .19 (=7%)   Weighted Average 2.77

If d_1 is the diameter at height h_1 and d_2 that at h_2, measured up from $Y = 0$ to $X = 1$, then consideration of similar triangles [cf. 12, p. 307] yields $\lambda = \ln[d_2h_1/d_1h_2]/\ln[d_1(1-h_2)/d_2(1-h_1)]$. If height 1 is that of T and height 2 that of levels A through F in turn (whose diameters we abbreviate by same letters), then weighting by 4,2,1,-,1,2,4 gives $\lambda = \ln4A/\ln(7/4A)$, $\ln2B/\ln(3/2B)$, $\ln(4C/3)/\ln(5/4C)$ at heights A,B,C, and reciprocally for D,E,F, ending with $\ln(7/4F)/\ln4F$ as equations for the six nomogram curves [loc. cit., pp. 308-309].

It was my privilege to be editor of the quarterly *Math.-Phys. Correspondence* from 1973-'83 in which Edwards' geometry and biology papers first appeared, since collected and expanded in book form [10 and 12 — 11 is out of print]. May the present summary and background help to bring that work to the attention of the wider readership it so richly deserves.

Bibliography

[1]   Stephen Eberhart, "Some Briggsian Bridges between Sense Perception, Postage Stamps, and Stylized Proportions," *Proceedings* of 2nd Bridges Conference in Winfield, KS, 1999.

[2]   Albrecht Dürer, *The Painter's Manual* (Nürnberg, 1525) translated and commented by Walter L. Strauss, Abaris Books, New York, 1977.

[3]   Albrecht Dürer, *Vier Bücher von menschlicher Proportion*, facsimile of 1st edition (Nürnberg, 1528), G.M. Wagner, London, 1970.

[4]   D'Arcy Thompson, *On Growth and Form* (1917), abridged edition, Cambridge University Press, 1961.

[5]   Louis Locher-Ernst, *Raum und Gegenraum* (1957), 2nd edition, Philos.-Anthropos. Verlag am Goetheanum, Dornach, 1970.

[6]   Louis Locher-Ernst, *Geometrische Metamorphosen* (1947-1955), Philos.-Anthropos. Verlag am Goetheanum, Dornach, 1970.

[7]   George Adams, *Strahlende Weltgestaltung*, Math.-Astron. Sektion am Goetheanum, Dornach, 1934.

[8]   Felix Klein, *Vorlesungen über nicht-euklidische Geometrie* (1892), 1928 edition, reprinted by Springer Verlag, Berlin, 1968.

[9]   Felix Klein, *Vorlesungen über höhere Geometrie* (1893), 1926 edition, reprinted by Springer Verlag, Berlin, 1968.

[10]   Lawrence Edwards, *Projective Geometry*, Rudolf Steiner Institute, Phœnixville, 1982.

[11]   Lawrence Edwards, *The Field of Form*, Floris Books, Edinburgh, 1982.

[12]   Lawrence Edwards, *The Vortex of Life*, Floris Books, Edinburgh, 1993.

[13]   Stephen Eberhart, "Grecian Amphoræ as Path-Curve Shapes," *Math.-Phys. Correspondence*, No. 27, 1979.

Exploring Art with Mathematics and Computer Programming

Alberto López-Santoyo
Fac. de Filosofía y Letras
UNAM
Arquitectura 19
04360 México, D.F., MEXICO
E-mail: fylgeo@servidor.unam.mx

Abstract

Even with simple concepts of pre-university mathematics and the aid of technology using computer programming one can explore the power of mathematics in art. Mainly distance between two points, the parabola, line segments, interpolation and random numbers were employed here. The images shown were obtained with very limited facilities using a very modest computer and a dot matrix printer. Perhaps teachers should encourage students to explore art using their mathematical knowledge and the aid of modern technology.

1. Introduction

Pre-university mathematics provides a wide range of material to explore from the point of view of art. One can start with simple concepts such as the distance between two points and then go to more complex ones. The aid of a computer is of course of great importance, but no matter how modest a computer is it is powerful enough to allow any one with some knowledge in mathematics and some experience in computer programming to get complex and artistic images. Students are able to explore art using some mathematics. Teachers may encourage them to do this since most students like to play with computers and produce works of art.

GW BASIC was used as programming language in the examples illustrated here. The outputs were obtained with a very modest computer and a dot matrix printer.

Duality is present in the four images made by computer due to the influence from real life experience.

The main mathematical items employed were as follows:

a)   Distance between two points.
b)   The parabola.
c)   Line segments, interpolation and random numbers.

In order to get a satisfactory result it was necessary to program the printer in text mode. The characters were the smallest available (upper captions) and the distance between characters and lines were reduced from the standard settings in order to get a compact set of lines and columns.

Some characters had to be designed to get the expected appearance according to what they had to represent and also because square characters were needed. Standard ones are larger in height than in width. This makes a big problem if one tries to obtain circles. Using the printer manual it was possible to program in GW BASIC some characters used. Of course this demanded patience and time in order to get the desired results.

With the aid of algebra it was possible to get complete flexibility for the size of the graphs, although it is not shown in the examples presented here. All the images can be printed in any size starting from a very small one to the biggest that can be obtained in a standard printer.

The four images made by computer are reduced 80 per cent. This means that 80 units of length in the reduced image correspond to 100 units in the original one.

2. Concentric Circles

For representing two sets of concentric circles (fig. 1) the concepts of distance between two points and the circle were used but only the equation for the distance between two points was needed:

$$D = [(X1\text{-}X2)^2 + (Y2\text{-}Y1)^2]^{1/2}$$ where D is the distance between the two points. X1,Y1 are the coordinates of one point and X2,Y2 are the coordinates of the other point. The obtained distance was used as the radius of a certain circle.

In the example illustrated the circles meet in the central part and give a sense of struggle between equal forces. The center at the lower left corner is white and the other on the upper right corner is black. From each center concentric circles radiate in a grading sequence from white to black or black to white.

Figure 1: *Concentric circles created using the distance between two points as radius*

A matrix was chosen with the number of characters (columns) to be printed along a sheet of paper and the number of rows necessary for the desired graph. In the matrix the position of two points were given as centers for the circles and the coordinates of those points were determined in relation to the established matrix. The centers were chosen at the upper right and lower left corners of the rectangle.

To find the way any point or pixel had to be represented or what character had to be used, it was necessary for each pixel in the matrix to compute its distance to the nearest center and use that distance as radius. Then select automatically the corresponding character chosen beforehand for that radius.

3. Parabola and Concentric Circles

A parabola was drawn over a set of concentric circles that grade from black to almost white, from the center outwards. The axis of the parabola was defined horizontally at the middle part of the rectangle. The center of the circles was chosen on the same line on the left edge of the rectangle.

The equation of the parabola was used together with the distance between two points in the case represented in figure 2. The parameters for the parabola were established in such a way as to get a satisfactory appearance for it. The center for the concentric circles was chosen outside the parabola on its axis. The axis of the parabola is here horizontal and therefore its equation is:

$$Y^2 = 4aX$$ where "a" is the parameter that gives the distance from the vertex to the focus. The parabola had preference over the concentric circles when printed. Therefore the circles act as a background for the parabola.

Figure 2: *Parabola over a set of concentric circles*

Here again as it was done on the concentric circles of figure 1, the rectangle of the chosen size was scanned pixel by pixel or point by point to allow the computer define or choose the corresponding character to be used. For each point or pixel there are three possibilities:

a)   The point is outside the parabola.
b)   The point is on the parabola.
c)   The point is inside the parabola.

If the point is outside the parabola then by using the equation of the distance between two points (the pixel to be printed and the center of the concentric circles), the corresponding character for a concentric circle was chosen.

If the point is on the parabola the character chosen beforehand for that locus was printed.

If the point is inside the parabola its distance to the axes was calculated. The computer chose the corresponding character in order to get straight lines, parallel to the axes, in a sequence from white to black.

4. Line Segments and Random Numbers

Line segments were drawn between 14 selected points, using linear interpolation, in order to get a terrain profile representing the two main volcanoes near Mexico City (Ixtaccíhuatl and Popocatepetl). The points were selected from photos and pictures of the two mountains. Some cosmetics were applied when doing this. For example it is very common to see in some paintings that the two mountains appear nearer together than they actually are.

The photo illustrated can give an idea of what they look from a field just outside Mexico City.

Photo 1: *View of Ixtaccíhuatl and Popocatepetl volcanoes*

Mexico City was originally founded on a lake. This lake has been drying out with time and only some small parts are left. It is now a bit difficult to find a place where one can see what the view was like in older times. Therefore what is depicted in these images is more a memory of a past landscape that people used to see.

The altitude of these mountains is enough for them to have permanent snow. For most of the year they look as it is shown in the photograph and in the images made by computer.

4.1. Landscape at night. In a first try, the image as shown in figure 3 intends to represent a night scene of the volcanoes and their reflection on a lake.

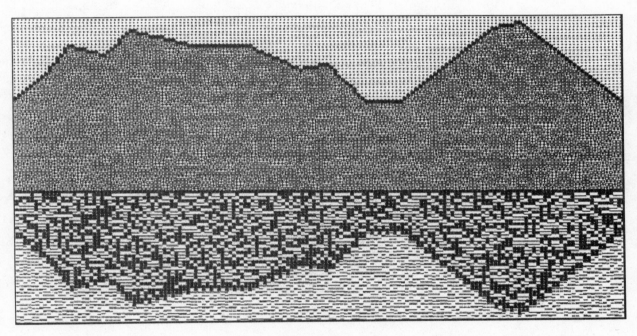

Figure 3: *Simulation of a view of the two volcanoes and its reflection on a lake at night*

Several layers had to be defined for this image. The main one is the profile of the terrain. Above it a character was chosen to fill the space representing the sky. Then a black character depicts the profile. Bellow the profile two gray characters printed at random simulate the mountains. Then a straight black line marks the boundary between land and water.

For the reflection of the mountains and for the reflection of the sky a pair of characters were selected for each one and printed at random using random numbers. These characters were designed to simulate a slight movement of the water surface. In order to represent the reflection of the terrain profile it was printed head down and with a double line so that it also simulates the slight movement of the water surface.

4.2. Landscape at daylight. When the image shown in figure 3 was finished, something a bit more complex was planned. The same landscape had to be represented but as it would be seen in daylight. This required to addition of two features: the snow layer on the upper part of the mountains, and the vegetation cover below it. The snow layer should appear on the profile of the volcanoes as well as on their reflection on the water.

Figure 4 shows the result of this attempt. The layer of snow is above a certain height. One character printed at random and having a low density simulates the snow cover on the highest parts of the volcanoes.

In order to represent the vegetation cover on the rocky surface below the snow layer, two characters were employed. The darker one simulates the vegetation cover. This character was printed at random and gives the idea of being upon the other one. It grades in density from low to medium as the vegetation cover increases with lower altitude.

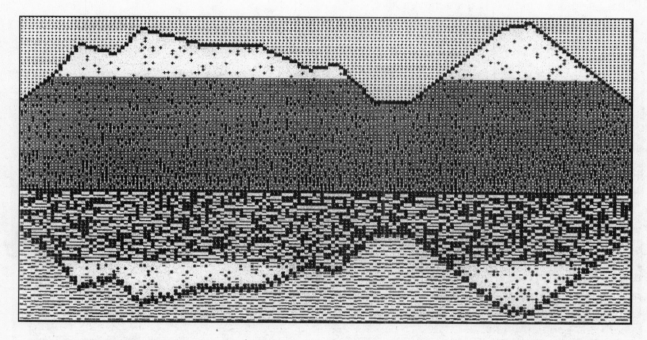

Figure 4: *Simulation of a view of the two volcanoes and its reflection on a lake at daylight*

Conclusions

It is possible to take advantage of mathematics in the field of art. Even pre-university students should be able to develop their artistic abilities or reinforce them by creating images with the aid of mathematics and computer programming. Teachers should try to encourage their students to explore art with the aid of mathematics.

Modern technology give us each time more options and software support to be used in several fields. This paper is an attempt in exploring art with mathematics and computer programming. Something similar might also be done with other tools of modern technology –software– and mathematics which students can handle.

Self-similar Tilings Based on Prototiles Constructed from Segments of Regular Polygons

Robert W. Fathauer
Tessellations
Tempe, AZ 85281, U.S.A.
E-mail: tessella@futureone.com

Abstract

Two infinite families of self-similar tilings are described which have apparently not been reported before. Each tiling is based on a single prototile that is a segment of a regular polygon. Each tiling is also edge to edge and bounded in the Euclidean plane, by means of the tiles being reduced in size by a fixed scaling factor. This results in self similarity. Tilings are constructed from these prototiles that are of a rich visual complexity, and an example is given of an Escher-like design based on one of these tilings.

1. Introduction

Tiling is a topic in which math and art combine wonderfully. While the rigorous mathematical treatment of tiling is beyond the understanding of the average person, anyone can appreciate the beauty of tilings. While the best-behaved tilings fill the infinite Euclidean plane, infinite tilings can also be finite in area by means of the tiles becoming infinitesimally small at the boundaries. In many cases, the boundaries of these tilings are fractal, which adds another element of beauty to them.

The sorts of tilings dealt with here are not to be confused with hyperbolic tilings. In the latter, hyperbolic geometry is used to depict an infinite tiling in, for example, the Poincaré disk. In such tilings, the individual tiles appear distorted in Euclidean space, which is how we perceive the world. In the tilings discussed in this paper the tiles are not distorted, but are all of the same shape (in mathematical terminology "similar").

The Dutch graphic artist M.C. Escher explored bounded tilings, but he only executed a single finished print of the general sort described here, namely "Square Limit" [1]. His other prints depicting bounded tilings, the four "Circle Limit" prints, are based on hyperbolic geometry. He also made several prints based on unbounded tilings in which the tiles become infinitesimally small at the center.

Others have worked on non-hyperbolic bounded tilings since Escher [2-5]. In almost all of these tilings, the prototile has two long edges, so that three or more of these tiles meet in a point at the center of the tiling. The resultant tilings have n-fold rotational symmetry, with $n \geq 3$.

In this paper, two families of tilings are presented in which, in their simplest form, the center is a line shared by two identical tiles that possess a single long edge. The tilings thus possess 2-fold rotational symmetry. The prototiles are constructed from segments of regular polygons. In the first family, the prototiles are isosceles triangles, while in the second they are trapezoids. The simple forms of these tilings are combined and modified to make more complex tilings of greater esthetic appeal. These families of tilings do not appear to have been reported before.

Figure 1 illustrates the construction of the prototiles from regular polygons. In the first family, denoted "s" tilings, the tiles are triangular and are generated by connecting two vertices encompassing two adjacent edges. In the second family, denoted " u" tilings, the tiles are trapezoidal and are generated by connecting two vertices encompassing three adjacent edges. The tilings are constructed by reducing these prototiles by a scaling factor given by the ratio of the short to long edges of the prototile, and then matching the long edges of the next smaller generation to the short edges of the larger generation. The simplest starting point for constructing a bounded tiling is to place two of the largest-generation tiles back to back. This process is illustrated in Figure 2.

2. s Tilings

Overlapping of tiles is not allowed in any reasonably well-behaved tiling, which restricts the allowed values of s. This is clarified in Figure 3 for two different values of s. It is readily apparent through experimentation that this condition restricts the allowed generating regular polygons to those with number of sides n = 4, 6, 8, 10, ... The triangular prototile generated from a polygon with n sides is simply referred to as the "s = n prototile". These triangles are isosceles, with the two identical angles given by π/s. The linear reduction factor from a given prototile to the next smaller generation of prototile is given by $1/(2(\cos(\pi/s)))$.

The simplest bounded infinite tiling for a given prototile is denoted the "pod" for that prototile. The pods for s = 4, 6, 8, 10, and 12 are shown in Figure 4. (In all of the figures, the tilings are constructed through a finite number of generations adequate to illustrate the general appearance of the infinite tiling.) All of these have fractal boundaries except for s = 4, for which the boundary is an octagon in which four sides are of one length and the other four of a second length $\sqrt{2}$ times the first length.

Note that there are two distinct types of vertices that are intrinsic to these tilings. The first type, marked "A" in Figure 4 is the point at which the tip (the point at which the two short edges meet) of a triangle meets the bases of n + 2 smaller triangles. The second type, marked "B" in Figure 4 is the point at which the tips of two triangles meet the bases of 4 smaller triangles. The point marked " C " indicates a third type of vertex that results from the choice of putting two large tiles back to back. It is possible to construct tilings without vertices of type C, as well as to create other types of vertices by construction choices. Only types A and B are always present, and there is an infinite number of each type in any tiling. Each pod possesses 2-fold rotational symmetry about its center and two axes of mirror symmetry.

Pods can be joined to form more complex tilings. The two basic configurations for joining pods that result in bounded tilings are denoted "rings" and "stars". Some examples of rings and stars based on s pods are shown in Figure 5. These have been restricted to those tilings that are edge to edge. This means that only certain numbers of pods can be combined in rings and stars for a given value of s. Pods can also be joined in periodic fashion to extend over the infinite Euclidean plane.

An even greater degree of complexity can be introduced in these tilings by "perforating" the pods. This refers to removal of the largest generation of tiles in the pods, followed by filling in of the resultant hole with tiles of the same and smaller size as the remaining tiles. This process is illustrated in Figure 6. A tiling obtained in this way is shown in Figure 7. Note that perforated stars and rings are equivalent to more complex combinations of unperforated pods.

An Escher-like design based on the s = 6 tiling is shown in Figure 8, where the triangular tile has been modified and detailing added to suggest a bird motif. The prototiles have been arranged in such a way that half the birds are flying left to right, and half right to left.

3. u tilings

Again, the tilings are restricted to those values of u for which the tiles do not overlap. It is apparent from experimentation with u tilings that this condition restricts the allowed generating regular polygons to those with number of sides n = 6, 10, 14, 18, ... The trapezoidal prototile generated from a polygon with n sides is similarly referred to as the u = n prototile. The prototiles have three sides of equal length, with the two small angles in the trapezoid given by $2\pi/u$. The linear reduction factor from a given prototile to the next smaller generation of prototile is given by $1/(1 + 2(\cos(2\pi/u)))$.

The pods for u = 6, 10, and 14 are shown in Figure 9. All of these have fractal boundaries except for u = 6, for which the boundary is a regular hexagon. There are two distinct types of intrinsic vertices for u tilings as well. Type "A" is a point at which a tip of a trapezoid meets the bases of u/2 +1 smaller tiles. Type "B" is a point at which the tips of two trapezoids meet the bases of 2 smaller tiles. Each pod again possesses 2-fold rotational symmetry about its center and two axes of mirror symmetry.

The u pods can also be combined to form more complex tilings. Two examples of rings and stars based on u pods are shown in Figure 10. Perforation of u pods is carried out in a similar fashion to that used for s pods. A perforated star is also shown in Figure 10.

Additional prototiles can of course be constructed from regular polygons by connecting vertices encompassing more than three edges. However, it can be shown that none of these allow well-behaved tilings.

4. Conclusion

Two new families of self-similar tilings have been demonstrated in which the prototile is a segment of a regular polygon. In the first family the prototiles are isosceles triangles, while they are trapezoids in the second family. The tilings constructed from these prototiles look quite different from most bounded tilings presented previously. There are an infinite number of distinct prototiles in each family, and tilings can be constructed from these prototiles that are of a rich visual complexity.

References

[1] *The Magic Mirror of M.C. Escher*, by Bruno Ernst (Ballantine Books, New York, 1976).

[2] Robert W. Fathauer, *Extending Recognizable-motif Tilings to Multiple-solution Tilings and Fractal Motifs*, to be published in the Proceedings of the Centennial Escher Congress, held in Rome and Ravello in June of 1998.

[3] Peter Raedschelders, private communication. Examples can be seen on Mr. Raedschelders web site, *http://home.planetinternet.be/~praedsch/index.htm.*

[4] Chaim Goodman-Strauss, private communication. Prof. Goodman-Strauss has constructed a number of tilings in which each tile is a fractal object related to the Koch snowflake.

[5] Robert W. Fathauer, presented at the 1999 Bridges Conference (Winfield, Kansas, August 1999).

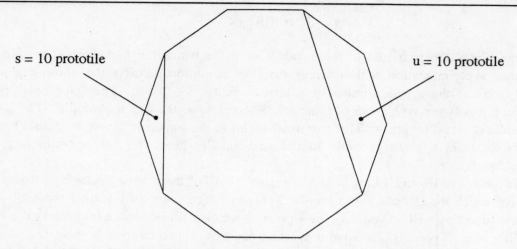

s = 10 prototile   u = 10 prototile

Figure 1: *Construction of the s and u prototiles from a regular polygon, for the case of a regular decagon.*

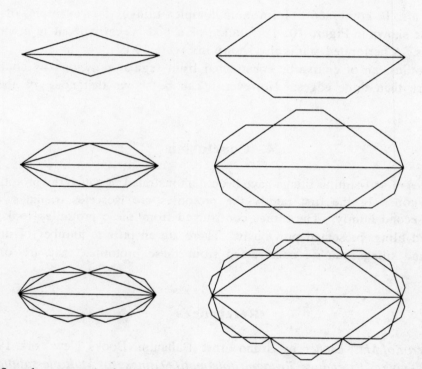

Figure 2: *The first three steps in the construction of the simplest bounded infinite tilings based on the s and u prototiles.*

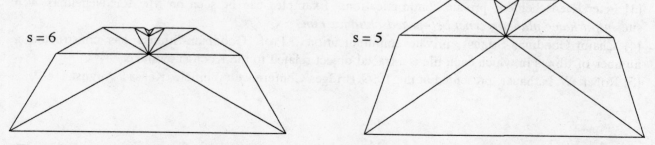

s = 6   s = 5

Figure 3: *A portion of two s tilings illustrating how the tilings fold in upon themselves. Note that the first case, s = 6, is well behaved, while the second case, s = 5, is not. The tiles overlap in the second case, which is not allowed.*

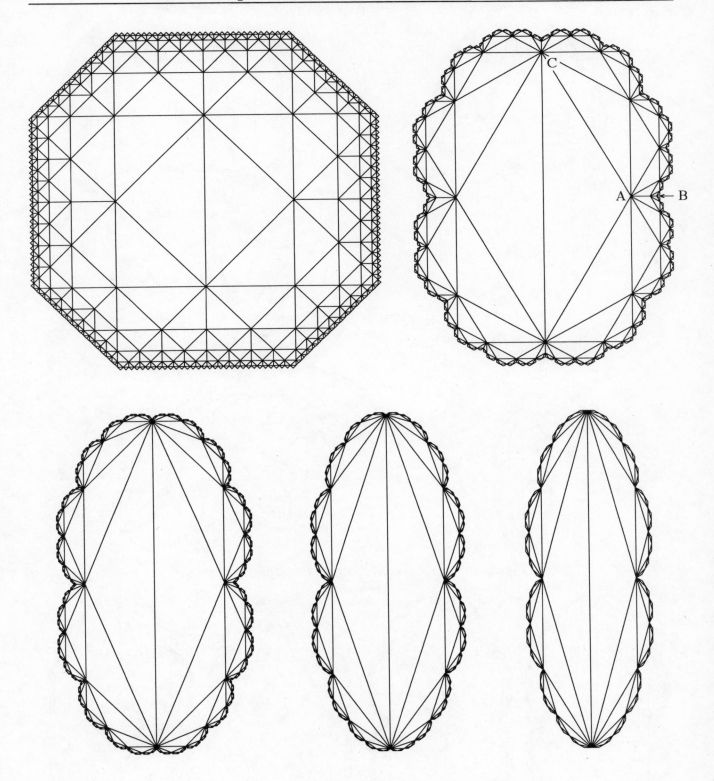

Figure 4: *Pods for s = 4, 6, 8, 10, and 12. The points A, B, and C on the s = 6 pod mark the three different types of vertices that occur.*

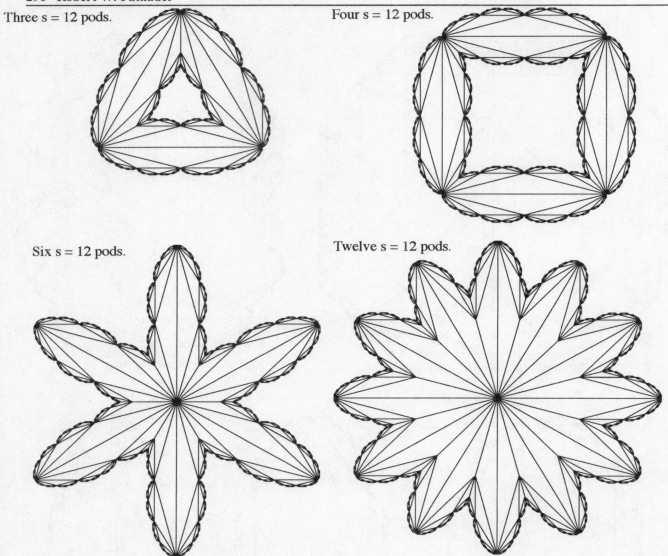

Three s = 12 pods.

Four s = 12 pods.

Six s = 12 pods.

Twelve s = 12 pods.

Figure 5: *Two rings and two stars based on s = 12 pods.*

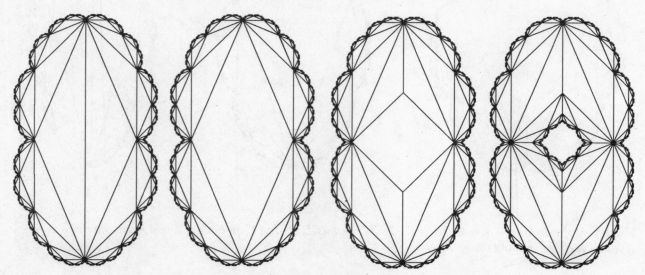

Figure 6: *The process by which a pod is perforated.*

Figure 7: *A ring of four s = 8 pods, perforated three times.*

Figure 8: *An Escher-like design based on an s = 6 tiling.*

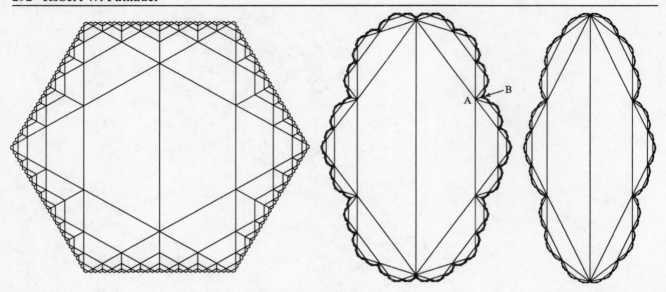

Figure 9: *Pods for u = 6, 10, and 14. The points A and B illustrate the two types of vertices intrinsic to u tilings.*

Figure 10: *A ring formed from six u = 18 pods, a star formed from five u = 10 pods, and the same star perforated once.*

Polyhedral Models in Group Theory
and Graph Theory

Raymond F. Tennant
Department of Mathematics, Statistics, and Computer Science
Eastern Kentucky University
Richmond, KY 40475
Email: tennant@acs.eku.edu

Abstract

A longstanding method for understanding concepts in mathematics involves the creation of two or three-dimensional images which describe a particular mathematical idea. From our earliest learning experiences, we are taught mathematics by appealing to our strong visual and tactile intuition. For students studying mathematics at the college or university level, the use of polyhedral models and graph theoretic constructions may be a valuable tool for gaining insight into abstract areas such as group theory and topology.

This investigation focuses on the use of Platonic and Archimedian solids to describe ideas in abstract algebra and to understand the concepts such as duality and symmetry subgroup. The reasoning behind several proofs of Euler s Formula are explored with the use of models. For the most part, planar graphs of polyhedra are used in place of actual three-dimensional models. This has the advantage of allowing for all of the vertices, edges, and faces to be viewed at the same time.

Planar Graphs

The notion of graph in graph theory is simply a diagram consisting of points, called vertices, joined together by lines, called edges. Each edge joins exactly two vertices or in the case of a loop, joins one vertex to itself. A graph is called planar if it can be drawn in the plane in such a way so that no two edges meet each other except at a vertex to which they both connect. The regions bounded edges are called faces.

Each of the Platonic Solids may be expressed as drawn as planar graphs. The faces of the polyhedron correspond to the faces of the graph including the unbounded region, called the face at infinity. In fact, any convex polyhedron may be drawn as a planar graph. To visualize this, imagine a convex polyhedron with glass faces. Placing your eye close to one of the faces and peering inside will give you a clear vision of all of the vertices and edges of the polyhedron. This image you see could be projected onto a planar graph.

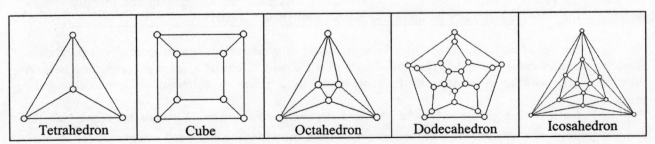

| Tetrahedron | Cube | Octahedron | Dodecahedron | Icosahedron |

Figure 1

Duality

The concept of duality of polyhedra may be understood through a sequence of graphs. If P is a connected planar graph and $P*$ then the dual graph of P, call it $P*$, can be constructed from P in the following manner. First, choose one point inside each face, including the face at infinity, of the planar drawing of P. These points are the vertices of $P*$. Next, for each edge of P, draw a line connecting the vertices of $P*$ which lie on each side of the edge. These new lines are the edges of $P*$. That a tetrahedron is dual to itself is seen in the graphs below.

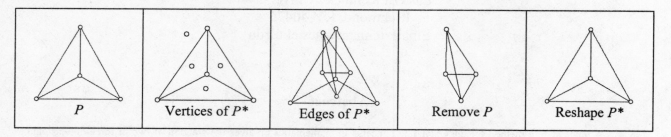

| P | Vertices of $P*$ | Edges of $P*$ | Remove P | Reshape $P*$ |

Figure 2

The process works for any convex polyhedron as is shown in the case of the truncated cube.

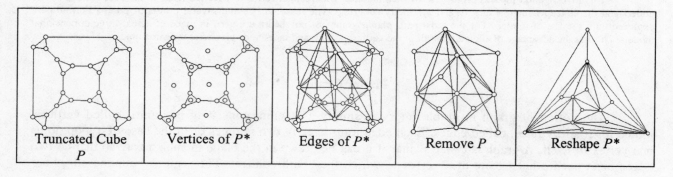

| Truncated Cube P | Vertices of $P*$ | Edges of $P*$ | Remove P | Reshape $P*$ |

Figure 3

Symmetry Groups

Definition   A symmetry of a polyhedral model is a rotation or reflection, which transforms the model so that it appears unchanged. The rotational symmetries along with the identity transformation form the symmetry group of rotations of a polyhedral model.

A number of classical groups may be represented as symmetry groups of rotations of polyhedra. A cyclic group of order n may be represented by a pyramid with a regular n-sided polygon for a base. A dihedral group with $2n$ elements may be represented by a prism or antiprism with n-sided regular polygons.

The alternating groups, A_4 and A_5, and the symmetric group, S_4, may be viewed as the symmetry group of a tetrahedron, icosahedron (or its dual, the dodecahedron), and octahedron (or its dual, the cube), respectively. Historically, the groups A_4, S_4, and A_5 were referred to as the tetrahedral, octahedral, and icosahedral groups.

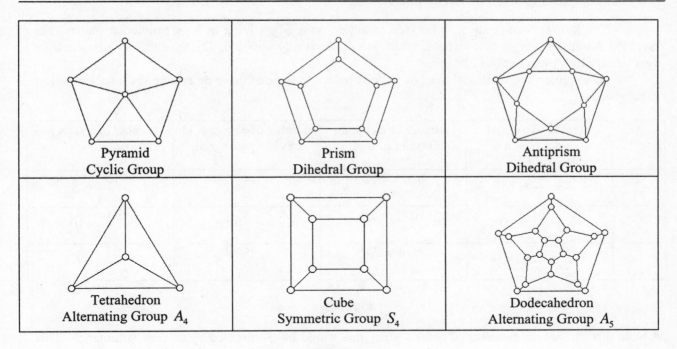

Figure 4

It turns out that Z_n, and, D_n, for any positive integer n, along with A_4, S_4, and A_5 make up a complete list of the groups which can be described as the rotational symmetry group of a convex polyhedron.

Sylow p-Subgroups of Symmetry Groups

Subgroups of symmetry groups may be described by looking at substructures of the polyhedron. For example, the group of rotations of the dodecahedron is A_5, group with $60 = 2^2 \cdot 3 \cdot 5$ elements. By this factorization, we see that A_5 has Sylow p-subgroups for $p = 2$, 3, and 5. Each of these subgroups may be thought of as acting on a substructure of the dodecahedron.

Definition   Let G be a finite group and let p be a prime which divides $|G|$. If p^k divides $|G|$ and p^{k+1} does not divide $|G|$ then any subgroup of G of order p^k is called a Sylow p-subgroup of G.

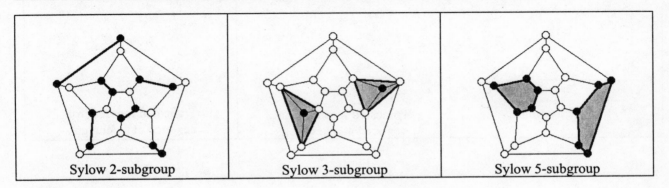

Figure 5

The Sylow 2-subgroup may be seen as acting on 6 edges lying in 3 perpendicular planes. The Sylow 3-subgroup acts on two vertices, which are diametrically opposed. The Sylow 5-subgroup acts on two faces, which lie in parallel planes.

By counting the rotational axes of various orders, it is possible to determine the numbers Sylow p-subgroups.

p	number of axes of order p on a dodecahedron	number of elements of order p in A_5	number of elements in a Sylow p-subgroup	number of Sylow p-subgroups
2	$\dfrac{e}{2} = \dfrac{30}{2} = 15$	$15 \cdot 1 = 15$ *	4	$\dfrac{15}{4-1} = 5$
3	$\dfrac{v}{2} = \dfrac{20}{2} = 10$	$10 \cdot 2 = 20$	3	$\dfrac{20}{3-1} = 10$
5	$\dfrac{f}{2} = \dfrac{12}{2} = 6$	$6 \cdot 4 = 24$	5	$\dfrac{24}{5-1} = 6$

Figure 6

* Note that A_5 has no elements of order 4 since this would be represented by an odd permutation. This means the Sylow 2-subgroups are isomorphic to $Z_2 \oplus Z_2$.

Students may easily verify that this calculation is correct by applying Sylow s Third Theorem.

Sylow s Third Theorem

Let n_p denote the number of Sylow p-subgroups.

Then $n_p \equiv 1 \pmod{p}$ and n_p divides $|G|$.

Orbit-Stabilizer Theorem

Creating new regular polyhedra from old ones by truncating or stellating may result in the new polyhedron having the same rotational symmetry as the original. For example, consider the icosahedron and the truncated icosahedron.

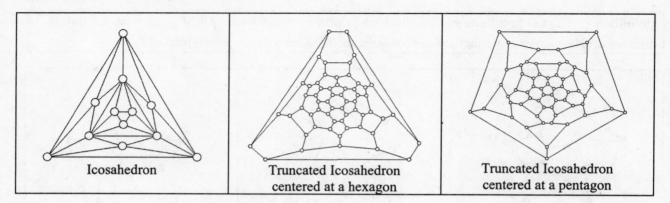

Icosahedron

Truncated Icosahedron centered at a hexagon

Truncated Icosahedron centered at a pentagon

Figure 7

A useful tool for determining the size of a finite symmetry group involves looking at orbits and stabilizers as the symmetry group acts on the polyhedron. The group of rotations of a polyhedron may be thought of as permuting around some geometric set of the polyhedron. We say the group is acting on the vertices, edges, faces, or some other set of components.

Definition   Let G be a group of rotations acting on the set I of components of a polyhedron.
For each $i \in I$, the orbit of i under G is defined by $orb_G(i) = \{\varphi(i): \quad \varphi \in G\}$.
For each $i \in I$, the stabilizer of i in G is defined by $stab_G(i) = \{\varphi \in G: \quad \varphi(i) = i\}$.

Orbit-Stabilizer Theorem
      Let G be the group of rotations acting on the set I of components of your model. For any $i \in I$,
$|G| = |orb_G(i)| \, |stab_G(i)|$.

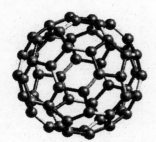

3-Dimensional
truncated icosahedron Also
known as a Buckyball,
or a soccerball

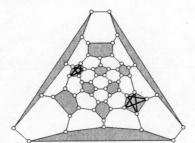

Acting on faces
i = a pentagonal face
with a star
orbit(i) = 12 pentagons
stabilizer(i) = 5 rotations

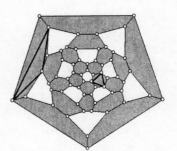

Acting on faces
i = a hexagonal face
with a triangle
orbit(i) = 20 hexagons
stabilizer(i) = 3 rotations

Figure 8

Whether viewing from pentagonal faces or hexagonal faces, the Orbit-Stabilizer Theorem gives us that there are $12 \cdot 5 = 20 \cdot 3 = 60$ elements in the rotational symmetry group of the truncated icosahedron. Since no new rotations have been introduced in truncating, the rotational symmetry group is A_5, same as the icosahedron.

Proofs of Euler s Formula

Euler s Formula (for Convex Polyhedra)
      Let P be a convex polyhedron, and let v, e, and f denote, respectively, the numbers of vertices, edges, and faces of P.
      Then $v - e + f = 2$.

Euler s first strategy for a proof (c. 1751) of his formula involved starting with a convex polyhedron and removing a vertex along with all of the edges and faces, which adjoin it. New triangle faces are added over the hole that has been created. With each step in this process, the value for $v - e + f$ stays the same. The desired result is to continue the process until a tetrahedron is reached and since $v - e + f = 2$ for a tetrahedron, then this must be the case for the original polyhedron. This process has a flaw in that at a given stage you may not be left with a polyhedron.

In 1813, Cauchy gave a proof of Euler s formula, which involved projecting a convex polyhedron onto the plane in the manner used in this discussion. He argued that the value of $v - e + f$ is the same for both the original polyhedron and its projection in the plane. Further, it is possible to add edges to the planar graph so that all the faces are triangles and $v - e + f$ remains the same. Finally, he showed that $v - e + f = 2$ for the planar graph with triangle faces.

This method is exhibited for the icosahedron below.

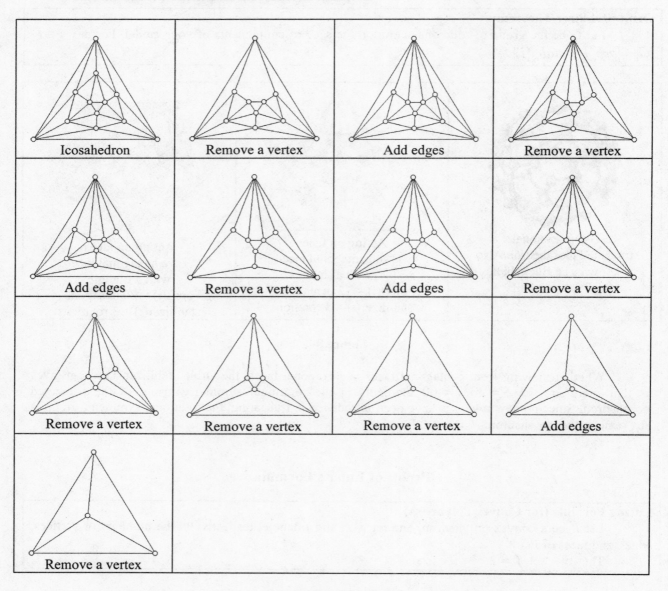

Figure 9

A novel proof of Euler s Formula, which is suitable for middle school students, describes a planar graph whose edges are dams, vertices are posts holding the dams together, the bounded faces are dry chambers and where the face at infinity is the ocean. The idea is to remove a dam (edge) so that the ocean rushes in and fills a chamber (face) with water. In this fashion, one dam (edge) is removed and one chamber (face) is flooded so that $v - e + f$ remains the same. We continue removing one dam (edge) so one chamber (face) is flooded until we have the ocean filling all of the chambers. At this final stage, we have all v posts (vertices) intact and 1 ocean (face). By this process, the posts have stayed connected by dams so there are v-1 dams (edges).

Since $v - e + f$ has stayed the same, we have $v - e + f = v - (v-1) + 1 = 2$.

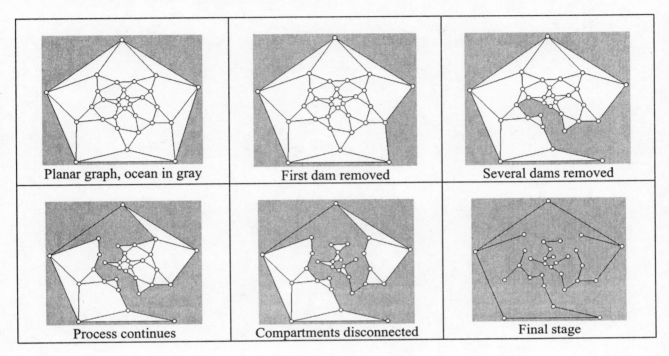

Planar graph, ocean in gray	First dam removed	Several dams removed
Process continues	Compartments disconnected	Final stage

Figure 10

One final note is in order. Verifying that a property holds for a particular polyhedron obviously is no guarantee that the property holds for all polyhedra. Also, a student s verification that a conjecture holds with models should not replace a rigorous proof but rather help in gaining confidence that a conjecture is true while the student struggles with a proof.

Questions for Students to Ask when Testing a Conjecture with a Model

1.	Is the model on which I am focused represent the properties of all or most of the polyhedra in the category of the conjecture?

2.	Can the steps involved in verifying the conjecture is true for the model be translated into logical steps in a rigorous proof for all polyhedra in a particular category?

References

[1]	Peter R. Cromwell, *Polyhedra*, Cambridge University Press, 1997.
[2]	Joseph A. Gallian, *Contemporary Abstract Algebra*, Houghton Mifflin College Publishers, 4th edition, 1998.
[3]	Robin J. Wilson, John J. Watkins, *Graphs: An Introductory Approach*, John Wiley and Sons, 1990.

Generalized Koch Snowflakes

Cheri Shakiban, Janine E. Bergstedt
Department of Mathematics
University of Saint Thomas
Saint Paul, Minnesota 55105
E-mail: c9shakiban@stthomas.edu

Abstract

In this paper, we show how a new procedure based on vector calculus and modular arithmatic is used to generate the Koch Snowflake. The procedure is then applied to create more generalized snowflakes. We also use a new method of calculating their fractal dimension based on the scaling factor.

1. Introduction

The fascinating topic of fractal geometry which was popularized by Benoit Mandelbrot, [1], in the seventies remains an intriguing subject. The most intuitive example of a fractal is the boundary of the Koch snowflake which was constructed by the Swedish mathematician Helge von Koch in 1904. The Koch snowflake begins with an equilateral triangle of unit side: the *initiator*. The first step in the construction replaces each side by adjoining a broken line made up of $m = 4$ equal sides of length $r = 1/3$. This broken line is referred to as a *generator*.

Initiator   Generator

Figure 1. *The Koch Curve's Initiator and Generator.*

Each subsequent stage of the construction proceeds similarly, replacing each line segment with a copy of the generator, reduced in size so as to have the same end points as those of the interval being replaced. The snowflake consists of the curve obtained by continuing the construction indefinitely.

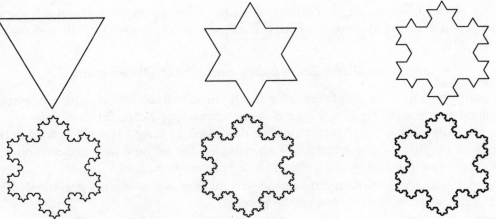

Figure 2. *Koch snowflake.*

Each side of the snowflake is self-similar, that is if we were to zoom in on the side once, twice, or an infinite times, the resulting pattern would be identical to the original generator. As we are using a recursive procedure, we can create a computer program to draw them. However, since it is impossible to run a program indefinitely, we cannot create an actual fractal but only a good approximation. Even though we only generate a fractal to a certain stage of self-similarity, it is not necessarily any less beautiful or interesting than its infinite counterpart. In fact, in most fractals, it would be difficult to tell the difference between the stages after a certain point.

2. Fractal Dimension:

One way in which we can try to "harness" the essence of fractals is by studying their dimension. The dimension of a fractal is a way of describing how densely a fractal occupies a space. Using dimension, we can objectively describe and compare any number of fractals. One way of defining fractal dimension is the method of self-similarity.

Definition: [2] For a self-similar figure subdivided into m similar pieces, each of which can be magnified by a factor of f to yield the whole figure, the *fractal dimension* is

$$D = \frac{\log(m)}{\log(f)} = \frac{\log(\text{number of pieces})}{\log(\text{magnification})}$$

The fractal dimension of a fractal exceeds its topological (Euclidean) dimension. The dimension of the Koch snowflake can be calculated using the self-similarity formula. We use the sides of the snowflake because no piece of the snowflake may be magnified to look like the whole object; however, pieces of the sides are self-similar. If we proceed to the kth stage of the construction, there are 4^k pieces used, but the magnification factor is 3^k. Therefore

$$D = \frac{\log(4^k)}{\log(3^k)} = \frac{k \log 4}{k \log 3} = 1.2618$$

Now what does this mean for the dimension of the Koch snowflake in its entirety? Well, one might jump the gun and assume since the snowflake is composed of three Koch curves as described above that the dimension of the snowflake would be three times our calculation of 1.2618. However, thinking of dimension as a sort of "density", we can apply the following theorem to find the dimension of the Koch snowflake.

THEOREM: [3] If $dim(B) \leq dim(A)$ for sets A and B, then $dim(A \cup B) = dim(A)$.

Since the Koch snowflake in its entirety is really a composite of three Koch curves as described above, each with the same dimension, $dim(A) = dim(B) = dim(C)$, the snowflake's fractal dimension is 1.2618.

3. Generalized Snowflakes and Their Dimension:

In this paper we will focus on generalizing the Koch snowflake; by creating a new routine that uses simple modular arithmetic and vector algebra to produce beautiful snowflake type fractals. Manipulating the process initiator/generator we described to create the Koch snowflake, we can create an infinite number of new generalized snowflakes. Let us now fully concern ourselves with the initiator/generator method by addressing three crucial questions:

(a) How does the method work on a mathematical (as opposed to simply graphical) level — i.e., what are the main components of the process?

(b) How can we utilize the components of the process to quickly and easily graph a snowflake (without having to graph each generation)?

(c) Can the initiator/generator method make calculating dimension of a snowflake purely routine, based solely on the basic components of the process?

By answering these three questions, in the end, we will be able to more fully understand the beauty that exists between mathematics, art and nature. To begin with, we can break down the initiator/generator process into two main components: vector algebra and modular arithmetic. These two simple subjects of mathematics will give us all that we need in order to understand the process, and later graph many generalized snowflakes created by the process. First, let us consider how vector algebra is used on the definition of the initiator. The only restriction here, for simplification purposes, is that the initiator be a regular polygon - an object of three or more sides of equal length. For the Koch snowflake, our initiator was an equilateral triangle.

3.1. Vector Algebra. For those who have no experience in vector algebra, or for those who are a bit rusty the basic idea centers around the idea of a vector. What is a vector? Think of a line segment with direction, i.e., an arrow. The key is "direction." Since up is entirely opposite from down, and left from right, there needs to be a way to tell one from the other. That's where the vector (arrow) comes in. How do vectors apply to an initiator of a snowflake (a polygon)? The graph must start at a point, which for simplification, we call the origin of the graphing axis. Also, let us have the first side be drawn to the right and be parallel to the horizontal axis. After drawing the first side, one would continue on to the second, third, fourth, etc. Since each side is added by drawing in a certain direction, they are considered to be a number of vectors and not line segments. If one were to never lift the pencil, the resulting graph would really be a sequence of vectors whose direction changed in a clockwise fashion. Consider the examples below.

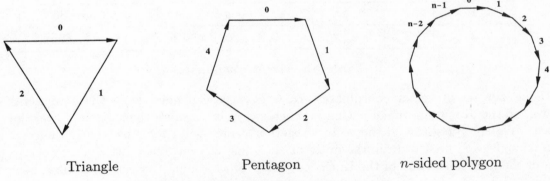

Triangle　　　　　Pentagon　　　　n-sided polygon

Figure 3. *Vectors and Polygons.*

Now that we know how vectors are used in defining the initiator of a snowflake, how are they used to define the generator? As can be seen in Figure 3, each vector is given a reference number according to the order in which it was "drawn." Note that we start at zero, although there would be no drastic consequences if we had started at one. If there are n sides in our polygon, our vectors will be designated with numbers ranging from zero to $n - 1$. Using these side indications, we can easily define a generator using the sides of the polygon. For example, let us return to the familiar Koch snowflake. The initiator is the triangle shown in Figure 3, with sides zero to two. We can use the negatives of each of those numbers, including negative zero! What this means for the graphical representation of our generator, is the negative of a certain side will have a direction completely opposite from the original side. The vector will in fact pivot one hundred and eighty degrees. So, if we were to consider the vector **-0**, its graphical representation would be identical to that of **0**, but would be pointing to the left instead of the right. We can see from Figure 4, that its generator is the set of sides **0,-1,-2,0**. This means that for every generation of the Koch snowflake, each line segment will be replaced by a scaled-down sequence of the original sides **0,-1,-2,0**.

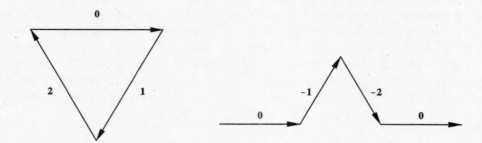

Figure 4.   *The Koch Snowflake Generator's Vector Sequence.*

To graph the entire snowflake, we must also know the coordinates of each vector. The vectors are numbered zero to $n - 1$. Using those numbers, the coordinates of the vector i equals $A_i = (r \cos \theta_i, r \sin \theta_i)$, where r is the scaled vector (1/scaling factor), and $\theta_i = -2\pi i/n$ for $i = 0, 1, \ldots, n - 1$. Vectors with a negative corresponding number, $-i$, have coordinates such that $A_{-i} = -A_i$. For the Koch snowflake, we start with an equilateral triangle ($n = 3$). If we assume that each side of the triangle has length one ($r = 1$), then:

Vector	i	θ_i	A_i
0	0	0	$A_0 = (\cos \theta, \sin \theta) = (1, 0)$
1	1	$-\frac{2}{3}\pi$	$A_1 = \left(\cos \frac{2}{3}\pi, -\sin \frac{2}{3}\pi \right) = (-.5, -.866)$
2	2	$-\frac{4}{3}\pi$	$A_2 = \left(\cos \frac{4}{3}\pi, -\sin \frac{4}{3}\pi \right) = (-.5, .866)$

Table 5.   *Vector Coordinates.*

Using our set of vector coordinates A_0, A_1, A_2, we can now graph the equilateral triangle initiator of the Koch snowflake. This is where we put to work the "algebra" part of "vector algebra." We are going to add the vectors together, which will enable us to find the corner points of the triangle, and finally graph the initiator. It is just as easy, but a little more work to find the graph of the first generation of the Koch snowflake. Since the pattern set of the first generation is $\mathbf{0, -1, -2, 0, 1, -2, -0, 1, 2, -0, -1, 2}$, all we need to do is go through the same procedure as above while remembering to scale down the new vectors. Since the length of the generator is three times that of the original initiator side, our scaling factor is three and we must reduce our new vectors by $r = 1/3$. Therefore, we can still use our old calculations for our A_i's, but we must multiply each one by 1/3. Once we know how to graph the Koch snowflake's initiator and first generation, it is routine to graph any generation of any snowflake if it weren't for two things. First of all, a lot of calculations are involved, even for first generation curves, secondly, you may be wondering where the pattern set for the first generation of the Koch snowflakes; $\mathbf{0, -1, -2, 0, 1, -2, -0, 1, 2, -0, -1, 2}$ came from. This is the part where our second basic mathematical subject comes into play. Of course, we're talking about modular arithmetic!

3.2. Modular Arithmetic. We can use modular arithmetic "$(x + y) \bmod z$". Our x, y, and z come from the initiator side presently being worked on, the generator pattern set, and the number of sides in the initiator. The results of a set of equations like this will give us our snowflake pattern set.

If we consider a generator with m elements that is applied to a n-sided polygon, we will produce a pattern of $n \times (m^k)$ elements for the kth generation. For the Koch snowflake, which has a three-sided initiator and a four-element generator, the pattern set will contain twelve elements for the first generation, forty-eight elements for the second generation and so on. The pattern set of a kth generation snowflake depends directly on the pattern set of the $(k-1)^{st}$ generation.

If we let the initiator be generation zero, we know that the first generation of a snowflake must depend on the pattern set of initiator. Thus, the first generation of the Koch snowflake builds on the equilateral triangle initiator, which has three sides. First we apply the generator to side 0. In our modular arithmetic equation $(x+y) \bmod z$, x is the side we are presently applying the generator to. For now, $x = 0$, but it will later also equal 1 and 2. We let y vary, in that it can equal each element of the generator. Finally, z is the number of sides in the initiator ($z = 3$). We need to keep track of all of the negative elements, since those will ultimately influence our pattern set. Remembering to do so is easiest if we adjust our modular arithmetic formula to become: $[(\text{sign of } x) \times (\text{sign of } y)] \times [(|x| + |y|) \bmod z]$. In this way, we first would calculate the absolute value of x plus the absolute value of $y \bmod z$, and then multiply that answer by the signs of x and y. We start with side 0 and the first four of twelve pattern elements.

G_0	G_1	G_2	G_0	G_1	G_2	G_0	G_1	G_2
0	0	0	1	1	1	2	2	2
		-1			-2			-0
		-2			-0			-1
		0			1			2
	-1	-1		-2	-2		-0	-0
		2			0			1
		0			1			2
		-1			-2			-0
	-2	-2		-0	-0		-1	-1
		0			1			2
		1			2			0
		-2			-0			-1
	0	0		1	1		2	2
		-1			-2			-0
		-2			-0			-1
		0			1			2

Table 6. *Sequence of Pattern Sets for the Koch Snowflake's Initiator G_0, First Generation G_1 & Second Generation G_2.*

Due to the construction of the generator, for any generation, a positive side i will return a pattern set of $i \bmod 3, -(i+1) \bmod 3, -(i+2) \bmod 3, i \bmod 3$. On the other hand, a negative side $-i$ will result in a pattern set of $-i \bmod 3, (i+1) \bmod 3, (i+2) \bmod 3, -i \bmod 3$. Using this shortcut, it is easier to calculate the second generation, which is summarized in the table on the preceding page.

A more efficient method of graphing involves finding the orientation of each vector in any generation of a snowflake individually, i.e., doing the calculations independently at each stage of a snowflake. In order to do this, we must first express the vector in question, w, in terms of a base m. As before, m is the number of elements in the generator, n is the number of vectors in the initiator, and k the generation of snowflake. Remember that for any generation k, where k ranges from one to infinity, the number of vectors in that particular snowflake equals to $n \times (m^k)$. To express the vector in question in base m we have to fill in the blanks "_" in the following equation:

$$w = \text{"\_"} \times m^k + \text{"\_"} \times m^{k-1} + \ldots + \text{"\_"} \times m^1 + \text{"\_"} \times m^0$$

Beginning with the space to the far left and working to the right. The solution for a space is the greatest number that does not force the solution to exceed w. For example for the 35th vector in the second generation of the Koch snowflake, since $m = 4$, $n = 3$, and $k = 2$, we get

$$35 = \text{"2"} \times 4^2 + \text{"0"} \times 4^1 + \text{"3"} \times 4^0$$

The actual representation of our vector 35 in terms of base 4 is in the form of the three integers that filled the spaces written together. Thus, 35 in base 4 equals 203_4. The subscript 4 indicates the base. Using the number 203_4, we can now find the orientation of vector 35. The first digit describes the component vector of the initiator on which vector 35 lies. Thus, the digit two tells us that vector 35 lies on vector two of the initiator, as shown in Figure 7(A) by the arrow. The second digit indicates which part of the generator applied to the previous component vector (two), vector 35 lies on. Thus, as the second digit is zero, we know that the second component describing vector 35 is the very first vector in the generator applied to initiator vector two. This vector is shown by the arrow in Figure 7(B). Lastly, our third digit is three and according to above, describes the vector in the generator applied to zero (the previous component vector) on which vector 35 lies. Three, indicating the last part of the generator (see the arrow in Figure 7(C)), gives us our final component for "building" the 35th vector of the second generation of the Koch snowflake.

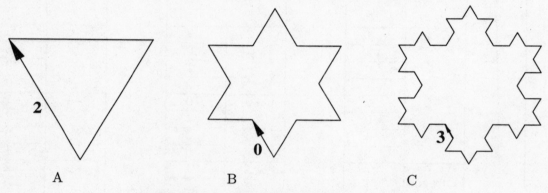

Figure 7. *Composing Side 35.*

This method was applied to create a Mathematica program "Fractastic", available at

(http : //webcampus3.stthomas.edu/c9shakiban/).

It uses three inputs — the number of sides in the initiator, the pattern set in the generator, and the number of desired generation — to compute and then output a graph of the initiator and a sequence of snowflake generations.

C. Snowflake Dimension:

As it turns out, we can also find the fractal dimension of the snowflake as a straightforward calculation, which comes in the form of a formula:

$$D = \frac{\log m}{\log f}$$

where m is the number of elements in the generator, and f is the scaling factor. If the generator is not listed, it is certainly possible after inspection of the first generation of the snowflake to determine the generator and continue with calculating the dimension. The variable f, on the other hand, is not as simple to find. It must be calculated and follows the formula below where $v_1, v_2, \ldots, v_N$ are the vector coordinates of the generator: $f = v_1 + v_2 + \ldots + v_N$ This procedure for finding snowflake dimension is far from being a perfect tool for all snowflakes. It works perfectly if the next generation of the snowflake does not overlap with the previous generation. When the snowflake generator crosses itself, goes back along the same line segment more than once, or lacks continuous "forward motion", the chances are that the snowflake, in its limit, is space filling. There is still innumerable details we don't know or understand about their fractal dimension.

4. Conclusion

In closing, we want to leave you thinking about the beautiful way in which mathematics and art are connected and though they seem so different, yet are often the same. In the following pages you can find a number of beautiful generalized snowflakes for your viewing. We would like to encourage you to try to look at these snowflakes not only as subjects of mathematics, but also as works of art, and in particular, as part of the mystifying beauty of the universe that appears even its smallest elements.

5. Examples

Initiator: square   Generator Pattern: $\{0, 3, 0, 1, 1, 0, 3, 0\}$
Snowflake dimension $= \log 8 / \log 4 = 1.5$

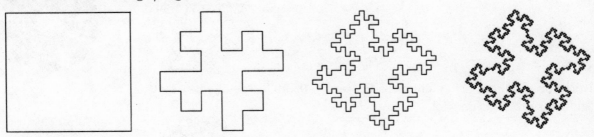

Initiator: pentagon   Generator Pattern: $\{0, -4, -3, -2, -1, 0\}$
Snowflake dimension $= 1.6309$

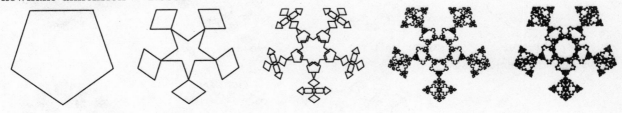

Initiator: hexagon   Generator Pattern: $\{0, -1, -2, -3, -4, -5, 0\}$
Snowflake dimension $= \log 7 / \log 3 = 1.77124$

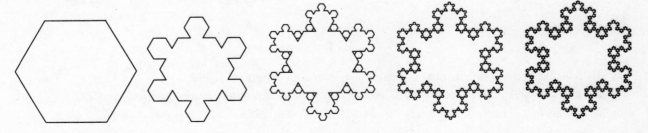

Initiator: heptagon   Generator Pattern: $\{0, -5, 3, -2, 1\}$
Snowflake dimension = 2; Space filling curve.

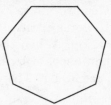  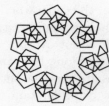 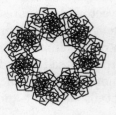 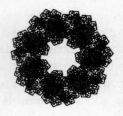

Initiator: octagon   Generator Pattern: $\{0, -7, -6, -5, -4, -3, -2, -1, 0\}$
Snowflake dimension = $\log 9 / \log 3 = 2$; Space filling curve.

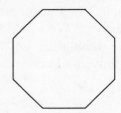 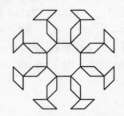 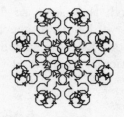 

Initiator: nonagon   Generator Pattern: $\{0, 2, 2, 2, 7, 7, 7, 0\}$
Snowflake dimension = 1.8692

 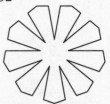  

Initiator: decagon   Generator Pattern: $\{0, -9, -7, -5, -3, -1, 0\}$
Snowflake dimension = 2

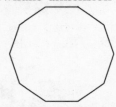 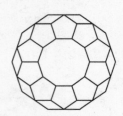 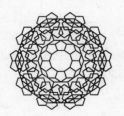 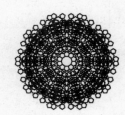

Initiator: 11–gon   Generator Pattern: $\{0, -9, -7, -5, -3, -1, 0\}$
Snowflake dimension = 2

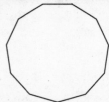 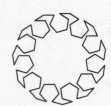  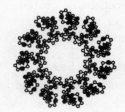

References

[1] B. Mandelbrot, *The Fractal Geometry of Nature*, Freeman, 1982.

[2] R. Devaney, *A First Course in Chaotic Dynamical Systems*, Addison-Wesley, 1992.

[3] M. Barnsley, *Fractals Everywhere*, Academic Press, 1988.

Visualization: From Biology to Culture

Brent Collins
90 Railroad Ave.
Gower, MO 64454
(816) 424-3436

Abstract

The capacity to comprehend the world through visualization is a product of our evolution in environments where it was the critical difference for the survival of our otherwise vulnerable species, and with exponentially increasing effect as it was invested in the symbolism of language where it could be shared in speech and eventually in writing. Viewed in unbroken continuity with these origins in prehistory, it was to become the sine qua non of science and the visual arts (music also has a visual dimension in its architectonics). The author specifically discusses the affinities between an art based on geometric visualization and model making in the physical sciences, noting that both share a motive force of aesthetic economy, and a reliance on the same suite of natural abilities in a nervous system adapted to function perceptually in three dimensions, acting upon them in particular through prehensile manipulations and technology. To exemplify these suggested "bridges" four recent sculptures are described in relation to their genesis in a visualization process similar to scientific thought in its analytic approach.

Geometric Visualization and Science

In this paper I hope to make four recent works intelligible in terms of the geometric visualizations which underlie them, and in doing so to demonstrate how unifying paradigms emerge in my work to the extent I am able to inventively articulate these visualizations into multi-layered grammars. Visualizing geometry does not require formal mathematical training (I'm a non-mathematician in that sense). It exists as a natural potential of our nervous system's evolution in three dimensions, though choosing it as an aesthetic focus will no doubt have been profoundly influenced by culture (archeologically dating from the appearance of spiral motifs in Paleolithic cave art), often through an osmosis which has been largely unconscious like so much learning.

Another reason an exegesis of the geometric thought in my sculptures may be interesting stems from the extent to which it otherwise tends to be elusive, apart from an immediate impression of their visual coherence, which is why they may sometimes seem familiar and strange at once. Actually if geometrically designed sculpture is successfully complex in its integration of harmonic elements, dissecting them reductively doesn't diminish its aesthetic resonance. The whole objective is a mystique of subtle analytic rigor whose significance as a revelation is simply the logic of the natural world which evolution has made permeable to our animal intelligence.

Imaginatively projecting the possibilities of molecular schematics is an analogous visualization process in science, which, though respecting different rules, has common ground with an approach of visual geometry in sculpture, each creatively searching for what will tenably work in three dimentions. The one reveals patterns present or potential in natural phenomena. The other creates patterns for aesthetic effect which is in turn a phenomenon of perceptual affect experienced by the neo-cortex (perhaps in that of chimpanzees quietly rapt before a sunset as well). Scientists often allude to the aesthetic dimensions of theory successfully distilled to elegant economy, while a paring to the essentials integral to design coherence characterizes geometric sculpture subject to minimalist constraints.

We might view scientists and mathematicians as having their own aesthetic aspirations, or at least as finding aesthetics a useful guide sometimes, which obviously implies a "bridge" of aesthetic concerns between the arts and sciences. And while it isn't infallible, we might surmise that aesthetic appreciation (beginning in a preference for symmetry as a fitness marker in mate selection?) is telling the neo-cortex what has the appearance of being or is at least provisionally right, hinting at the definitive reality beneath the provisionality of conscious experience, and coaxing the mind to a more lucid gestalt of self-recognition as a process of comprehension...such reflections can only have the poetic cast of an artists' perspective, and I voice them with some ambivalence preferring the refuge in concreteness sculpture has always held for me to the problematic uncertainties language holds.

First Sculpture

This sculpture (Figure 1; also see front cover), a ribbon deployed in an asymmetrically enriched trefoil pattern, is the fourth and only knot in a series of ribbons with negative Gaussian curvature along their entire length to the degree their other design features permit. The zero mean culture which I try to approximate as an aesthetic optimum, results in the simplest case when upward curves are negated by downward curves in the double arch formation of the familiar equestrian saddle. This is achieved in these ribbons because in cross section they are a crescent whose concave profile faces away from either the underlying curvature of the spherical surfaces they pass over or from the sharp U-turns in their paths over these surfaces. A twist becomes necessary to accommodate the back-and-forth transition in orientation from the sphere's curvature to the U-turns on its surface. In this particular sculpture the underlying geometry of the ribbon's global circuit consists of three smaller spheres ensconced within a larger sphere at 120-degree angles in relation to its center. The ribbon traverses the interior of the sculpture over the three smaller spheres to accomplish the three crossings of a trefoil, and transits the surface of the larger enclosing sphere when defining the three-lobed exterior of a trefoil. The sculpture's title, "Music of the Spheres," has an obvious descriptive aptness.

The trefoil pattern of the sculpture is asymmetrically enriched by a maneuver deforming the edge patterns of its ribbons front-to-back into a systematic deviation from the perfect symmetry of a classically configured trefoil. This deformation does this with a cogent economy in which the edge patterns of each face continue to have a graceful threefold rotational symmetry. Essentially the curvatures on one face sharpen (see front cover), while those on the other becomes gentler (see Figure 1), a relationship which possibly conserves the total curvature the trefoil would have restored to a classical configuration of perfect front-to-back symmetry. The deformation also displaces the pinch point where the crossing ribbons come closest together from the center of the trefoil to a locus nearer the face whose curvatures are sharper. This intuitive maneuver has a parallel in the topology of systematic grid deformations D'Arcy Thompson found evolutionary changes in morphology following in related species.

Second Sculpture

This untitled sculpture (Figure 2) is a toroidal band one by five inches in rectangular cross section. Its circuit encompasses six adjoining vertices of a cube, a geometry that resolves into a non-planar hexagon whose successive segments zigzag at right angles to each other. (For a more interpretable view of the same underlying geometry see the sculpture in Figure 4 where it is less disguised than in the sculpture under discussion.) At each of the six vertices the band twists 180 degrees like a Mobius strip. Matters, however, do not rest here with an even number of 180 degree twist increments and an apparently orientable surface, but become more subtle because integral to the object's global geometry the ribbon or band must always move across two contiguous sides of the cube in going from vertex to vertex around the hexagon. This confers an additional twist increment of 90 degrees to the segments between the vertices, and obviously occurs six times for a cumulative 540 additional degrees of twist, or three further 180 degree twist increments, making the total an odd number of nine and the ribbon consequently non-orientable. When the actual sculpture is viewed, the working in concert of these discrete maneuvers forms a seamless holism which dissolves them and becomes the focus of perception.

Figure 1: *"Music of the Spheres," laminated wood, 28 x 28 x 20 inches, 1999*

Figure 2: *Untitled, laminated wood, 33 x 33 x 21 inches, 1999*

The ribbon must furthermore twist in three different ways to achieve these maneuvers with a consistently unidirectional cheirality:

A) At each vertice of the hexagon, the ribbon twists 180 degrees over the exterior of the original toroidal column I laminated to carve the sculpture from.

B) In three of the connecting segments between the vertices, the ribbon twists 90 degrees through the interior of the column around its central longitudinal axis.

C) In the other three it also twists through the interior of the column, but with one enge hinged to an axis on the surface of the column.

These three twist patterns succeed each other in tandem as follows: AB AC AB AC AB AC, and in doing so create a convincingly dynamic evocation of natural turbulence, their most precise resonance being perhaps with the stabilized periodic motion of controlled chaos: correlations not entirely foreseen but completely welcome in their aesthetic freshness.

Rotating this object around its vertical axis from one hexagonal face to the other, and then inverting it, will fit it back within itself. Actually I think this is true of all such 180 degree rotations along the radial axes connecting vertices across from each other.

While the object's features are easy to summarize, they nonetheless are in a configuration of significant originality in which symmetry and seeming asymmetry dance in counterpoint. In particular the twist which emerges integrally from the object's global geometry, rather than in the more usual way as localized feature over a given length of ribbon, has a pleasingly unfamiliar subtlety. I wanted to somehow suggest the ultimately enigmatic coherence of the natural world the mind is a part of...an estimable aspiration likely leading to only modest achievement, yet even a hint of non-linear motion can quicken our appreciation of life's unpredictable dynamics: the enormous complexity out of which comprehension, compassion and apparent choice are born.

Figure 3: *Untitled, laminated wood 34 x 34 x 17 inches, 1999*

Third Sculpture

To create this sculpture (Figure 3) I initially laminated a triangular toroid. Subtracting from the trunk of this wood lamination, I consecutively produced a series of geometries, each within the enclosure of the preceding. First, a sculptural column, in cross-section a circle six inches in diameter, which spirals clockwise around the longitudinal axis of the original lamination: this linear and planar axis is simply an equilateral triangle with rounded vertices. Continuing to work through removal, I next created a twisting band, in cross section a three by six inch rectangle with rounded corners, within the previous spiraling column: the axis of this band's clockwise twist is the spiraling line which passes through the center of the column. The band twists 180 degrees over each side of the triangular toroid to form a triply twisted Mobius surface at this stage. The final subtractive step was to resolve this band into a single, double-orbiting column intersecting itself, a geometry which can be visualized in cross section as two circles three inches in diameter overlapping so their respective centers are two and three-quarter inches from each other. The continuous cleft of self-intersection which results creates a surprisingly strong resonance of organic muscularity for such economy of means. This is further enhanced by the curvaceous richness the fused columns have in embodying cumulative spiral and twist patterns.

An interesting topological transition has occurred as well. The single continuous edge of the non-orientable Mobius strip becomes the single continuous cleft of self-intersection between the column's two circuits. Altogether this piece exemplifies how multi-layered grammars can be progressively articulated by unschooled geometric visualization alone.

Figure 4: *Untitled, laminated wood,*
29 x 29 x 20 inches, 2000

Figure 5: *Untitled, laminated wood,*
29 x 29 x 20 inches, 2000

Fourth Sculpture

This sculpture (Figures 4 and 5) merges design features from the paradigms of the second and third. Like the second it is a toroid following the hexagonal zigzag pattern formed by the segments connecting six adjoining vertices on a cube. And at one stage in its production it was also a band twisting 90 degrees along each segment of the hexagon, as it must by moving across contiguous faces of the cube to keep the same orientation when entering successive vertices. Unlike the second sculpture, however, it does not twist an additional 180 degrees at each vertex Instead it is like the third sculpture in twisting only the 540 degrees (six times 90 degrees) of a triply twisted Mobius strip. It is also like the third in that its band, which intermediately was a three by six inch rectangle with rounded corners in cross section, is further resolved into a single, double-wound self-intersecting column. This again gives a purely geometric conception a surprisingly sensuous resonance of muscular anatomy, while creating the triplet of over-under crossings which define a trefoil knot.

Two views of it are shown to highlight how dramatically its aspect changes from different perspectives. It is highly symmetric, yet as dynamic as serene, and its instantly perceivable coherence nonetheless somehow seems to have a counterintuitive center of gravity, particularly on first encounter...the hope being that taking aesthetic perception to a place of unfamiliar coherence can potentially be an experience of renewal, perhaps the attraction of sanity felt when some new depth is glimpsed.

What Do you See?

Nathaniel A. Friedman
Department of Mathematics
University at Albany
Albany, New York 12222
artmath@csc.albany.edu

Abstract

In this paper I first describe a kindergarten workshop experience which opened my eyes to the fact that children in kindergarten already possess a certain visual literacy. The giant error in education is that explicitly developing this visual literacy is not considered, except in art courses. An example in the form of a checkerboard is then discussed as a source of developing children's visual literacy in mathematics with applications to the arts. Thus one can teach children to see both as a mathematician and as an artist.

1. Kindergarten and Visual Literacy

I recently taught artmath workshops in kindergarten classes. I started by passing around a variety of seashells that included both complete shells as well as sliced shells and broken shells where the inner structure was visible. The two-dimensional spiral at the end of the shell was pointed out as well as the three-dimensional spiral structure of the shell itself. Thus the children experienced the spiral structure as well as the form-space sculpture of these ingenious early examples of architecture.

The second exercise was making Geos (geometric sculptures) using foam core shapes such as squares, triangles, and rhombi joined by inserting round wooden toothpicks in the styrofoam edges, as discussed in [1]. Of course each sculpture was different although each child was given the same set of six shapes.

For the third exercise I had brought a variety of wire knots made from clothesline wire covered with blue plastic so that the knot shape is clearly visible. I first discussed the various knot shapes, which may have been slightly beyond them. However, we then dipped the wire knots into a soap-water solution to obtain the soapfilm minimal surfaces. They thoroughly enjoyed seeing the beautiful minimal surface sculptures, which is always the case in all the children's workshops I've taught, as well as with adults.

The kindergarten workshop experience was invaluable because it opened my eyes to the fact that the children can appreciate the above exercises because they touch, see, and feel as they experience the exercises. In particular, they already have a certain visual literacy. The giant error of education is that it does not consider developing this visual literacy in K-12, except in art courses. Hopefully due to the efforts of Bridges, ISAMA, Nexus, and others, this will change.

My purpose in this paper is to present an example that can be introduced to children in order to develop their visual literacy in mathematical concepts, as well as have applications to art. The operational question to ask the children Is What Do You See? This gives them the opportunity to look and to try to see, which is what it is all about.

2. Checkerboard Example

This is a variation and extension of an example that I discussed in [2]. Consider a 4x4 checkerboard, as shown in Figure 1. What do you see? Take your time - it's all yours.

Figure 1

There are at least three ways to see the checkerboard. In [2] I only saw it numerically. However, one can also see it from a geometric viewpoint as well as a topological viewpoint. The geometric and topological viewpoints can usually be understood by kindergarten children whereas the numerical viewpoint requires addition and multiplication.

To begin, it is helpful to rotate the checkerboard 45□ - say clockwise - so that it is on point, as in Figure 2.

Figure 2

From a geometric viewpoint, one sees that there is vertical as well as horizontal symmetry. Using a mirror, this can be explained as mirror symmetry by placing the mirror on the horizontal and vertical diagonals. One can also point out the half-turn symmetry which is rotational symmetry by a rotation of 180□ about the center point.

To extend this exercise, one can image a checkerboard extended infinitely in all directions. Why not introduce them to infinity early in life? Thus one has an infinite checkerboard, which is a first wallpaper pattern for them, as approximated in Figure 3. Now what else do you see.

Figure 3

For the infinite checkerboard, there are infinitely many lines of reflection. There are the previous diagonal lines of reflection. There are also horizontal and vertical lines of reflection dividing the squares in half. There are centers for half-turns at the corners of squares. There are also now centers for half-turns at the center of each square.

One can now also see translation symmetry. There is diagonal translation symmetry, where the translation distance is □2 units. There is also horizontal and vertical translation symmetry, where the translation distance is 2 units.

Finally there is also glide-reflection symmetry. For the diagonal line axis one translates □2 units and then reflects. For horizontal and vertical lines through the centers of squares, one translates 2 units and then reflects. For horizontal and vertical lines along the sides of squares, one translates 1 unit and reflects.

This example introduces children to symmetry. In fact one can visually present all frieze patterns and wallpaper patterns to children. Exercises are then to classify frieze patterns and wallpaper patterns. This is strictly a non-trivial visual exercise. For example, in the following wallpaper patterns, can you see centers of half-turns and glide reflections.

Figure 4

We will now consider the checkerboard in Figure 1 from a topological point of view. If we regard the small squares as countries, then the black-white coloring distinguishes the countries. Thus we can consider the checkerboard pattern as a simple two-color map. Note that the checkerboard pattern can be generated by equally spaced lines connecting opposite sides of the square. Can you see a general Two-Color Theorem? Can you see a generalization that would lead to a Three-Color Theorem?

A possible Two-Color Theorem is any map generated by lines that connect points on different sides of the square is two-colorable. The proof is inductive. Each time a new line is drawn, the colors of the regions on one side of the new line are switched. Note that this distinguishes any two regions having a boundary on the new line. Also regions previously distinguished will remain distinguished. Thus the two-coloring is preserved at each stage. The first step is obvious.

From an artistic point of view, it is interesting to present the development of the map one stage at a time. For example, if the map is generated by four lines, we consider four squares and add one line at a time, as shown in Figure 5.

Figure 5

We refer to the four images in Figure 5 as an example of <u>sequential art</u>. Here the artwork consists of all four stages so that one can appreciate the development from minimal to more complex. Note that the sequence of 4 lines can be introduced in 4! = 24 possible permutations.

One can also introduce curved lines as well as closed curves such as squares, triangles, and circles. In this case one switches the colors inside or outside the closed curve in order to preserve the two-color map. An example of sequential art is shown in Figure 6, starting with a circle.

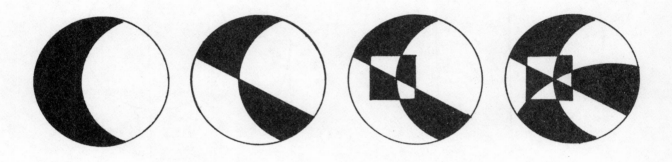

Figure 6

One can also ignore the 2-color map restriction and simple use the grid to generate a 2-color design. Moreover, in the work of Douglas Peden [3] he uses a grid of sinusoidal lines as a generalization of the usual grid. In this case he is producing optical art, where the surface appears to oscillate. An example is shown in Figure 7 (courtesy of Douglas Peden).

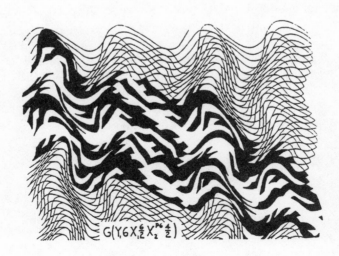

Figure 7

For a three-color map, one allows the line to stop at a previously drawn line. Two examples are shown in Figure 8. The proof of the three-color property was described to me by Richard Steinberg [5].

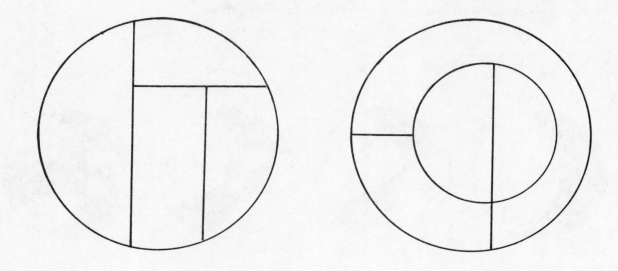

Figure 8

A large number of examples of geometric art appear in [4]. In particular, there are excellent examples of optical art in black and white by Victor Vasarely and Bridget Riley.

We will now look at the checkerboard in Figure 1 numerically. Since the checkerboard is 4 by 4, we see 16 squares. There are 2 white and 2 black in each row and each column so we see $2 \square 4 + 2 \square 4 = 8 + 8 = 16$. What other numerical relationship do you see? Consider the position in Figure 2. Take your time - it's yours.

The diagonals in Figure 1 appear vertically in Figure 2 and are more obvious in this position. The number of squares in each vertical diagonal starting from the left are 1,2,3,4,3,2,1. Thus we see the symmetric sum

(1) $$1 + 2 + 3 + 4 + 3 + 2 + 1 = 16 = 4^2 .$$

If we consider a 5 by 5 square, we will se

(2) $$1 + 2 + 3 + 4 + 5 + 4 + 3 + 2 + 1 = 25 = 5^2 .$$

Now can you guess the following symmetric sum?

(3) $$1 + 2 + 3 + \cdots + 9 + 10 + 9 + \cdots + 3 + 2 + 1 .$$

What is the general formula?

In an n by n square we see the general formula

(4) $$1 + 2 + 3 + \cdots + (n - 1) + n + (n - 1) + \cdots + 3 + 2 + 1 = n^2 .$$

From (4), one can derive the formula

(5) $$1 + 2 + 3 + \cdots + (n - 1) + n = n(n + 1)/2 .$$

Lastly, here is the formula for the sum of the first n odd integers.

(6) $$1 + 3 + 5 + \cdots + (2n - 1) = n^2 .$$

Can you find (6) in an n by n square? Consider n = 3 to start.

Lastly, we will view the checkerboard in Figure 1 from one other geometric viewpoint, which in a sense is the most obvious. Namely, a tiling of the plane by squares. A first generalization is to present children with cardboard four-sided shapes, as in Figure 9. An exercise is to have them trace the shape on paper in order to construct a tiling.

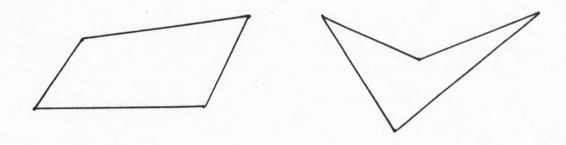

Figure 9

One can then tell the children that any four-sided shape will tile. For older children, one might even attempt to explain a proof.

The main point is that a checkerboard is a simple example of tiling that leads to the vast subject of tiling in general. In particular, there are many beautiful examples of colored tilings such as Islamic tilings.

The MAA publication Mathematics Magazine contains a large number of visual examples under the title Proofs Without Words. These are excellent exercises in developing visual literacy. It all comes down to What Do You See? There is a lot out there to look at and when you look long enough, you start to see.

References

[1]   N. Friedman, "Geometric Sculpture For K-12: Geos, Hyperseeing, and Hypersculptures, Conference Proceedings, 1999 Bridges: Mathematical Connections in Art, Music, and Science, Reza Sarhangi, Editor, Winfield, Kansas, USA.

[2]   N. Friedman, "Hyperseeing, Hypersculptures, and Space Curves", Conference Proceedings, 1998 Bridges: Mathematical Connections in Art, Music, and Science, Reza Sarhangi, Editor, Winfield, Kansas, USA.

[3]   D.D. Peden, "Bridges of Mathematics, Art, and Physics", Conference Proceedings, 1998 Bridges: Mathematical Connections in Art, Music, and Science, Reza Sarhangi, Editor, Winfield, Kansas, USA.

[4]   G. Rickey, Constructivism, Origins and Evolution, George Braziller, Inc. New York, 1967, revised 1995.

[5]   R. Steinberg, private communication, 1993.

Persian Arts: A Brief Study

Reza Sarhangi

Mathematics and Computer Science

Southwestern College

Winfield, Kansas, 67156, USA

sarhangi@jinx.sckans.edu

Abstract

Persia has left numerous marks on the civilizations and cultures of human beings, dating back to ancient times. The purpose of this article is to glance through the Persian arts in different eras and also to present a more detailed visual analysis of a ceramic design.

1. Introduction

Persia (now Iran) was one of the nations which were invaded by Muslim Arabs in the seventh century. This brought to an end the *Sasanian* dynasty in Iran, and it was an end to the last phase of ancient Middle Eastern civilization there. The interchange of cultures and combinations of arts among nations living in a vast area, which included North Africa, a part of Europe, and the Middle East, created a type of art, known as Islamic Art. Even though there are common elements in the arts of included nations, each region and nation has kept its own identities in the presentation and performance of it. The greatest flourish of Islamic art was in the period of 800-1600 AD. Not all artists or nations involved in or influenced by the Islamic art were Muslims. Persians, because of their background in art and architecture were one of the most influential among these Islamic nations. "Iran, and Persian influence on literary taste becomes apparent in Arabic literature from the mid-8th century onward. Many stories and tales were transmitted from, or through, Iran to the Arab world and often from there to Western Europe" [1]. This influence also included visual art, architecture, music, laws, and medicine. In this article we focus our study of Islamic art to what has been performed under Persian cultures and beliefs.

2. Persian Pre-Islamic Art

Architecture and art were at their peak during the *Achaemenian* dynasty (549-325 BC), later falling, yet regaining strength during the Sasanian dynasty (224-642 AD). The ruins of *Persepolis*, the massive ceremonial palace complex built in the Achaemenian era, and the ruins of the Sasanian structures are evidence of a rich background in architecture (figure 1). Zoroaster's hymns in the *Gathas*, which date back to the second millennium BC, and other Zoroastrian items all present examples of their literary art. Metal goblets, glassware, ceramic vessels, stone and metal engravings and sculptures show sophistication in visual arts in ancient Persia (figure 2). Phillips Stevens of the Anthropology Department at the State University of New York at Buffalo points out "We cannot overstate the influence of the religion and culture of Persia on the development of Christianity. The sack of Babylon by Cyrus the Great in 539 BC ended the exile. Cyrus aided the Jews in their return to Israel, contributed to rebuilding the Temple, and was hailed as a messiah ... and Persian culture became ennobled and exemplary in Israel. New concepts of angels, ... and the structure of the whole Apocalypse ... were among the direct influences of Persia's religion, Zoroastrianism" [2]. There are older examples of arts from this region, dating back to 6000 BC, which present friezes with various symmetrical constructions and rosettes of numerous cyclic and dihedral groups [3]. Maspero writes that the Chinese achievements in the third and fourth centuries BC, especially in astronomy and geometry, were due to information that they received from India and especially from Persia [4]. The art of the Persian carpet has its roots in ancient times.

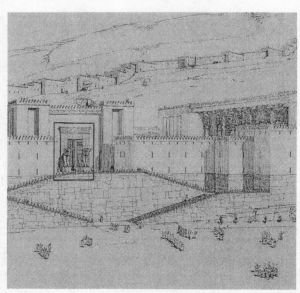

Figure 1: *Two 18-foot limestone bulls flank the doorway of the Xerxes Gateway at Persepolis. Persepolis was started by Darius the Great of the Achaemenian dynasty, and continued by his successors, notably Xerxes. It was a spiritual rather than an administrative capital. Noruz, the Persian New Year according to the vernal equinox, was one of the main events celebrated there.*

Figure 2: *(a) Lion Goblet   (b) Gold Goblet with four winged oxen design, late second millennium and early first millennium BC. Lions and bulls were symbols of two opposite powers of summer warmth and winter coldness respectively.*

The ancient Persian arts included details in design and construction, which were carried to the Islamic era. Even though some branches such as ceramics and tiling improved drastically, the traditions of some of the other Persian arts (such as sculpting) received serious damage or were completely destroyed.

3. Persian Arts after Islam

In Persian Islamic art, similar to other regions of the Islamic world, representation of living beings for decoration of mosques and theology schools was prohibited. (This faithfully obeys both the ancient Zoroastrian traditions of Persia and the Judeo-Christian commandment in *Exodus* 20.) Therefore, the center of the artistic tradition lay in abstract geometrical designs, in calligraphy, and in floral forms. Figures 3 and 4 are examples for the calligraphic and for the floral-inspired designs, respectively. In calligraphy, words, as the medium of divine revelation, were written or carved in the walls or around the domes using geometrical rules. In floral-based forms, one spiral-shape branch with some leaves and flowers, without beginning or end, grows out of another, and with the application of mathematical symmetries, it may have numerous variations.

Figure 3: *A calligraphy design with two written lines. The lower line is written in Arabic and is from the Qur'an. The top line is in the Persian Language (Farsi). Using geometrical shapes, it is written that "the key of the treasure is in the hand of the scientist-philosopher." The script is in a stylized fashion that is difficult to read, even for a native user of the language.*

Figure 4: *A floral design from a part of the dome interior of the Marble Palace in Tehran. The dome is closely modeled on the Sheik Lotfolah Mosque in Esfahan, built in the Safavied period.*

3.1. The Art of the Miniature. In Persia, a highly refined art of miniatures developed. Iran was connected to China through the commercial exchanges and trading routes commonly referred to as the Silk Road. This road was instrumental, not only in the expansion of trade, but also the exchange of ideas and thoughts in the arts. After the invasion of China and then Iran by the Mongol dynasty, the relationship between Chinese and Persian artists expanded. The cooperation of artists improved the art of ceramics and miniatures. The art of the miniature became popular for the decoration of books and creating pictures of mythological figures and royal families and scenes of their hunting, fights, and celebrations (figure 5a). Miniature artists also represented the imagery world of the Persian poems (figure 5b):

> "A rose without the glow of a lover bears no joy;
> Without wine to drink the spring brings no joy."
>
> Hafez (1325-1390 AD)

Figure 5: *(a) A prince is hunting ,(b) Lovers served with wine.*

In China, porcelain designs were inspired by Persian verses of poetry and by maxims demonstrating perfection in the art of calligraphy. In Persia, Chinese paintings were mentioned in Persian books and manuscripts and influenced the Persian miniature. In buildings built by the side of the Silk Road used by caravans, there are wall paintings by Chinese artists depicting Persian faces of the Sasanian period. In return, most faces and dresses in the traditional Persian miniature had oriental origins.

In the traditional Persian miniature, the principles of perspective are not applied. The miniature artist believes that perspective is an optical illusion and in order for him to reach the divine milieu, he tries to refrain from observing the principles of perspective. Another characteristic of this art is that the artist uses dots to create shades, as in some of the Impressionists' works, and in the pop art of Roy Lichtenstein and others in the 1960s and 1970s.

Persian miniatures, however, never found a place in the architecture and decoration of religious buildings. The world of geometry acted as the main source for inspiration of the artist-geometer for decorations of mosques, theology schools, mausoleums, and students' residential buildings.

The traditional Persian visual artist avoids emptiness in his art. Each part of the art carries details that have their own symmetries independent from the entire work. Inside a flower or a star design, there are other self-similar smaller designs that have their own identities. This is also true in the imagery world of a poem.

3.2. The Literary Arts. The most sophisticated part of Persian poetry is deeply involved with the ambiguity in which the reader is oscillating between the worldly and the divine. The *Ghazal* was considered the most appropriate genre form for this type of poetry. The Persian poet *Hafez*, undoubtedly, is the most recognized among all other poets in this style. *Johann Wolfgang von Goethe* introduced Hafez to the German-speaking readers in his enchanting poems book, *West-östlicher Divan*, published in 1819. This book was the first aesthetic appreciation of Persian poetry in Europe, and was done by one of the highest European literary intellectuals of the century.

> The rose has flushed red, the bud has burst,
> And drunk with joy is the nightingale—
> Hail, Sufis! Lovers of wine, all hail!
> For wine is proclaimed to a world athirst.
> Like a rock your repentance seemed to you;
> Behold the marvel! of what avail
> Was your rock, for a goblet has cleft it in two!
> Hafez

Classical Persian poetry often mentions knights and kings from Iran's history alongside those from Arabic heroic tales. The cup of wine offered by the "old man of the Magis (Zoroastrian sage)" is comparable to the miraculous cup owned by the Iranian mythical king Jamshid or to Alexander's mirror, which showed the marvels of the world; the nightingale may sing "Zoroastrian tunes" when it contemplates the "fire temple of the rose." … Minute arabesque-like descriptions of nature, particularly of garden scenes, are frequent: the rose and the nightingale have almost become substitutes for mythological figures. The versatile writer was expected to introduce elegant allusions to classical Arabic and Persian literature and to folklore and to know enough about astrology, alchemy, and medicine to use the relevant technical terms accurately. Images inspired by the pastimes of the grandees — chess, polo, hunting, and the like — were as necessary for a good poem as were those referring to music, painting, and calligraphy [1].

The writer was also expected to use puns and to play with words of two or more meanings. He might write verses that could provide an intelligible meaning even when read backward. He had to be able to handle chronograms, codes based on the numerical values of a phrase or verse, which, when understood, gave the date of some relevant event. Later writers sometimes supplied the date of a book's compilation by hiding a chronogram in its title. A favorite device in poetry was the "question and answer" form, employed in the whole poem, or only in chosen sections [1].

3.3. Persian Architecture and Ceramic Art. There were two types of buildings that survived for centuries, religious buildings, and royal residences and fortresses.

In regards to their architecture and structure, each religious school was generally one or two stories in height and was comprised of many rooms, called *Hojreh*, with each room belonging to separate individuals to reside. These buildings generally had a rectangular, open courtyard in the middle, with large rooms for studying and gathering, and normally were constructed near a mosque or shrine of a religious scholar.

The first mosques were designed as a square or rectangular space, which included a small ablution pool in the middle. The design came from the Zoroastrian temples [5]. Throughout time, religious centers remaining from the Sasanian dynasty rebuilt and changed to become mosques, and the aesthetics of these renovations were improved by the talent of Persian architects, along with a combination of artists from other cultures. The primary mosque in *Esfahan*, *Masjede Jame Esfahan*, has been estimated to be around 1200 years old, even though the main parts of this mosque had been destroyed and rebuilt throughout time. According to some sources of information, this mosque was first a Zoroastrian temple [5]. The oldest date carved in this mosque shows a date of around 1000 years ago, which is carved on the dome. This mosque included a large library with a three-volume bibliography of the library books and collections, however, a major part of this library was lost in a fire.

Palaces were not as secure as the mosques and schools were. Each conqueror, in general, destroyed the old structures. Nevertheless, there are palaces and fortress that have been kept in their original shapes in some older cities.

The decoration of walls using bricks also became important. Brick in its original shape is not flexible, but after cutting and carving it can be used everywhere in the surface of buildings for decoration, including round surfaces, which at that time was a very new idea. In the twelfth century, it became popular to use bricks to cover a surrounding area and then to design the area inside using ceramics with four colors of light blue, navy blue, brown and black. The color scheme improved rapidly by increasing the number of colors made through different combinations of metals.

During the fourteenth century, designs using tiling intensified, and the component tiles became smaller and smaller and the designs more complex. The sophisticated designs also involved curved tiles. Due to this transformation, the most complicated designs of Persian carpets could be carried out in ceramics by cutting and arranging ceramics composed of solid colors (figure 6).

Figure 6: *A ceramic pattern from the Safavid era, incorporating carpet-like details.*

In the *Safavid* era, especially during 1524-1629 period, there was a renaissance for architecture and art. The art of ceramics changed considerably throughout time, and every day new ideas were developed. The way in which ceramics were composed was also improved, introducing different sizes and colors.

The geometric designs and shapes used in architecture, based on their materials and the methods of their execution, can be divided into the following categories:

1. Lay brick designs.
2. Combination of brick designs for the border with ceramics for the main part.
3. Cutting and pasting ceramics.
4. Wooden designs created by using wooden strips which have been cut in different angles and have been put together in order to make a geometrical design and also function as a type of protection and blind for windows.
5. The art of Stucco for creating three-dimensional designs, that resemble stalactites, which were placed as decoration for corners and interiors of domes.

Although decorative designs were constructed using a wide range of materials, their mathematical layouts were quite similar. There were a number of hidden mathematical problems in the design of the entire project with respect to the local decorations and designs, which needed to be studied before the execution. Besides planning for the structure of the building, the architect needed to study the balance and harmony among designs performed for the decoration within an entire project, which is called rhythm of design. The movement of this rhythm from the floor to the wall, and finally to the interior dome design should make a harmony and gives us a feeling of balance and proportion for the entire execution. In fact, in each design we can find the pure mathematics of the division of space and the construction of shapes using essential geometric tools. In order to have an appropriate design and its execution, whether on ceramics, wood, stucco or other mediums, another important item to consider was color.

Before the use of ceramics, Persians used painting as decoration for walls. The use of painting continued until around the eleventh century, when ceramics grew increasingly popular, replacing painting.

4. Study of a Ceramic Design

Figure 6 is a ceramic design, which includes carpet-like details made from solidly colored, small, curved tiles (figure 7). The division of space of the layout creates geometrical pieces. Most of the time these pieces are standard and can be used for creating innumerable patterns. Therefore, each piece found its own name and identity known to the designer and ceramic maker. The artist-geometer learned how to construct each piece geometrically for a certain size without creating the entire pattern. Therefore he could communicate with the ceramic maker without referring to the entire design. Figure 8 provides the geometrical constructions of pieces in figure 6.

Figure 7: *The details of a design used in the ceramic pattern in Figure 6.*

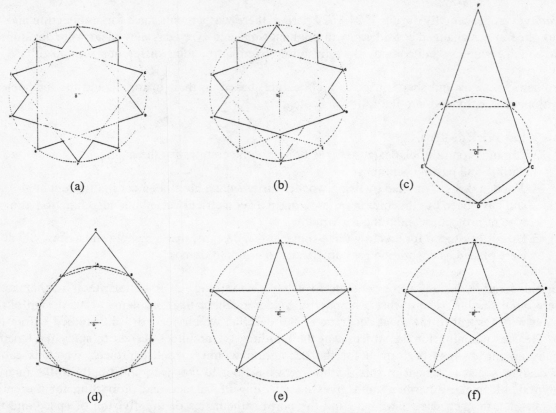

Figure 8: Ceramic designs: *(a) Sharp Ten Shamseh (Sun-like), (b) Cut Shamseh, (c) Sharp Toranj (d) Sharp Shesh (six), (e) Sharp Setareh (star), (f) Se Pari (three-wing). All the above designs have been created based on the division of a circle to five (that easily can be extended to ten). This division has fascinated artists and mathematicians for centuries [6].*

5. Conclusion

From the first scientific notions of astronomy and the measurement of the celestial sphere and determination of the new year according to the vernal equinox, to the mathematics of shapes, solids, and numbers, Persian arts have played important roles. Persian workers have combined skills in theoretical mathematics and practical techniques such as ceramics, along with artistic ideas from their own history and around the world. Putting these together, they have expressed several appealing forms, which bridge mathematics and the arts.

Acknowledgment: I would like to thank Bruce Martin for his helpful comments to complete this article.

References

[1] *Islamic arts*, Encyclopædia Britannica Online

[2] P. Stevens, *Further Thoughts on the Star of Bethlehem*, Skeptical Inquirer magazine, March 2000.

[3] S. V. Jablan, *Theory of Symmetry and Ornament*, Matematički Institut, Beograd, 1995.

[4] H. Maspero, *China in Antiquity*, translated by F. A. Kierman, Jr. Kent, Eng., Dawson, 1978.

[5] M. MaherAnNaghsh, *Design and Execution in Persian Ceramics*, Reza Abbasi Museum Press, Tehran, 1984.

[6] J. Kappraff, *Connections, The Geometric Bridge Between Art and Science*, McGraw-Hill, Inc, 1990.

Polyhedra, Learning by Building: Design and Use of a Math-Ed Tool

Simon Morgan
Rice University School Mathematics Project
Rice University, MS-172, 6100 Main Street, Houston, Texas 77005
email: smorgan@math.rice.edu

Eva Knoll
EK Design and Consulting
1236 St. Marc no.1, Montreal, Quebec, Canada, H3H 2E5
EvaKnoll@netscape.net

Abstract

This is a preliminary report on design features of large, light-weight, modular equilateral triangles and classroom activities developed for using them. They facilitate the fast teaching of three dimensional geometry together with basic math skills, and create a lasting motivational impact on low achievers and their subsequent performance in math and science.

In directed discovery activities, lasting from 20 to 90 minutes, large models of basic polyhedra are made, enabling their properties to be explored. Faces, edges and vertices can all be counted and tabulated, providing opportunities to see number patterns and inter-relationships, to plot graphs, to extract algebraic relationships and to look for proofs of those relationships. These building activities can be kept central, under the teacher's control for large classes with limited time, or building can be split out into groups of children where co-operative problem solving skills are also developed.

In interviews, children have stressed the effectiveness of learning by building the shapes themselves. In classroom activities, it is clear to see that these triangles make children excited. Learning by building gives a concrete, active, authentic and personal experience of mathematics to children and teachers enabling them to feel the full excitement of the subject.

Introduction

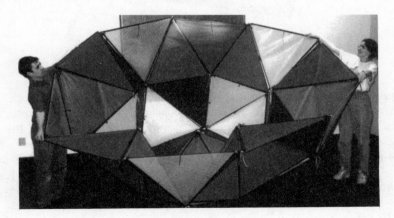

Figure 1: Eva Knoll and Simon Morgan holding half an endo-pentakis-icosi-dodecahedron., Montreal, March 1999

This is a preliminary report on an educational project which begin as an exercise in the combining of mathematics and art. Eva Knoll developed a way of folding an origami grid of equilateral triangles from a disc of paper, and then constructing polyhedra from this grid [2]. With the help of Simon Morgan and Roger Tobie the same technique was used to make a 1 m diameter 80-sided shape. The experience of this construction process convinced us to go ahead and build larger triangles (see figure 1) so that a whole group could participate in a barn-raising event [1] and [3]. To take the project further, and with the support of Rice University, these ideas and materials were taken into a Houston middle school to be tried out in mathematics lessons.

The first thing that struck us was how the size, colors and construction of the triangles excited the children and made them to want to build three dimensional shapes. In the classroom, this translated into the triangles' effectiveness as teaching and learning tools for geometry and a way of reaching and motivating low performers. The simple modular construction of equilateral triangles makes them an extremely versatile tool for teaching different concepts in a short space of time and creates endless possibilities for building shapes of a scale which really appeals to children. The educational applications have grown steadily since then, including workshops for teachers as well as deeper mathematical investigations in geometry and topology for the classroom.

Description and design features of the triangles

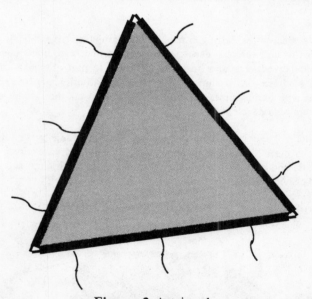

Figure 2 A triangle

The patented equilateral triangles are made of kite materials: carbon fiber rods, nylon and clear Mylar sheeting. Laces are attached to each edge so the triangles can be fastened to each other (see Figure 2). Each of three rods is enclosed in a black seem running along an edge of the triangle. The rods are hollow allowing a cord to be passed through each rod and tied fast to give the triangle its rigidity. This design was motivated entirely by the need for modular triangles which could be used in group participation sculpture making events (see phase 2 below, [1] and [3]). Here are the key design features which were identified as important for educational effectiveness in the classroom:

a) Size is the big attraction. We made the triangles as large they could be built, 1 m edge length. Children were immediately excited by that size because it enabled them to get inside the shapes. This experience of being able to see polyhedra you have built yourself from the inside and the outside stimulates the mind in a way that cannot be achieved on a hand-held scale or by looking at pictures or computer simulations. The size also adds versatility to classroom activities so that either the whole class can watch shapes being made, seeing every detail, at the front of the class, or children can work separately in groups where it is easy for the teacher not only to see what they are doing, but also to show the whole class the different results.

b) No right angles! The departure from right angles takes children away from the familiar rectangular world of cuboid rooms, buildings and objects. Every shape made looks unusual in an unpredictable way and therefore provokes curiosity.

c) Light weight materials give versatility in the classroom. The light weight hi-tech materials enable the polyhedra to be built safely, easily and quickly using lace ties. They also ensure that the polyhedra are attractive and rigid with a clear geometric shape. Fully assembled polyhedra can easily be picked up and moved around by a teacher or a child. This helps children develop a tactile feel for their shapes as well as allowing the shapes to be held up and spun around to show how they look from different angles, and to demonstrate their symmetries.

d) Modularity gives simplicity, speed, versatility, full utilization and demands problem solving. Having just one equilateral triangle modular component coming in a range of colors makes for straightforward construction of an unlimited number of fascinating shapes, with no time consuming setting up of materials. Any child can continue building any shape, using any of the triangles, until they are all completely used up. The flexibility of the how triangles can be put together forces children to make choices and to decide how to build a shape, either in free form experimentation or in trying to build a polyhedron with specified properties.

e) Lace ties enable speed and discovery learning. These are great for a fast moving class. They keep the instructions simple as everyone knows how to tie shoe laces. Three ties are needed for one edge, so it does not take too long for a group of children to put a large shape together. They just want to keep building more and bigger shapes. Speed of untying is also critical for this classroom tool. It enables many shapes to be built and modified, one after the other, re-using the same triangles and enables fast dismantling and storage at the end of the class which saves teacher time.

f) Rainbow colors attract interest and dispel fear. Using deep saturated rainbow colors adds to the triangles' immediate appeal and emphasizes the contrast between adjacent faces on polyhedra. Any fear of mathematics tends to be forgotten as children decide which is their favorite color of triangle to start building with.

g) Clear triangles show the inside and the outside at the same time. Clear triangles are like windows into another world, the inside of a polyhedron. If a polyhedron is made entirely of clear triangles it is possible to see all the faces, edges and vertices at once, which helps emphasize its three dimensionality.

h) Black ribbon edging gives geometric definition to shapes. There is a 1/2 inch black ribbon bordering each triangle. This marks out clearly the edges and vertices of polyhedra which in turn demarcate the distinct faces and their orientations.

i) Rigid rods in each edge give tactile and conceptual understanding. Rods inside each triangle edge under the black ribbon have a tactile quality, they are straight, rigid and what you feel and hold when you pick up a triangle. This direct tactile experience of edges adds to the child's understanding of polyhedra. The rods are useful when explaining what happens to edges when triangles are assembled into polyhedra. A triangle by itself has one rod per edge, but in a polyhedron two triangles come together along an edge giving two rods per edge. If we pick up a polyhedron by an edge we can actually feel the two rods in that edge, giving a concrete experience that precedes or reinforces the abstract concept (lesson 4).

Classroom activities

We explain the lesson plans developed so far in chronological order. Additionally we have often allowed children to build whatever they want with the triangles. This free form work has involved a lot of variety, learning and problem solving which encourages children to think for themselves. Phases 1 and 2 took place together in one semester culminating a big event [3]. Phase 3 followed as an attempt to develop more mathematical content for lessons with the triangles. The work with low performers in phase 4 was done in response to seeing successes with individual 'at risk' children who used the triangles and to address the special needs of this group. Free form work and lesson 3 make good starting points for students at all levels who are going to use the triangles.

Phase 1: Platonic solids. These early activities were designed to help the children experience the geometry of solids through building, seeing and physically getting inside shapes. They took place in a high school geometry class and with regular 8th grade students in Lanier Middle School, Houston, TX.

Lesson 1: Filling tetrahedra and the stella octangula. This lesson involved constructing a large tetrahedron with 2 m edge length and four triangles making up each face. This gave a chance to review how volume and surface area scale with length. The children were asked how many 1 m edge length tetrahedra could fill the inside. After four tetrahedra were placed inside, a space was left that was filled by another mystery shape that the children constructed with triangles.

Figure 3 Tape reveals the cube

Figure 4 Building icosahedra

Finally, to continue the space filling theme the middle triangle of each face was stellated to produce a stella octangula (see figure 3). Another stella octangula was made by stellating an octahedron, and then the two stella octanguli were placed so that they enclosed an octahedron, demonstrating how to fill 3 dimensional space. The key excitement points in this activity were trial and error with filling the large tetrahedron and looking inside the stella octangula during construction.

Lesson 2: Building icosahedra. The children built two icosahedra, one hand held by folding a paper disc [2], and a scaled up version using the triangles. The large version icosahedron had to be colored with 5 colors so that no two triangles at a vertex had the same color. The children achieved this by planning their coloring on the paper nets they had used for the small icosahedron, laying out big triangles in a net following that plan, and then closing up the giant net just as they had done with the paper models (see figure 4). The key excitement points were having a paper icosahedron that unfolded into snowflake shape that they could take home and getting inside their full size icosahedra.

Phase 2: Geraldine. Barn-raising an endo-pentakis-icosi-dodecahedron (see website video [3] and [1]).

Phase 3: Theorems and proofs by directed discovery. For group work, a set of step by step instructions was prepared, each of which required some problem solving e.g. build a closed shape with four triangles. For each step there were some tabulated records to be made, such as writing down the number of edges or vertices, or drawing the shape. The adult instructors went round and checked that each building step had been done correctly and that each child had written down the observations. For low performers we sometimes adapted this plan in phase 4. At various stages the class can be brought together to examine the different shapes made and to look for trends, to plot graphs, to extract algebraic relationships and to discuss proofs of formulas or theorems derived from experiment. As much as possible the children are asked to suggest things and come up with their own reasons rather than just being told the answers.

Lesson 3: Making n-gon based pyramids. This lesson was designed as a preliminary lesson to introduce children to the difference between the flat plane and a pointed vertex of a polyhedron. They count six triangles that would come together flat on the floor, and then see different pyramids produced as they reduce the numbers of triangles coming together at a vertex. Then they are asked to make closed up shapes with a fixed number of triangles at each vertex, i.e. three to make a tetrahedron, four for an octahedron and five for an icosahedron. For gifted children this lesson can combine with lessons 4 or 5. See phase 4 for enriching this lesson for low performers.

Lesson 4: Combinatorial topology (5th grade and above). The simplest formula involving a polyhedron with all triangular faces is $3 \times F = 2 \times E$ where F is number of faces and E is number of edges. This can be discovered by building shapes with a fixed number of triangles and counting the edges. Tabulating and graphing the number of edges for the different number of faces can enable the children to find a formula relating the quantities. Children also discover by experiment that you cannot make a polyhedron with an odd number of triangles. Finally we ask why is the formula true? Then we observe that for each triangle there are three rods in the edges of one triangle, and in each polyhedral edge there are two rods. So we see we are counting the total number of rods in two ways, as three times the number of triangles or two times the number of edges in the polyhedron (Rods = 3 x Faces = 2 x Edges). Then we see that if we have an odd number of faces, we would get a half edge, thus proving (by contradiction) that it is impossible to make a polyhedron with an odd number of triangular faces.

Lesson 5: The angle deficit theorem (7th grade and above). The Euler formula for a polyhedron with no holes (faces - edges + vertices = 2) has a geometric counter part, the angle deficit theorem. Angle deficit at a vertex can be measured by adding all the angles at the vertex and subtracting that from 360 degrees. For example, a vertex of a cube has three 90 degree angles, giving a total of 270 degrees. Subtract this from 360 and we get a deficit of 90 degrees. The theorem states that the sum of all angle deficits at each vertex on a polyhedron adds up to 720 degrees, or two whole turns. A proof of this (not part of the lesson) can be derived from the euler characteristic of a sphere and the fact that the sum of the interior angles of a triangle is 180 degrees. Also see lesson 6 for another proof.

The class activities involve building tetrahedra, octahedra and icosahedra, counting and tabulating the angle deficit at each vertex (recorded in fractions of 360 degrees) and the number of vertices. The children then are asked to find a number pattern (see table 1).

Angle deficit (Pointiness)	no.of vertices	Polyhedron
1/2	4	Tetrahedron
1/3	6	Octahedron
1/6	12	Icosahedron

Table 1

They notice that the number of vertices is always twice the denominator in column 1. They are shown that this means the product of angle deficit and number of vertices is always two. We also discussed why the number of vertices increases as the angle deficit decreases. Basically angle deficit measures how pointy or sharp a vertex is, which in turn determines the angles between faces meeting at that vertex. To close up the shape it is necessary for the faces to go all the way around the shape. Pointy vertices (e.g. on tetrahedra) turn the faces around more than less pointy vertices (e.g. on icosahedra). So if you have vertices that are less pointy, or flatter looking, you need more of them to get the faces all the way around the shape. These explanations become more meaningful when the teacher is holding up and comparing giant sized tetrahedra and icosahedra. Physically pointing out the shapes of vertices and the angles between faces around them as the explanation progresses makes it completely clear. Then the angle deficit formula is stated, explained and verified where possible by the children for other shapes such as pyramids, prisms, the cube and the dodecahedron.

Lesson 6: Non-euclidean geometry (9th grade). This lesson starts with the children investigating how a geodesic or shortest path goes over an edge of a tetrahedron. Then three points (a,b,c on the left of figure 6) are chosen around a tetrahedral vertex and connected by masking tape geodesics so that a vertex of the tetrahedron is enclosed within this 'geodesic triangle'. The children measure the sum of the internal angles of the triangle and find that they add up to 360 degrees instead of the usual 180 degrees for triangles in the plane. The enclosed tetrahedral vertex is then opened up by untying along one edge of the tetrahedron. See figure 6 where three triangles around a tetrahedral vertex are opened up and laid out flat to examine the angles. The geodesic triangle going around through points 'a', 'b' and 'c' on the left (and also 'e' and 'f' when flattened out on the right) is marked by the children with masking tape.

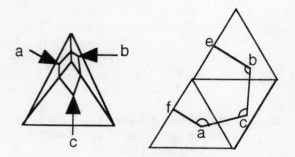

Figure 6 Tetrahedron with a geodesic triangle 'abc' (left) becomes pentagon 'facbe' after the enclosed tetrahedral vertex is opened up by untying (right).

One group of children then tried to prove mathematically that their protractor measurements were accurate. One child proved this himself and his proof was shown to the rest of the class on the triangles that had been unfolded, as in the right hand side of figure 6. Pentagon 'abcdef' has a sum of interior angles of 540 degrees (as it can be decomposed into three triangles) and as 'eb' is parallel to 'af', the angles at 'e' and 'f' must add up to be 180 degrees, leaving 360 degrees for the sum of the angles at 'a', 'b' and 'c'.

The general formula for the sum of the interior angles of a triangle on a polyhedron is 180 degrees plus the angle deficits of the enclosed vertices. This can also be used to prove the angle deficit theorem by considering two complementary triangles bounded by a given set of three geodesics. This lesson can be varied by repeating the exercise using octahedral or icosahedral vertices and by allowing a geodesic triangle to enclose more than one vertex.

Low performing children

Phase 4: Remedial work. We have carried out work with remedial 6th grade through 12th grade groups taking from all the above lesson plans and allowing free form work. One 6th grade remedial teacher used the triangles exclusively with free form work to encourage independent thinking and group work. The children discovered a wide variety of three dimensional shapes for themselves.

Additionally we used the triangles for teaching and reviewing two dimensional geometry and use of protractors. When making pyramids (lesson 3), we place masking tape around their bases on the floor, then remove the pyramids and see what polygon emerges. Then we can measure and calculate the interior angles. Masking tape is used to triangulate the hexagons and pentagons that emerge, and the properties of the triangles can be discussed, measured and proven. Doing geometry 'life-size' on the triangles and on the floor with masking tape and chalk is an effective strategy to elicit student participation in the geometry proofs and discussions. Children are more likely to make suggestions and ask for something to be explained again until they understand it when working with large manipulatives. It is valuable to be able work a proof or derivation in different ways, so whenever students make suggestions the teacher can help them to develop the line of reasoning. If a class then sees many different proofs of the same thing based on student suggestions, then all the better.

With low performing children, in contrast to high performers, it is best to ask each group to build a different shape and keep a table of results on the blackboard, rather than having each child keep their own record as each group builds every shape. Low performers can complete the whole of lessons 1, 3 and 4, but more time is needed when examining number patterns in data, plotting graphs and seeing how to derive equations. When lesson 3 is done with protractor work, low performers can then go on to do part of lesson 5.

Two key pedagogical aspects of working with low performers in mathematics are diagnosis of deficiencies in mathematical basics and then teaching with 'hands-on' manipulatives. These activities certainly provide the hands on manipulatives that these students need, and in addition provide teachers a chance to see what basic math skills their low performers have. We have found with the number work that teachers can be surprised by their students, not just by seeing gaps in ability, but also by finding that some students had abilities that had never previously emerged. Just as important as these pedagogical aspects are long term motivational improvements in the children that have been observed.

Discussion

Using this design of triangle gets children excited about mathematics and teachers have found that these activities can completely change the attitude of children who had been low performing and described by the teachers as 'disconnected from school'. The children have stressed how their understanding was helped by building shapes themselves. In the video [3] from Phases 1 and 2, teacher Jackie Sack says:

"when kids are exposed to something of this size that they can feel and touch and move around then they truly begin to understand it, understanding with a sense of excitement and curiosity and this is impossible to achieve on a tiny hand held scale. Once you have this excitement and curiosity you do and discover all kinds of things."

This comment seems also to apply when we later broadened the scope of the mathematics covered in the lessons. The concrete real world nature of these shapes makes mathematics real and relevant to children in a way working on paper may not. As the children are working with triangles, seeing how to put them together, having ideas, trying things out, being mathematically creative and innovative (particularly in free form work), and getting immediate results all the time, they see a very different view of mathematics. This is not a view constructed step by step by a teacher or a book, it is not seeing someone else's explanation of mathematics, it is seeing by doing it oneself. Learning by building gives a concrete, active, authentic and personal experience of mathematics to children and teachers enabling them to feel the full excitement of the subject.

These building and discovery learning activities that go beyond the syllabus have proven to give benefits to the children's learning and use of basic syllabus material that more than justify the time spent on them. These activities are full of opportunities to review and apply arithmetic and algebra skills and two dimensional geometry. The scientific method is also integrated into this process by use of experiment, making, tabulating and graphing observations, seeing patterns in data, trying to define and explain them mathematically, and then going on to discuss mathematical proofs of observed trends.

Results attained so far justify large scale educational research into the advantages of using these materials and activities, both in terms of geometry and basic math skills and also in terms of motivating underachievers in mathematics in a lasting way. For school classrooms, we recommend having a central project for each district which has a set of triangles and one or two people to go out to schools to use them with children. Schools that use the triangles a lot can buy their own set. For museum displays, we recommend having shapes you can get inside and organized shape building activities for visitors school groups. The authors are available to supply triangles and lesson plans.

References

[1] Eva Knoll and Simon Morgan. *Barn Raising an Endo-pentakis-icosidodecahedron.* In Bridges: Mathematical Connections in Art, Music, and Sciences. Conference Proceedings 1999. Reza Sarhangi, Editor.

[2] Eva Knoll. *From Circle to Icosahedron.* In Bridges: Mathematical Connections in Art, Music, and Sciences. Conference Proceedings 2000. Reza Sarhangi, Editor.

[3] Website video: http://math.rice.edu/~rusmp/

Symmetry and Beauty of Human Faces

Teresa Breyer[1]
Department of Mathematics
University of Wisconsin
Madison, WI 53706, U.S.A.
And Technical University of Vienna, Austria
E-mail: breyerteresa@hotmail.com

Abstract

In this project we study the correlation between symmetry and attractiveness of faces through statistical pattern recognition and neural networks. We design an intelligent agent that learns to use a given set of standards to distinguish between beautiful and ordinary faces. We represent the faces as feature vectors in a vector space in order to measure the degree of symmetry. To implement a larger data set, we use 2-dimensional wavelet representation to improve the performance.

1 Introduction

There have been many studies trying to figure out whether there is a correlation between facial symmetry and attractiveness. Below we briefly mention some related results:

In 1878 [9] Francis Galton found out, that composite faces (faces that are a combination of several faces) were considered more beautiful than the original faces. In 1990 Langlois and Roggman confirmed this. They argued that increasing averageness also increases attractiveness. Of course these composite faces were also more symmetric.

Conversely the findings of Grammer and Thornhill in 1994 [8] suggested that there was a negative correlation between attractiveness and fluctuating asymmetry (a measure for the deviation from bilateral symmetry used by biologists).

Rotem Kowner did a study on "Facial Asymmetry and Attractiveness Judgement " in 1996. In one of his experiments he took pictures of young children, young adults and old people. He used the original images as well as symmetrical composites, formed by taking one hemisphere and flipping it vertically. But symmetry only had a positive influence on older people. Kowner [1] suggests "that facial asymmetry has a curvilinear effect on facial attractiveness, as both extreme…asymmetry …and extreme symmetry…somewhat diminish attractiveness". The proposed neural network in essence learns from examples to approximate Kowner's conjectural curvilinear function.

Our approach to discuss the correlation between symmetry and attractiveness of faces is new in the sense that we use an intelligent agent for the study of a cultural issue. Our learning-machine, a multi-layer perceptron network, which provides a simplified imitation of information processing in the human brain, can learn to distinguish between beautiful and ordinary faces. This is the first attempt to explore the underlying mathematical structure and patterns that are related to human emotion, preferences, perception of beauty, etc.

[1] This research was done with supervision of Amir Assadi as part of the NSF Project Symmetry supported in part by NSF-HER-DUE-CCD and UW-Madison Office of the Provost

One approach to conceptualizing the human brain, where the basic units, the neurons, are connected in average to 10 000 other neurons, is via artificial neural networks. In some models for artificial neural networks, one models parts of the cortex by layers of neural nodes, each node, also called perceptron, being connected to all nodes of the two adjacent layers, as in Figure 1. In the human brain, the relevant cortical neurons respond to input stimuli by "spikes," that is, they pass on signals, if the input they have accepted exceeds a certain level, called threshold. An artificial network is modeled as a graph whose nodes correspond to neurons and whose edges measure neuronal connectivity strength. We assign values, called *weights*, to the edges connecting nodes. These pass the signals received from the nodes of the preceding layer on to all nodes of the following layer after multiplying them with the weights of the corresponding edge. At the initial stage, these weights are set to random values. Learning means giving the input-layer some input, comparing the output calculated by the network to the desired output. The objective is to construct an error function from measuring the difference between the desired output and network's output as a function of the network weights, and to modify the weights so as to minimize the error function. One achieves this by propagating the error back and adjusting the weights according to the so-called "back-propagation algorithm." By repeating this procedure with a large set of data, the error made by the network will consistently diminish, provided that the data members form clusters or any other appropriate configuration in the given n-dimensional space. This is related to the convergence properties of the network. A network with only one layer of weights would only be able to separate classes with linear hyperplanes. Our network consists of two layers of weights. This second layer makes it possible to classify with convex hulls [4].

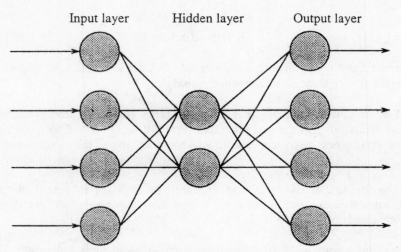

Figure 1: *A Multi layer Artificial Neural Network [4]*

According to Kowner [1], perfectly symmetric faces as well as extremely asymmetric ones are more likely to be evaluated as less attractive. So if the information representing the level of symmetry is presented to the perceptron network, one expects that the algorithm should converge. We distinguish only between two classes: the class of data members that we assume or simply agree as representing "beautiful faces" and the class of data members representing ordinary faces, by some reasonable standards. The points in pattern space representing the latter class should result in a cluster for perfectly symmetric faces, surrounded by an area of less density of points of this class, where the points representing beautiful faces outweigh. Moving farther away from the center there should be an increase in the density of members of the class of ordinary faces caused by the points representing faces with rather extreme asymmetries again. In this project we run a test to validate Kowner's hypotheses.

As mentioned before this artificial network needs a set of data to learn, i.e. to adjust its weights. So the data would include information about the face itself as well as whether it is considered as beautiful. For this purpose we have to rely on the perception of beauty of a comparatively small group of people. Consequently the intelligent agent will learn to judge according to the opinion of these people. What they consider beautiful will again strongly depend on their cultural and ethnic background. Still we can assume that this group's judgment, determined by averaging the individual judgments, is a good representative of what the part of society they belong to consider beautiful.

This artificial neural network will be fast at adding up and multiplying all the inputs and at giving us its judgment whether a face is beautiful or not, but it still depends on us providing it with the applicable type of data. Even if we use more sophisticated methods to represent faces than simple measurements of the degree of symmetry, at some point the machine will fail to recognize a face as such. We just have to think of smiling and crying faces, images of faces taken from the front and profiles, faces covered by a scarf or smoking a cigarette. Our brain won't have any difficulty recognizing these images as faces, but training a computer to fulfill this task is a much more complex issue. Trying to understand better how our brain works, in our example how it perceives images, will help achieve these goals. But up to now, many more questions regarding the human brain remain to be fully explored.

Remarks. [2] 1. The reasons for using only female faces in our study are: (a) mixing female and male images in one study complicates the analysis due to differences in weights of parameters entering in judgements of beauty in different genders. There are more subtle correlations between the two types that are as yet not well-understood. (b) It was simply easier to collect appropriate data for female faces.
2. We have not included any data that would compare systematically the response of human subjects versus the network, simply due to lack of time and resources, such as special permission for human subject studies. However, it is important and interesting to conduct such a study.
3. There is considerable anatomical evidence to support the hypothesis that the brain of human and other primates has specialized networks for face recognition (localized in the inferotemporal cortex IT). This ties the computational studies of symmetry in human faces to the biology more readily than symmetry of other objects. For us, this is another reason to favor faces versus other objects.

2 Measurements of facial degree of asymmetry

I wanted to find out if there is a correlation between symmetry and attractiveness of faces. So I used Kowner's [1] measurements of the facial degree of asymmetry, which is a modification of the method used by Grammer and Thornhill [8].

There are more sophisticated ways to represent faces. We will come across 2 other methods using deformable templates respectively wavelet analysis later on. These are much more complicated and I wanted to concentrate on the aspect of symmetry first.
First Krowner [1] defines 12 specific points in a face:
"The outermost and innermost eye corners",
"The right and left junctions where the lower part of the ear touches the head"
"The rightmost and leftmost points of the nose in the lower nose region"
"The top leftmost and rightmost points of the chin"
"The rightmost and leftmost points of the mouth"
He connects each pair of points with a straight line and calculates a midpoint with the following formula: "([left point-right point] /2)+right point"[1].

---

[2] We thank the referee for helpful comments. These remarks are in response to referee's suggestions.

To represent the faces as vectors, I chose these midpoints as coordinates and the distance between the outermost left and right eye corners for standardizing.

3 Another method to describe faces

Alan L. Yuille [2] describes "an approach for extracting facial features from images and for determining the spatial organization between these features using the concept of deformable templates".
He describes two different approaches. In the first one he uses global templates consisting of the basic features connected with springs. He adjusts these to the faces by changing the parameters of the template, which are the locations of the features. He defines a function containing of two parts: The first part is a measure for how well the single features actually fit. The second part is a cost function for the springs and is responsible for the spatial relations. Yuille maximizes this function and uses the values of parameters, at which the maximum is achieved, to represent the face.

In the second approach he uses templates for the eyes, the mouth, etc. He represents these features before concentrating on the spatial relations.

In both approaches the templates only explore the specified features and ignore the rest.
An advantage is that this method can be made robust, so that for example also a mouth smoking a cigarette can be recognized as a mouth.

4 Data and Principal Component Analysis

As data I took 16 faces from one ethnic group. I chose pictures of African American women from the December/January 2000 edition of the magazine Black Hair. All of them look straightforward; their mouths are closed and not smiling. Each face is represented by a vector, which consists of the measurements of the facial asymmetry degree as in Kowner[1].

I performed *principal* component analysis (PCA) [3] to find out which linear combination of the measured features carry the most statistically significant information and to exclude less important information. Principal components for a data set given by a collection of feature vectors are the unit eigenvectors of the covariance matrix of the data. PCA in this case could be interpreted as transforming the standard coordinate system into the one spanned by the *principal components* of the matrix consisting of the face vectors. The origin of this new coordinate system is the sample mean of the data. The covariance matrix is given by the formula $BB^T/(N-1)$ where B denotes the matrix consisting of the original vectors from which the sample mean is subtracted and N the dimension of the vector space. The first principal component is the eigenvector corresponding to the largest eigenvalue, which has largest variance. Up to now the data has not been compressed and no information is lost. In this new coordinate system the coefficients of the coordinates corresponding to the last few principal components will be negligible and we are going to suppress them.

This results in a new coordinate system of smaller dimension encoding the features that is more efficient than the original representation.

A Matlab function (prcoan.m) prepares the data as explained and then does the principal component analysis. First it calculates the covariance matrix. Then it carries out the singular value decomposition. After putting the eigenvectors in order according to the size of the associated eigenvalue, starting with the one belonging to the largest eigenvalue, it transforms the original data into the new coordinate system. Then it also calculates the percentage of the information contained by the principal components. In my example the first PC contains 87.7%, the second one 10.79 % and the remaining 3 each only less than 1% of the information. So I transformed my data into a plane spanned by the first two principal components.

I presented the network with these components, carrying 98.49% of the information, after adding another coordinate of the value 1 indicating that the faces used for the measurements and PCA were considered as beautiful. My analysis at this point is based on the assumption that the faces of models in fashion magazines are selected typically from beautiful women. If we change this assumption by having another collection of faces representative of perceived bias towards beauty, then the results may differ, while the procedures and algorithms remain the same.

5 Neural Network-Matlab implementation

The input layer and the first layer, which is a hidden layer, have as many nodes as the data dimensions. The second layer, which already is the output layer, consists of just one node. Only the values 1 and 0 are admissible. The target output tells if a face is considered beautiful (1) or ordinary (0).

The program consists of one main function neuraln.m. It calls the functions forward2.m, which multiplies for each node in each layer the inputs with the weights, sums these and passes these as new inputs to the next layer, and the function pback2.m. This function does the backwards calculation, which means it distributes the mistake made in one layer on the preceding layer using the weights and adjusts the weights according to this. I applied a modified version of the formula in Beale and Jackson [4] (Chapter 4.5, p.73-74) by simply using the value of the output of the corresponding node instead of the final output for the error terms for hidden units. A momentum term is also introduced. It controls the changes in the weights. As long as the error is large, bigger changes are made, but as the error decreases also the changes become smaller.

To test whether a face is considered beautiful by the network, you just have to transform the vector representing this face into the same coordinate system determined through PCA and rerun pforward2.m with the already found weight matrices. The output will be a number between 0 and 1.

6 Results

After the network learns from data, which means letting it adjust the weights, it should be able to recognize faces considered as beautiful by the hairstyle magazine as such. We presented it with the measurements of a completely symmetric face, the null vector transformed into the new coordinate system, and got an output of 0.9888. This emphasizes that a symmetric face is considered as beautiful by this network and consequently also following the trend of this magazine. But according to Kowner [4] perfect symmetry should have had a negative effect on facial attractiveness.

Analyzing the distribution of the points in the face vector space more carefully we might be able to detect a lower density around the point representing a perfectly symmetric face. The network's output value at that point still could be slightly smaller than the ones of surrounding points. In our example we only have 20 points in a plane, which won't display a very sophisticated pattern. Enlarging the data set might improve the result. At least it would narrow the ambiguity of possible interpretations. Using another way of representing faces than these measurements of the degree of asymmetry is another alternative.

We have to keep in mind that our method of representing faces only concentrated on a few measurements of bilateral symmetry; and so the sample size and the details of recording features are not representative of the realistic complexity of faces. It ignores a lot of other aspects as shading and proportions, just to mention a few.

When we look at a face we see much more than the features recorded above. We associate our experiences with people and their facial expressions, and these might have a positive or negative effect on the way we perceive them. If we meet a person that has similarities with somebody we didn't get along with in the past, we will probably judge him as less attractive. To mention another point, we performed our test on the ideally symmetric face with the measurements of a face that doesn't exist in real world. At least we haven't found it in our face collection and we probably haven't seen too many perfectly symmetric faces in our lives. Similarly, Kowner created synthetic symmetric faces by taking one hemisphere and flipping it vertically instead of using faces that really occur in nature, that we are confronted with in real life. In both cases, we have left unanswered the real question of how *natural* symmetry (as opposed to *synthetic* symmetry) and perception of beauty in the human society are combined.

7 Work in progress

To improve the results and to implement a larger data set, we are going to use 2-dimensional wavelet representation. Again we transform our data and, to compress it, we try to store only those coefficients that include most of the information about the images. These wavelet methods are similar to *Fourier* transformation.

In Fourier analysis [6] we describe an L^2-function with its Fourier series representation. Since the periodic functions $\{\sin(nx), \cos(nx), n \in N\}$ form a basis of L^2, we can represent the functions with the coefficients according to this new basis. The strength of this method lies in its ability to capture frequency and its weakness is in providing an efficient representation when the function varies non-periodically.

In wavelet analysis [6], we construct an orthonormal system spanning the whole space by taking one block-function, the mother wavelet, and translating and dilating it. The simplest one-dimensional mother wavelet is the *Haar function*,

$$\psi(x) = \begin{cases} 1, & 0 \leq x < \tfrac{1}{2} \\ -1, & \tfrac{1}{2} \leq x < 1 \\ 0, & \text{otherwise.} \end{cases}$$

L^2-functions can be approximated as precisely as necessary with a finite number of block functions of the derived orthonormal system.

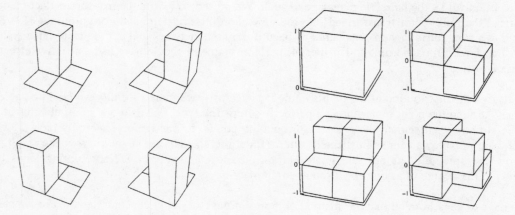

Figure 2: *The original block function basis [5]*   Figure 3: *Tensor-Product Haar Wavelet [5]*

In our case, we want to transform 2-dimensional matrices, whose entries measure the gray-scale of the pixels of our images. For simplicity of exposition, we mention only the simplest two-dimensional wavelets, namely, the *Tensor-Product Haar Wavelets*. They consist of the tensor products of the function $\psi(x)$ from above and the function $\varphi(x)$, which equals 1 on its support, the interval $[0,1)$.

The 4x4 matrix associated with a square-step function can be expressed in terms of the latter tensor products. Originally the entries of the matrix, the values of a pixel, represent the height of the block in the corresponding quadrant. The entries of the transformed matrix are a certain linear combination of all four blocks (see Figure 2 and 3) and allow the following interpretation:

The upper left-hand entry measures the average of the original entries.

The upper right-hand entry measures the horizontal change in the original entries.

The lower left-hand entry measures the vertical in the original entries.

The lower right-hand entry measures possible diagonal edges.

This transformation does not result in any loss of information.

To transform the matrix of our image we split it into 4x4 matrices and apply this method to them. Then we save the upper right-hand values of all the 4x4 sub-matrices in the upper right-hand corner of the original matrix, the upper left-hand values in the upper left-hand corner, etc. Then we continue with the upper left¼-hand quarter of the matrix, which now consists of the image with only ¼ of the pixel values. We can repeat this procedure, as in Figure 3, until we are down to one average value.

(a): Original Image

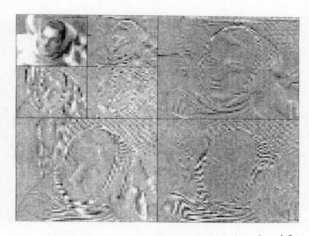

(b): Haar Wavelet Decomposition at level 2

Figure 4

After each step we gain a new matrix without any loss of information, but instead of giving only the values of the matrix we encode information about the "patterns of change" [5].

In our work in progress, we compare different wavelet decompositions for the images, to determine the criteria that selects the optimal wavelet family to save the most significant features, while it compresses the data through a minimal representation.

References

[1] Rotem Kowner, *Facial Asymmetry and Attractiveness Judgment in Developmental Perspective*, Journal of Experimental Psychology: Human Perception and Performance, Vol. 22, No.3, pp. 662-675. 1996.

[2]Alan L. Yuille, *Deformable Templates for Face Recognition*, Journal of Cognitive Neuroscience, Vol. 3, No. 1. 1989 – 1992.

[3]Lay, *Linear Algebra, Chapter 8: Symmetric Matrices and Quadratic Forms,* Addison-Wesley, 1997.

[4]Beale, *Neural Networks and Pattern Recognition in Human-Computer Interaction,* New York: Ellis Horwood. 1992.

[5]Yves Nievergelt, *Wavelets Made Easy*, Birkhaeuser, 1999.

[6]R. Todd Ogden, *Essential Wavelets for Statistical Applications and Data Analysis*, Birkhaeuser, 1997.

[7]David Beymer, Tomaso Poggio, *Image Representations for Visual Learning*, Science, Vol. 272, pp. 1905-1909. 1996.

[8]K. Grammer, R.Thornhill, *Human facial attractiveness and sexual selection: The role of averageness and Symmetry*, Journal of Comparative Psychology, Vol. 108, pp. 233-242. 1994.

[9]F. J. Galton, *Composite portraits*, Nature, Vol. 18, pp. 97-100. 1878

The Rubik's-Cube Design Problem

Hana M. Bizek
121 West Chicago Ave
Westmont, IL 60559
Email: hbizek@ameritech.net

Abstract

The design problem spans art and mathematics. Its subproblems make incursions into science. We begin by stating, what the problem is -- a set of algorithms that enable us to create a three-dimensional, composite, pleasant, geometrical design on a set of Rubik's cubes. We explain parity pairs, which influence design symmetry. We actually construct a simple design, the Menger Sponge. We conclude this article with some observations about fractals. The nature of the Rubik's cube makes it possible to create designs which mimic fractals.

1. The Name of the Game

1.1 Introduction. The goal of the design problem is to construct, by conventional cube manipulation, a composite, pleasant, geometrical design on a set of Rubik's cubes. Art enters the problem via these designs, which, if I may say so myself, are not unattractive. The math is brought into the problem by the cubes themselves, and by an acute necessity to develop certain algorithms based on mathematics. The idea of using the Rubik's cube as an art medium occurred to others on the World Wide Web. The major difference is that they create two dimensional, picture-like structures, whereas my designs are fully three dimensional, perhaps similar to sculptures.

1.2 The Basics. A solution algorithm for a Rubik's cube is independent of colors on its faces. This fact led David Singmaster, an English mathematician and a leading authority on the Rubik's cube, to devise a notation for its faces and rotations. He labeled faces according to their position on a fixed cube as F for front face, B for back face, U for up face, D for down face, L for left face and R for right face. A rotation of the up face, say, is labeled as U for a clockwise rotation by 90 degrees, U2 for a rotation by 180 degrees, and U' for a clockwise (counterclockwise) rotation by 270 (90) degrees. One may consult a number of books, e. g. [1], and [2].

There are three kinds of Rubik's cubes in a given three-dimensional design. The corner cubes form corners of a design and display three faces. For a rectangular-solid or cubical design there are always eight corners. The smallest object is made up of the corners only as a 2x2x2 larger design. The edge cubes are cubes, which form edges of a design and display two faces. Finally, the center cubes are cubes that form centers of a design and display just one face. My fellow artists, who create picture designs, work essentially with center cubes.

To summarize, the basic, unrefined, and fairly simple-sounding design algorithm is:
1. Construct patterns on individual cubes
2. Color-synchronize all cubes, so that they display some geometrical symmetry
3. Stack the cubes together in the manner of a three-dimensional jigsaw puzzle; the patterned cubes, created using the previous two steps, are the jigsaw pieces.

When refining the above steps, the mathematics of the Rubik's cube and the need to display geometrical symmetry on each and every face of the design, including the bottom face confront one. I tried to simplify the first step somewhat by breaking each pattern into a set of sequences. One needs a few sequences to construct quite a lot of patterns. Instead of memorizing each pattern one merely needs to know how individual pieces of the Rubik's cube (called cubies) transfer under the action of a given sequence. This method is described in Reference [3]. A color scheme is a fixed array of colors on opposite faces of a Rubik's cube. Once a color scheme is chosen, it remains the same throughout the design. All corner and edge cubes should have identical color scheme. For the center cubes it is sufficient to have the required colors somewhere on the cube.

1.3 Parity Pairs. Parity pairs play a major role in the design problem. Without them, the designs could not be of fewer than six colors and the design symmetry, as we know it, would not exist. Suppose we have two cubes of identical color schemes, such that the colors on the U, D, F and B face of one cube are identical to the colors on the U, D, F and B face of the other cube. If the color of the L/R face of one cube is identical to the color of the R/L face of the other cube, such a pair of cubes form a parity pair, as seen in Fig. 1 (a). The cubes in Fig. 1 (b) do not form a parity pair.

Let eight cubes form four parity pairs. Place two members of such a pair next to each other. Because a pair of opposite faces of one cube is switched relative to the other cube, the two internal faces of this 2-cube structure that touch, are colored the same, leaving only five colors on its six faces. We can form four such cube structures - one for each parity pair. Those four structures can be further combined to form two 4-cube structures; each has four colors on its combined six faces. They are the top and bottom layers of the 2x2x2 clean "design." Combining them will produce the 2x2x2 clean "design" of three colors only on its six faces. To see this, please obtain eight Rubik's cubes in four parity pairs and create this "design."

Such arrangement of cubes is used as corners in larger designs, leading to reflection invariant designs. Those are (usually) cubes that display the same design on their combined opposite faces. Reflection invariance is the simplest design symmetry induced by parity pairs. In Figures 2 and 3 we see two original designs with additional parity-pair requirements. The three-color Vasarely design, shown in Figure 4, could not be constructed without a judicious use of these pairs. Space limitations prohibit further discussion of this topic.

2. The Menger Sponge

2.1 Introducing the Sponge. The Menger Sponge, shown in Fig. 5, is not an original design, but is an adaptation of a classical fractal. The centers are clean. The color of the centers must be identical to the color of the center cubes of the adjoining edges and corners. Execute the F B' U D' R L' F B' sequence on suitably oriented Rubik's cubes and, if necessary, repeat it twice. The fact that we have created this fractal from Rubik's cubes leads to the following two questions:

1) Can other fractals be created from Rubik's cubes as designs?
2) Can we learn something new about fractals from the design problem?
The answer to the first question is yes. Two other fractals are shown in Figs. 6a and 6b. The answer to the second question is that I am going to try in the next section.

3. Fractal geometry

3.1 Fractal Properties. Fractal objects are of considerable interest. There is a lot of literature on fractals on the Web. But besides showing pretty computer-generated graphics, there is little else. No real mathematics surrounds fractals. Yet I believe time is ripe for such exploration. The mathematics developed over centuries deals with stable or nearly stable systems. To complement our understanding of natural phenomena we need to consider systems that are characterized by fractals.

Fractals have two main features, fractional dimensions and self-similarity. If one constructs a smaller and smaller version of a fractal, the smaller version bears a strong resemblance to the original fractal. This property is known as self-similarity. The notion of dimension is described in Reference [4]. The formula arrived at is:

$$D = \log (\text{number of pieces})/\log (\text{magnification})$$

where "number of pieces" is the number of constituent pieces into which an object has been subdivided. We divide the object into pieces which, when magnified by a certain factor called "magnification," give the original object.

3.2 Fractals and Rubik's-cube Designs. Some of the designs closely resemble fractals (the Menger Sponge). A real Menger Sponge [5] has holes, which we cannot drill into the Rubik's cubes without ruining them. We pretend the holes are there by defining a "hole color" and the "background color." For the Menger Sponge the color occupied by holes is the hole color, while the surrounding pieces constitute the background color. A gasket is a "solid" fractal. A carpet is a "flat" fractal. All fractals, shown in the figures, are gaskets. A corresponding carpet is readily obtained as a picture design. The zeroth iteration is the clean cube. The cube with a suitable pattern on it is the first iteration, or seed. Define a general rule of self-similar iteration. To go from nth iteration to n + 1st iteration, we do the following:

1) If the n-th iteration cubie has the color of the background, replace it by the seed.
2) If the n-th iteration cubie has the color of the hole, replace it by a clean cube having the hole color. In other words, "enlarge the hole."

This rule has self-similarity built into it. Let us apply it to the case of a Menger Sponge. To go from first to second iteration, we examine the cubies on the seed. We replace the surrounding cubies by the Rubik's cube with seed pattern on them. The center cubie is a hole so we replace it by the clean cube. The Menger Sponge Design, shown in Fig. 5, is the second iteration. To go from second to third iteration we proceed in exactly the same manner. The seeds will replace all background-color cubies. Nine Rubik's cubes replace nine cubies of the center cube. The center cubies of the surrounding eight cubes will be replaced by clean Rubik's cubes.

3.3 The Real Set. Just as rational and irrational numbers combine to give real numbers, one should combine a dimension of fractals and spaces with integer dimension. Call this combined set a real set. A unique dimension, which is a real number, characterizes each member of the real set. One does not think of ordinary spaces as self-similar, but they are. Let us apply a rule of self-similar iteration sketched above. For a point, we invert the seed for the Menger Sponge. Let the centerpiece be background and the surrounding 8 pieces be holes. Then the dimension is $\log(1)/\log(3) = 0/\log(3) = 0$. For the line let us choose the seed to be a Rubik's cube and do U2 D2 on it. The background color is taken by the cubies in the middle layer, while the holes are occupied by the colors in the U and D layer. Using this seed we get $\log(3)/\log(3) = 1$. A second iterartion would get nine cubies in the middle layer, but the magnification also increases to 9, yielding $\log(9)/\log(9) = 1$. By placing integer-dimension spaces and fractals into a real set, we can utilize what we know about real spaces to learn about fractals and vice versa. Fractals are not esoteric objects, but merely a part of the whole picture.

3.4 Fractal Processes. The box fractal shown in Ref. [4], p. 132, is a carpet. The seed is obtained on a face of a Rubik's cube by doing, say, L2 R2 F2 B2 U2 D2 to obtain a checkerboard pattern. By choosing the corner cubies and a center cubie as background colors, and the edge cubies as hole colors, we obtain the seed of the box fractal. Using the above rule for iteration (replace background cubies by seeds and hole cubies by clean cubes) we obtain the second iteration as shown in Ref. [4]. A box fractal gasket is shown in Fig. 6a. It is called box fractal A therein. The dimension is $\log(5)/\log(3)$ for a box fractal A carpet. A second box fractal in Fig. 6b is called box fractal B. Its seed is obtained by switching the hole colors and the background colors of the box fractal A. The second iteration of this fractal gasket is shown in Fig. 6b. The dimension of the box fractal B carpet is $\log(4)/\log(3)$.

Instead of iterating, we combine the second iterations of both box fractal A and box fractal B to produce a third fractal, the so-called checkerboard pattern (Fig. 7). The corresponding carpet would be a fractal of dimension $\log(41)/\log(9)$. This is a seed of the combined fractal, or its first iteration. The second iteration would be obtained the usual way, by taking each cubie of the background color and replacing it by the seed. Not by a single Rubik's cube, but by the whole 9-cube structure. Nine clean Rubik's cubes should accordingly replace each hole cubie. To sum up: a fractal process is a way to combine two simple fractals. The result of the combination is a third fractal which is self-similar and which has different fractional dimension. A second iteration could readily be constructed, either as a carpet or gasket, but would require 6561 and 531441 cubes, respectively! A computer programmed to take care of such iterations best carries out the second iteration. I myself would be curious to see the result.

What we will probably see is this: the lower the dimension, the less background there will be, and therefore less stability. On the other hand, the spaces with integer dimensions, too, have more stability with increasing dimensions. This is a pretty intuitive statement, but it stands to reason that a point will have a much smaller region of stability than a volume. We can combine these two observations and make the following statement: the lower the dimension (both integer and fractal) of a system is, the less stable that system will be. So, if we want to improve the stability of a system, we should strive for fractal processes whose dimension is as close as possible to an integer dimension, and translate those fractal processes to physical processes through, perhaps, the use of Mandelbrot sets. The case I have illustrated here is an idealized case, perhaps not suitable to any real physical system. But the possibility of doing that exists. One needs to utilize the computer and investigate these fractal processes.

I do not wish to imply that investigating fractal processes will be easy. This should probably be a task for the next millennium. Fractals occur around us in nature. How did they get there? Through some fractal process? Or through some process that may not be fractal? Do such processes exist? And what is the mathematics to handle them?

4. Conclusion

4.1 The Last Few Words. The design problem is multidisciplinary. It has a bit of everything: art, mathematics and science. Other subproblems to consider are to develop an algorithm and computerize the design problem for others to do. Some expert programmers should implement their skills and code the pertinent algorithms. Rubik's cube is sometimes cited as an example of cellular three-dimensional automata. Perhaps those problems can be studied with such a code.

Multidisciplinary problems shoud receive more attention in cutting-edge research. After all, Nature does not compartmentalize itself, man has compartmentalized Nature. Thus separate disciplines like physics, chemistry, biology, etc, etc, have eolved over the centuries, with their own specific rules. An attempt to cross those boundaries has been made with combination of fields, such as physical chemistry, biophysics, biochemistry and others. This attempt should be further widened. Study of fractal systems, for example, should cover many fields. Experts in those fields should benefit.

5. Acknowledgments

Thanks are due to Zdislav V. Kovarik, professor of mathematics at McMaster University, for telling me about this forum and giving me the pertinent web site. Ms. Bonnie S. Cady helped me with the formatting and introduced me to the intricacies of MS Word and picture handling. Mr. Josef Jelinek put together http://cube.misto.cz, a URL containing scanned photographs of 12 designs. And last, but not least, I would like to thank those who visited this URL. They kept the faith and kept me going. ☺

References

[1] D. Singmaster, *Notes on Rubik's Magic Cube*, Enslow Publishers, 1981
[2] D. Taylor and L. Rylands, *Cube Games*, Holt, Rinehart and Winston, 1981
[3] Hana M. Bizek, *Mathematics of the Rubik's Cube Design*, Dorrance, 1997
[4] Robert L. Devaney, *Chaos, Fractals and Dynamics*, Addison-Wesley, 1990
[5] Heinz-Otto Peitgen, Hartmut Jurgens, Deitmar Saupe, *Chaos and Fractals New Frontiers of Science*, Springer-Verlag, 1990

List of Figures

Figure 1: (a) These cubes form a parity pair   (b) These cubes do not form a parity pair.

Figure 2: Marie Design   **Figure 3**: Jaroslav Design

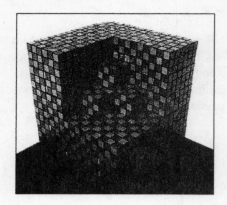

Figure 4. Vasarely Design

Figure 5. The Menger Sponge Design

Figure 6a. Box Fractal A

Figure 6b. Box Fractal B

Figure 7: The Checkerboard

Mathematics and Art: Bill and Escher

Michele Emmer
Dipartimento di matematica
Università di Roma "La Sapienza", 00185 Roma, Italy
email:emmer@mat.uniroma1.it

Introduction

Mathematicians pay considerable attention to the aesthetic qualities of their discipline; this explains why many of them think that mathematical and artistic activities are to some extent very similar, or comparable. Such an attitude also explains in part why mathematicians have periodically not only attempted to explain the fascinating beauty of their discipline, but also to interpret and analyze art through mathematics. On the other hand, there are many cases of artists, from different periods, who have been fascinated by mathematics and have tried to use its ideas and techniques in their work. And some artists have indeed been forced to become mathematicians, as happened during the Renaissance.

Renaissance painters turned to mathematics not only because they had the problem of depicting the natural world realistically on canvas, of producing scenes in three dimensions with depth, but also, as Morris Kline has pointed out in his important book on mathematics in western culture [1], they were profoundly influenced by the rediscovery of Greek philosophy. They were wholly convinced that mathematics was the true essence of the physical world and that the universe was ordered and explainable in geometric terms. This great interest forced Renaissance painters to become – as Kline defined them – the best applied mathematicians of the period. Since the professional mathematicians of that time did not have the geometric instruments which the artists needed, they themselves also had to become learned and active theoretical mathematicians.

"Even though there still exist some pseudo-humanists who are proud of their non-comprehension of mathematics, the growing number of laymen who regret not being able to take part fully in this banquet of the Gods is quite reassuring." This is the opening of the preface to the book *Les Grands Courants de la Pensée Mathématique*, the preface of which was written by the mathematician François Le Lionnais [2]. The aim of the book was to attempt to show "not the immobile panorama of the sectors pertaining to mathematics, but first and foremost the direction towards which the diverse mathematical disciplines are moving." In the chapter on *Arts et Esthétique: Les Mathématiques et la Beauté*, Le Lionnais replies to those who would reduce the relationship between mathematics and art to a question of proportions and numbers: "In mathematics there exists a beauty which must not be confused with the possible influence of mathematics on the beauty of the works of art. The aesthetics of mathematics must be clearly distinguished from the applications of mathematics to aesthetics."

The words used by Le Lionnais to describe the beauty of mathematics are no different from those used by many other mathematicians: "And so, beauty shows itself in mathematics just as it shows itself in the other sciences, in the arts, in life and in nature. While at times comparable to pure music, to great painting or poetry, the emotions that mathematics arouse are of a different nature that cannot be understood unless one has had a sort of illumination within oneself. The beauty of mathematics is certainly no guarantee of truth or usefulness. But it enables some people to enjoy a unique experience, while for others it provides the certainty that mathematics will continue to be practiced for the benefit of everyone and for the glory of the human adventure."

I would like to mention two artists who have attracted my interest in the study of a possible relationship between art and mathematics: Max Bill and M. C. Escher. I have made documentary films about both artists' works. I worked directly with Bill, while in Escher's case, I used his works without ever having met him. I will try to point out several features that these two artists have in common [3], [4].

Max Bill's mathematical approach to art

There is no doubt that the clearest approach to the possibility of a mathematical approach to the arts has been formulated by the famous Swiss artist Max Bill. In 1949 he wrote: [5] "By a mathematical approach to art, it is hardly necessary to say I do not mean any fanciful ideas for turning out art by some ingenious system of ready-reckoning with the aid of mathematical formulas. So far as composition is concerned, every former school of art can be said to have had a more or less mathematical basis. Even in modern art, artists have used methods based on calculation, inasmuch as these elements, alongside those of a more personal and emotional nature, give balance and harmony to any work of art."

These methods had become more and more superficial, for the artist's repertory of methods had remained unchanged, except for the theory of perspective, since the days of ancient Egypt. The innovation occurred at the beginning of the twentieth century: "It was probably Kandinsky who gave the immediate impulse towards an entirely fresh conception of art. As early as 1912... Kandinsky in his book *Ueber das Geistige in der Kunst* [6] indicated the possibility of a new direction which would lead to the substitution of a mathematical approach for improvisations of the artist's imagination... It is objected that art has nothing to do with mathematics; that mathematics, beside being by its very nature as dry as dust and as unemotional, is a branch of speculative thought and as such in direct antithesis to those emotive values inherent in aesthetics... yet art plainly calls for both feeling and reasoning."

We must not forget that Max Bill was first and foremost a sculptor who believed that geometry, which expresses the relations between positions in the plane and in the space, is the primary method of cognition, and can therefore enable us to apprehend our physical surroundings, so, too, some of its basic elements will furnish us with laws to appraise the interactions of separate objects, or group of objects, one to another. And again, since it is mathematics that lends significance to these relationships, it is only a natural step from having perceived them to desiring to portray them. Visualized presentations of that kind have been known since antiquity, and they undoubtedly provoke an aesthetic reaction in the beholder. In the search for new formal idioms expressive of the technical sensibilities of our age, these borderline exemplars had much the same order of importance as the "discovery" of native West African sculpture by the Cubists.[7]

And here is the definition of what must be a mathematical approach to the arts: "It must not be supposed that an art based on the principles of mathematics, such as I have just adumbrated, is in any sense the same thing as a plastic or pictorial interpretation of the latter. Indeed, it employs virtually none of the resources implicit in the term "pure mathematics". The art in question can, perhaps, best be defined as the building up of significant patterns from the ever-changing relations, rhythms and proportions of abstract forms, each one of which, having its own causality, is tantamount to a law unto itself. As such, it presents some analogy to mathematics itself where every fresh advance had its "immaculate conception" in the brain of one or other of the great pioneers.

Thus Euclidean geometry no longer possesses more than a limited validity in modern science, and it has an equally restricted utility in art. To convince his readers, after having clarified his thoughts, Bill needed to provide some examples which pertained to his point of view as an artist – examples of what he called "the mystery enveloping all mathematical problems", "the inexplicability of space – space that can stagger us by beginning on one side and ending in a completely changed aspect on the other, which somehow manages to remain that self-made side; the remoteness or nearness of infinity – infinity which may be found doubling back from the far horizon to present itself to us as immediately at hand; limitations without boundaries; disjunctive and disparate multiplicities constituting coherent and unified entities; identical shapes rendered wholly diverse by the merest inflection; fields of attraction that fluctuate in strength; or, again, the space in all its robust solidity; parallels that intersect; straight lines untroubled by relativity, and ellipses which form straight lines at every point of their curves.

For though these evocations might seem only the phantasmagorical figments of the artist's inward vision, they are the projections of latent forces... Hence all such visionary elements help to furnish art with fresh content. Far from creating a new formalism, what these can yield is something far transcending surface values since they not only embody form as beauty, but also form in which intuitions or ideas or conjectures have taken visible substance.

It may be contended that the result of this would be to reduce art to a branch of metaphysical philosophy. ... I assume that art could be made a unique vehicle for the direct transmission of ideas because, if these were expressed by pictures or plastically, there would be no danger of their original meaning being perverted by whatever fallacious interpretations. Mental concepts are not as yet directly communicable to our apprehension without the medium of language; though they might ultimately become so by the medium of art.

Figure 1: *Max Bill, Endless ribbon, granite, 1935-1953.*

Thus the more succinctly a train of thought was expounded and the more comprehensive the unity of its basic idea, the closer it would approximate to the prerequisites of the mathematical way of thinking. The orbit of human vision has widened and art has annexed fresh territories that were formerly denied to it. In one of these recently conquered domains, the artist is now free to exploit the untapped resources of that vast new field of inspiration. And despite the fact that the basis of this mathematical way of thinking in art is in reason, its dynamic content is able to launch us on astral flights which soar into unknown and still uncharted regions of the imagination."

This is very clear praise by a great artist for the artistic quality of certain aspects of mathematics. When mathematicians think about the beauty of mathematics, they generally have in mind examples of a type that are not comprehensible to non-specialists. However, the artist states that not only does mathematics provide aesthetically relevant models, but also that it has deep cultural influence. And this is why a mathematical route to art is possible.

One of the visual ideas that Max Bill used without realizing it was the Moebius ribbon. Max Bill called some of his sculptures *Endless Ribbons* and in fact they are shaped like Moebius ribbons. In his article *How I began to make single-faced surfaces*, Bill recounts how he discovered the Moebius surfaces [8]: "Marcel Breuer, my old friend from the Bauhaus, is the real originator of my single-faced sculptures. This is how it happened: in 1935 in Zurich, together with Emil and Alfred Roth, I was building the Doldertal houses which in those days were much talked about. One day, Marcel told me he had been commissioned to design a house for an exhibition in London. It was to be the model of a house in which everything, even the fireplace, would be electric. It was obvious to all of us that an electric fireplace, which glowed without flames, would not be a particularly attractive object. Marcel asked me if I would like to make a piece of sculpture to be hung above it. I began looking for ideas – a structure that could be hung over a fireplace and which might even turn in the upward flow of hot air. With its shape and movement, it would in a sense act as a substitute for the flames. Art instead of fire! After many experiments, I came up with a solution that seemed reasonable."

The interesting thing to note is that Bill thought he had invented a completely new shape. Even more curious was that he discovered it by twisting a strip of paper, just as Moebius had done many years previously.

"Not long afterwards, people began to congratulate me on my fresh and original interpretation of the Egyptian symbol for infinity and the Moebius ribbon. I had never heard of either of them. My mathematical knowledge had never gone beyond routine architectural calculations, and I had no great interest in mathematics."

Max Bill's *Endless Ribbon* was put on display for the first time at the Milan Triennale exhibition in 1936. "Since the 1940s – wrote Bill – I had been thinking about problems of topology. From them I developed a sort of logic of shape. There were two reasons why I kept on being attracted by this particular theme: 1) the idea of an infinite surface – which was nevertheless finite – the idea of a finite infinity; 2) the possibility of developing surfaces that – as a consequence of the intrinsic laws implied – would almost inevitably lead to shapes that would prove the existence of the aesthetic reality. But both 1) and 2) also indicated another direction. If non-oriented topologic structures could exist only by virtue of their aesthetic reality, then, in spite of their exactness, I could not have been satisfied by them. I'm convinced that the basis of their efficacy lies in part in their symbolic value. They are models for contemplation and reflection."

Bill had the idea of setting up a room for his topological sculptures in the topological section of the permanent mathematics exhibition in London's Science Museum. Unfortunately, the project was never carried out.

Figure 2: *Max Bill, Kontinuität, granite, 1986; from the book W. Spies, Kontinuität. Granit-Monolith von max Bill, Deutsche bank, 1986.*

The graphic artist M.C. Escher (1898-1972)

The Dutch artist Maurits Cornelis Escher was born in 1898. Escher's tale is quite unusual: for a long time, his work was almost completely unknown and unappreciated. After almost 20 years spent in Italy, he had shown only a handful of times. At a certain point in the Sixties, however, his fame began to grow, among scientists, mathematicians, physicists, and crystallographers. The history of the relationship of Escher with scientists, mathematicians especially, is quite interesting in terms of understanding how the Dutch artist thought of his work. Escher defined himself, and not unreasonably, a graphic designer. The fundamental event in his life, he said, was in 1938, when the Escher family had already left Italy after a long stay. "In Switzerland, Belgium and Holland... I found the landscape and architecture much less interesting than in Southern Italy. I felt driven to distance myself ever more from more or less direct and realistic illustration of the surrounding reality. There is no doubt that these unusual circumstances were responsible for my having brought to light these 'interior visions.'"

All the illustrations in his first book, except for the first seven works, were done with the idea of communicating a detail of these interior visions. "The ideas that are basic to them often bear witness to my amazement and wonder at the laws of nature which operate in the world around us. He who wonders discovers that this is in itself a wonder. By keenly confronting the enigmas that surround us, and by considering and analyzing the observations that I had made, I ended up in the domain of mathematics. Although I am absolutely without training or knowledge in the exact sciences, I often seem to have more in common with mathematicians than with my fellow artists." [9]

These last two comments by Escher summarize perfectly the privileged relationship that the artist established with the scientific community. That Escher himself thought of scientists as his privileged audience is beyond doubt.

The title of the first monographs on Escher "The world of M.C. Escher" [10] is very appropriate. Escher created his own world of meticulously constructed images; a world at once fantastic, imaginative and magic but also realistic, coherent and detached; observed with an eye apparently lacking in emotion. Escher was an artist of minute details, tiny elements that create instability and are disturbing the apparent calm of the whole.

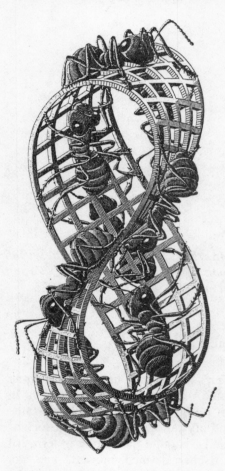

Figure 3: *M.C. Escher, Moebius Strip II, wood engraving, 1963;by the kind permission of the M.C. Escher Foundation. Baarn. Holland. All M.C. Escher works © Escher Heirs c/o Cordon Art, Baarn, Olanda. All right reserved*

The first reason for using a movie camera to film Escher's prints is that cinematic technique first of all lets us concentrate the attention on what the filmmaker wants to be looked at. The film forces us to look at Escher's world as if we were part of it. We are not distracted by anything. The second reason is that Escher's drawing technique is very precise and detailed. Many of these details are very minute despite being very precisely drawn. Cinema technique allows us both to isolate these details and also magnify them many times in order to appreciate from close up the precision of the artist's method. Many of Escher's works are like a story that develops. They must be observed in the sequence suggested by Escher himself. This is why the movie camera permits a very precise and accurate analysis of Escher's works.

A very significant example, which sums up the various aspects, is the "Metamorphoses"; in particular "Metamorphosis II" which immediately seemed to me a perfect cinematic sequence. I have used the animation of this Escher's work as the final sequence of the video.

Escher made the following comments to this work: [11]: "First the black insect silhouettes join; at the moment when they touch, their little white background has become the shape of a fish. Then figures and background change places and white fish can be seen swimming against a black background. A succession of figures with a number of metamorphoses acquires a dynamic character. Above I pointed out the difference between a series of cinematographic images projected on a screen and the series of figures in the regular division of plane. Although in the latter the figures are shown all at once, side by side, in both cases the time factor plays a role. "

In 1964 Escher was visiting his son George Escher in Canada. He was invited by several organizations in the USA, including the MIT and the Bell Laboratories, to give presentations on his work. Shortly after arriving in Canada, Escher had to be admitted to Saint Michael's hospital in Toronto for an emergency operation. All appointments had to be canceled, and Escher would never again have a chance to give his carefully prepared lectures.

Escher had written the complete English text of his lectures and these texts have been preserved. In a book published in 1989 "Escher on Escher: exploring the infinite"[12] all these materials have been published. The chapter is entitled: "The lectures that were never given". The final part of his talk was dedicated to "Metamorphose II": "I propose to round off this talk by showing you a woodcut strip with a length of thirteen feet. It's much too long to display in one or even in two slides, so I had it photographed in six parts, which I can present in three successive pairs and which you are invited to look at as if it were one uninterrupted piece of paper. It is a picture story consisting of many successive stages of transformations. The word "Metamorphose" itself serves as a point of departure. Placed horizontally and vertically in the plane, with the letters O and M as points of intersection, the words are gradually transformed into a mosaic of black and white squares, which, in turn, develop into reptiles. If a comparison with music is allowed, one might say that, up to this point, the melody was written in two-quarter measure.

Now the rhythm changes: bluish elements are added to the white and black, and it turns into a three-quarter measure. By and by each figure simplifies into a regular hexagon. At this point an association of ideas occurs: hexagons are reminiscent of the cells of a honeycomb, and no sooner has this thought occurred than a bee larva begins to stir in every cell. In a flash every adult larva has developed into a mature bee, and soon these insects fly out into space.

The life span of my bees is short, for their black silhouettes soon merge to serve another function, namely, to provide a background for white fishes. These also, in turn, merge into each other, and the interspacing takes on the shape of black birds. Then, in the distance, against a white background, appear little red-bird silhouettes. Constantly gaining in size, their contours soon touch those of their black fellow birds. What then remains of the white also takes a bird shape, so that three bird motifs, each with its own specific form and color, now entirely fill the surface in a rhythmic pattern.

Again simplification follows: each bird is transformed into a rhomb, and this gives rise to a second association of ideas: a hexagon made up of three rhombs gives a plastic effect, appearing perceptively as a cube. From cube to house is but one step, and from the house a town is built up. It's a typical little town of southern Italy on the Mediterranean, with, as commonly seen on the Amalfi coast, a Saracen tower standing in the water and linked to shore by a bridge. (It is the town of Atrani)

Now emerges the third association of ideas: town and sea are left behind, and interest is now centered on the tower: the rock and the other pieces on a chessboard. Meanwhile, the strip of paper on which "Metamorphose" is portrayed has grown to some twelve feet in length. It's time to finish the story, and this opportunity is offered by the chessboard, by the white and black squares, which at the start emerged from the letters and which now return to that same word "Metamorphose".

This was the end of the lecture of Escher. It is a story he was telling based on his woodcut "Metamorphose". When I was making my video I was not aware of the text of this lecture of Escher. But my idea was exactly the same described by Escher, perhaps more: not only a story but what in cinema is a story-board, a precise description of a sequence to be filmed, made usually by drawings and words. So I have used Escher's original engraving as a storyboard for making the animation of "Metamorphose". What Escher has to described by words, because it was impossible to show the complete work to the audience, it is described in the movie without any words: just the animation, following the time of changing of the various forms in the woodcut, and music, another suggestion of Escher himself! To paraphrase Escher we can say: In this film they are the images and not the words that come first. [13]

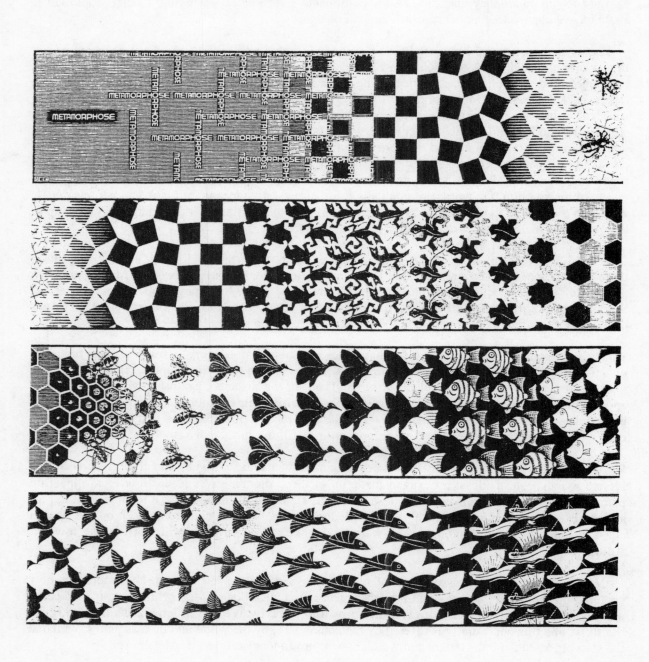

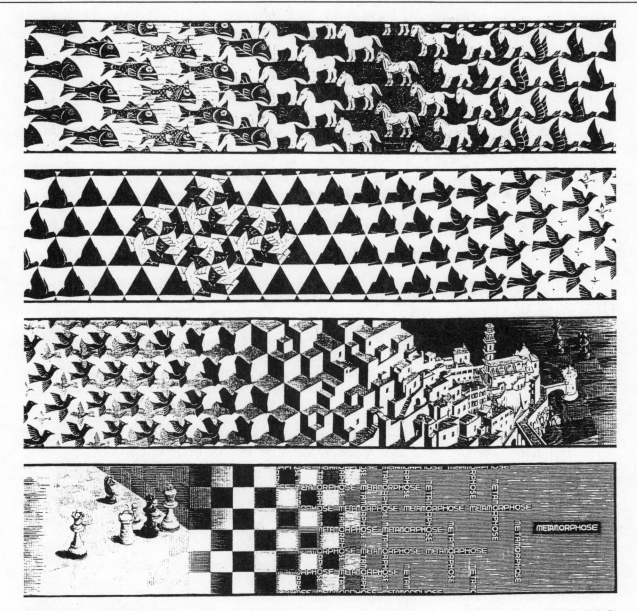

Figure 4: *M. C. Escher, Metamorphose, xylogravure, 1967/68. by the kind permission of the M.C. Escher Foundation. Baarn. Holland. All M.C. Escher works © Escher Heirs c/o Cordon Art, Baarn, Olanda. All right reserved*

Concluding remarks

A new phenomenon that appeared only a few years ago is modifying the quality of the relationship between mathematics and art. Even granting that mathematics is an art, and that it is possible to apply mathematical criteria in art, and thus to build a mathematical approach to art, art nonetheless remains something different from mathematics, and no mathematician expects to be remembered as a great artist. Thanks to the explosion of technology and the increasingly sophisticated techniques offered by computer graphics, over the past few years it has become possible to visualize mathematical phenomena whose existence could hardly have been otherwise supposed. "And since this science possesses these fundamental elements and puts them into meaningful relationships, it follows that such facts can be represented, or transformed into images...which have an unquestionably aesthetic effect." So wrote Max Bill in 1949. Do his words apply to the images created by mathematicians in recent years?

The new fact is that some mathematicians not only continue to claim that their discipline is an art, but also now wish to be thought of as artists in their own right. Over the past several years, there has been an enormous output of mathematical images, of objects resulting from extremely disparate mathematical studies. Some examples are highly interesting from both the strictly mathematical and aesthetic standpoints, at least according to the mathematicians who developed them. I do not mean at all to say that if it is possible to speak of progress in mathematical knowledge, then this must imply an idea of improvement in the domain of art, where an argument of this kind makes no sense. At any rate, we can expect further changes in the relationships between mathematics and art, and I may say so, most of the credit must go to the mathematicians. [14]

References:

[1] Morris Kline, *Mathematics in Western Culture*, Oxford Univ. Press, 1965

[2] François Le Lionnais, a cura di, *Les grands courants de la pensée mathématique*, Librairie Scientifique et Technique A. Blanchard, Paris, 1962.

[3] M. Emmer, *Moebius Band,* video, 27 minutes, color, sound, series *Art and Mathematiques*, Rome, Italy

_____, *Visual Art and Mathematics: the Moebius Band*, Leonardo, vol. 13 (1980), pp. 108-111.

[4] M. Emmer, *The Fanstatic World of M.C. Escher*, video, 50 minutes, color, sound, series *Art and Mathematics*, Rome, Italy.

[5] Max Bill, *Die mathematische denkweise in der kunst unserer zeit*, Werk, n.3 , 1949; reprinted in english with corrections by the author in M. Emmer, ed., *The Visual Mind*, Boston, MIT Press, 1993.

[6] Wassily Kandinsky, *Ueber das Geistige in der Kunst* , R. Piper & Co., Monaco, 1912.

[7] W. Rubin, ed., *Primitivism in 20th Century Art*, Museum of Modern Art di New York , New York,1984.

[8] M. Bill, *Come cominciai a fare le superfici a faccia unica,* 1972, in A. C. Quintavalle, ed., *Max Bill, catalogo della mostra*, Parma, 1977, pp. 23-25.

_____, *Endless Ribbon,* in *Mathematics Calendar*, June, Springer-Verlag, Berlin, 1979.

[9] M.C. Escher, *The Graphic Work* , Macdonald, London, 1967.

[10] J.L. Locher, ed. ,*The World of M. C. Escher*, H. N. Abrams, Inc., New York, 1971.

[11] M.C.Escher,*Regelmatige Vlakverdeling*, De Roos Foundation, Utrecht, 1958; reprinted in J.L. Locher, ed., *La vie et l'Oeuvre de M.C. Escher*, Chêne, Paris, 1981.

[12] M.C. Escher, *Escher on Escher: exploring the infinite*, H. N. Abrams, New York, 1989, pp. 48-53.

[13] M. Emmer, *Movies on M.C. Escher and their Mathematical Appeal*, in H.S.M. Coxeter, M. Emmer, R. Penrose, M. Teuber, eds., *M.C. Escher: Art and Science*, North-Holland, Amsterdam, 1986. pp. 249-262.

[14] M. Emmer, *La perfezione visibile*, Theoria, ed., Roma, 1991.

_____, Matematica e Estetica, Laterza ed., Roma, to appear.

_____, ed., *The Visual Mind 2 :Art and Mathematics*, MIT Press, Boston, in preparation.

Bridges, June Bugs, and Creativity

Daniel F. Daniel
Integrative Studies Program
Southwestern College
Winfield, KS 67156
ddaniel@sckans.edu

Gar Bethel
212 Iowa
Winfield, KS 67156

Curiosity under Control?

As we contemplate a third gathering for the Bridges Conference, I am delighted to provide a brief commentary on Gar Bethel's poem about a four year old friend who is an occasional charge of his.

If we could begin to ground the conference in something that even approaches the wonder and awe of the world we experience through children, we would have an exciting first step onto the bridge that is before us. This is of course what those who continue to work in these fields, at once so various and wonderful yet for some slouching toward unsuspected convergences – have not lost.

Mathematics, sciences, and arts of all kinds continue to be awe-full and enchanted. Last year I recall a session in which a young artist proposed the building of a bridge as an icon for the conference. Within minutes, a half dozen of the participants began to graphically and verbally play with the idea. One of the members of that group went well beyond the initial play and his paper on a bridge for the Bridges Conference may be found in *Bridges' 2000*. (See "–To Build a Twisted Bridge –" by Carlo Séquin)

The poet (the maker or craftsman) comments on the child as he confronts wonders like the coffee maker, the coffee grinder, a microwave oven, and a flashlight. Educator and neurologist Howard Gardner reminds us that young children have always posed fascinating questions about the world. "Children in the first five to ten years have ample opportunity to let their imaginations roam, to raise questions about phenomena that inspire doubt or awe; and then, at least sometimes, to pursue these questions for a while as they walk in the fields, or fall asleep at night. (*Creating Minds*, 1993, p.88)

This observation, at once so mundane and so difficult to honor as we educators work with children, provides the introduction to Gardner's chapter on the physicist, titled *Albert Einstein: The Perennial Child*. Here Einstein muses,

How did it come to pass that I was the one to develop the theory of relativity? The reason, I think, is that a normal adult never stops to think about problems of space and time. These are things which he has thought of as a child. But my intellectual development was retarded, as a result of which I began to wonder about space and time only when I had already grown up. Naturally I could go deeper into the problem than a child with normal abilities. (*Creating Minds*, 89)

For those who might be regretting your lack of a parallel retardation as a child, I.I. Rabi, a contemporary of Einstein's arrives in Gardner's text to insist: " I think that physicists are the Peter Pans of the human race. They never grow up and they keep their curiosity. Once you are so sophisticated, you know too much – far too much."(*Creating Minds*)

The poet is content to offer space and to let the child give up on Dr. Seuss in favor of the wonders of a flashlight on the ceiling and in an ear at bedtime. He endures and appears to enjoy the anatomy investigation via the blunt instrument of the flashlight as the child moves from ceiling to wall to the tympanum of an inner ear, "where skin and bone make a tiny drum." In an effort to instruct and to lead the child to sleep, the poet sings a song. The song, from roughly the same era as inventor Nikola Tesla, is "The Band Played On." I introduce Tesla, inventor of fluorescent lighting, the bladeless turbine and much more, only to remind us that these experiments don't always bring wisdom or joy.

Once Tesla perched on the roof of the barn, clutching the family umbrella and hyperventilating on the fresh mountain breeze until his body felt light and the dizziness in his head convinced him he could fly. Plunging to earth, he lay unconscious and was carried off to bed by his mother. His sixteen-bug-power motor, was, likewise, not an unqualified success. This was a light contrivance made of splinters forming a windmill, with a spindle and pulley attached to live June bugs. When the glued insects beat their wings, as they did desperately, the bug-power engine prepared to take off. This line of research was forever abandoned, however, when a young friend dropped by who fancied the taste of June bugs. Noticing a jarful standing near, he began cramming them into his mouth. The youthful inventor threw up. (Cheney, *Tesla*: *Man Out Of Time*, 1981, p.7)

Physicist Sheldon Glashow echoes this curiosity when he recalls the moment that he "suddenly realized you *could* turn an inner tube inside out through a hole in its side. Things like that were so much more interesting to me than American history or English literature."(*Interactions*, 1988, p.35). In this review of contemporary physics, he says, "Science has been my life because it is the systemization of curiosity. When I was a child growing up in Manhattan, I wanted to know how a car or a clock worked, what a rainbow was, and why an uncooked egg could not be made to spin. My greatest discovery was that science can be more than a mere hobby. It could be my profession. People would actually pay me to do what I most wanted to do: to satisfy my own curiosity."(*Interactions*, p.*xi*) How would one measure the distance between the inner tube and the quark; the umbrella jump and the turbine; the passage of light through skin and the "schizoid nature" of light itself? (Or of light themselves?)

And finally the mathematician who seemed to many to remain a child well into his eighties, Paul Erdös. Famous for his eternal quest on behalf of mathematics and for the company of mathematicians, Erdös could not be bothered by the world's products. "Some French socialist said that private property was theft. . . .I say that private property is a nuisance." To the man for whom a child was an *epsilon* and those who had stopped producing mathematics were said to have died (those who had died were said to "have left"), Erdös was obsessed by the reality which mathematics revealed to him. He constantly sought out new and younger collaborators.

With 485 co-authors, Erdös collaborated with more people than any other mathematician in history. Those lucky 485 are said to have an Erdös number of 1, a coveted code phrase in the mathematics world for having written a paper with the master himself. If your Erdös number is 2, it means you have published with someone who has published with Erdös. If your Erdös number is 3, you have published with someone who has published with someone who has published with Erdös. Einstein had an Erdös number of 2, and the highest known number is 7. (Hoffman, *The Man Who Loved Only Numbers*, 1998, p.13)

It is this sense of play which continues to honor one of the most child-like of mathematicians we have seen.

While these are in many ways aberrant illustrations of something difficult to capture, we continue to believe that curiosity can also be democratized in adulthood as it is in childhood. We have a two-semester course in theoretical and applied creativity in our integrative studies program at Southwestern College. We continue to hope to find the button – which once pushed, or the metaphor — which once spoken, or the theorem — which once solved will unlock the wonder which seems often to have fled from the childless land of undergraduate life and work. Recalling the often repeated story of mathematician David Hilbert who, when told of a student who had left mathematics to become a poet, replied: "Good. He didn't have the imagination to be a mathematician." Someday the walls dividing all the academic fields represented at this conference will be understood as the human fictions they surely are. And we will all understand that we were each a scientist at age four. And more. So much more.

Science at Four

Before we go to bed, he helps
me make my morning coffee –
a paper filter in a plastic basket,

three spoons of beans ground up
in the dangerous *noise* machine,
and he helps me pour the water in –

but we can't turn it on till morning.
What happens inside, we can't see.
It's like a human brain thinking-

in goes a simple element
and out comes a dark stimulant –
no food value, just a mildly

addictive drug to get you started.
What happens inside is a mystery.
But that's just my way of thinking.

He's interested in the simple
mechanics, like those of the *micro wave*.
He doesn't drink. It's just pure science.

In bed the fantasies of Dr. Seuss
are given short shrift to the rechargeable
flashlight in the dark. He's in a hurry

to perform his experiments –
examining the tiny bulb and large
reflector, shining the light

on the ceiling with its concentric
Saturn circles and counting each one,
waving the light in squiggly lines

so, as he says, he can write his name,
though he can't yet write any name.
With closed fingers and positioned thumb

I make the shadowed head of a goose,
complete with the proper sound effects.
He moves the shadows of his fingers

on the wall – the size of a giant's,
and he makes a tent of the covers
and basks in the light's reflected glow.

Then he examines my ear hole
and hears of the dangers of poking there
where skin and bone make a tiny drum.

He shines in my nose and sees hairs.
He shines in my mouth and sees my tongue
and teeth and what seems like two holes.

I close my eyes, and he sees eyebrows
and tender lids and lashes that appear
to form a tear at one of the corners.

Then the struggle for control begins.
It's time to shut off his light and sleep.
But at four he's learning to assert

his will, and I am unwilling
to give up my benevolent force.
Of course, we both live under illusion.

No one in the long run has control,
and very little in the short haul.
The light off, we listen to the rain.

I tell him it's becoming so cold
in the morning the grass will be white.
He asks for his current favorite song,

one that was popular at the turn
of this century, just as this
century is about to turn-

and I give him its anapestic rhythm
in two separate tempos –
slow and deliberate, fast and lively –

Casey would waltz with the strawberry
blond, and the band played on.
He'd glide o'er the floor with the girl

he adored, and the band played on.
His brain was so loaded it nearly
exploded. The poor girl would shake

with alarm. He married the girl
with the strawberry curls,
and the band played on – without knowing

he was learning the tiers of metaphor –
from *strawberry* for a blond, and curls,
its variation, to exaggerated metaphors

like *loaded* and *exploded* for dramatic
effect, to a tiny philosophy for all our
lives – *and the band played on* –

all in a plot of courtship and marriage –
and life goes on and on and on.
And the universe had expanded

when we woke. Typical of his thinking,
even before he checked the cartoons
on TV, he wanted to know

if the grass was white, so he opened
the blinds, and I was both right
and wrong. It wasn't frost. It was snow.

He immediately wanted to go out
and feel, feel all that white,
which meant the production of helping him on

with his socks, so he could pull them up,
and tying his shoes with a knot,
and as he instructed – though he could not

do it himself – adding another knot,
a bow, and then helping him start
the zipper on his hooded jacket

which is, if you think, a series of tiny knots,
as we are one knot together,
in his life, in my life – in life,

though that's only a theory.

Saccades and Perceptual Geometry:
Symmetry Detection through Entropy Minimization

Hamid Eghbalnia and Amir Assadi
Department of Mathematics
University of Wisconsin – Madison
Madison, WI 53706-1338, USA

Introduction

The world surrounding us is composed of objects. Object recognition is an important factor in our survival and functionality in the world. Through our senses, we learn the properties of objects and distinguish them from one another. We learn, in particular, that objects occupy "volume" and are bounded by "surfaces", the type of entities whose existence and properties are learned through a combination of senses. Eventually, our visual perception of the external world relies on our ability to distinguish various pieces of surfaces, to integrate collections of surfaces into parts of an object, and to fill any missing information by inference and other mechanisms that develop as part of our survival strategy. Thus, a theory of visual perception of surfaces is at the heart of any comprehensive theory of human perceptual organization. First studied by Gestalt psychologists early in this century, perceptual organization concerns how retinal images are structured, how the various regions and elements are perceived as being related to each other in terms of part-whole relations, as well as various geometrical relations. Among the most pervasive and difficult problems in vision science are the nature of perceptual organization and the mechanisms responsible for it.

Perceptual geometry is an emerging field of interdisciplinary research whose objectives focus on study of geometry from the perspective of visual perception, and in turn, apply such geometric findings to the ecological study of vision. Perceptual geometry attempts to answer fundamental questions in perception of form and representation of space through synthesis of cognitive and biological theories of visual perception with geometric theories of the physical world. In our previous papers (Assadi 99), we have proposed a basic mathematical model for the Gestalt of surfaces, that is, the simplest phase of global perception of surfaces in the environment, as opposed to visual perception in laboratory and psychophysical experiments under controlled parameters. The result at this point is a computational model for perception of form (or Gestalt) of surfaces in natural environments, including those that have rough and non-smooth small-scale structure but with a perceived global (larger-scale) geometric form. Examples include grass and meadow, surfaces textured with sand-paper, natural scenes having rough texture such as the skin of crocodile, pine cones, a field of sea urchins, forests, etc.

The problems of figure-ground separation and scene segmentation in perceptual geometry could be formulated in terms of structural regularity of regions of images in statistical and information theoretic terms. Intuitively, as well as in psychophysical studies performed by cognitive scientists (e.g. Palmer, Rock, Tyler), perception of local structural regularity is fundamentally correlated with perception of local symmetry of surfaces, and under parallel projection of planar surfaces, with local symmetries of their images. In other words, such local symmetries distinguish prevalent regularity of common surfaces in the environment from randomness in arbitrary composition of colored dots, or what is the same, they distinguish between a meaningful image versus a generic pattern of a totally random selection of light intensities in matrices encoding local incoherence in optical properties. From a mathematical point of view, it can be shown that in the space of all possible patterns of light (i.e. all large matrices of same size with non-negative coefficients), the set of possible images of natural scenes is a very small subset. In the technical jargon of mathematical analysis, true images form a subset of measure zero in the space of all possible 2-dimensional patterns representing arbitrary light intensities.

We use the term symmetry in reference to all kinds of transformations that leave invariant some form of geometry, together with the related concepts such as harmony. Therefore, similarity in Euclidean plane geometry is a form of symmetry (in the so-called conformal geometry, where angles are preserved) although it is not necessarily a rigid motion. The term quasi-symmetry can be used for perceived regularity of structure that is compelling in its organization, but fails to be a strict symmetry in the mathematical sense above. This aspect of our research has a long-term history in psychology as the investigation of cognitive processes that underlie human perception of geometric forms, much the same way that Henri Poincare' posed in his 19-th Century treatise Science and Hypothesis and led to his construction of non-Euclidean geometry and optics [19].

Among many researchers studying symmetry, S. Palmer and I. Rock have contributed to our understanding of the role of Gestalt in perceptual organization in the context of symmetry, (e.g. Palmer, 1977; Palmer & Rock, 1994), and the formation of those aspects of our research that could belong to the domain of Gestalt Psychology, or mathematically speaking, passage of local-to-global properties in the context of geometric form. One of the main topics investigated by Palmer concerns the perception of internal geometrical structure in 2-D figures, including symmetry (Palmer, 1985; Palmer & Hemenway, 1978), part-whole relations form vs. texture (Kimchi & Palmer, 1982), figural goodness (Palmer, 1991), object-centered reference frames (Palmer, 1990), and grouping (Palmer, 1983; Palmer & Rock, 1994). In all these research areas, Palmer's approach has been to develop objective behavioral methods that can be used to study the stimulus factors that influence perceptual phenomena. In many cases, his theories have taken a transformational approach to problems of perceptual organization, where he analyzes phenomena in terms of the action of the similarity group (Palmer, 1982, 1983, 1991).

Biological Motivation

Our research has a long-term objective: to investigate the cognitive processes that underlie our perception of geometric forms. The time is ripe to address this question in the realm of cognitive neuroscience. While our proposed computational model serves to investigate and support the cognitive theories, it reveals the potential for a refined approach to cognitive and biological models of *visual information processing channels*. One could generally agree that the increasingly rapid pace of advances in our understanding of the biology of the brain and advances in computation power will open new ways for investigation of information processing in the brain. Thus, we are optimistic that in foreseeable future, we will have the scientific tools to understand neuronal substrates of low-level computations of visual, tactile, motor and auditory processes, which contribute to our perception of symmetry and regularity in structure.

This paper is a second computational attempt at modeling detection of repetitive patterns in visual perception, and potential reasoning about almost structural regularity in the presence of noise and natural imperfections of visual stimuli. Being as small a step as it may be, we propose a possible approach for visual perception of geometric form of surfaces endowed with symmetric patterns in their texture. The key biological observation is the dynamic nature of vision: visual perception of the physical world depends on saccades, the tiny, almost instantaneous jitters of the eyeballs (e.g. see pp. 78-80). Without such seemingly random motions of the eyes, the photoreceptor cells of the retina get chemically saturated from the steady invasion of photons, and the image on the retina ceases to exist! All neurons in the visual cortex have receptive fields (see Hubel 1995, page 41-43), defined, roughly speaking, as the cone-shaped region of the space measured in terms of the visual angle. This potentially stimulates a neuron through its variation of light intensity. The concept of the receptive field and its biological properties leads us to hypothesize that: there is a biologically realistic hybrid computational model of intermediate-level vision in which the process of neuronal detection of visual symmetry in the presence of repetitive patterns involves an adaptive series of comparison of patterns of spike trains. This is due to visual stimuli that are brought about by a sequence of saccades.

We refer to this statement as *the Adaptive Saccades Hypothesis* for detection of symmetry. The long-term goal of the present research includes formulation and verification of models that incorporate hypotheses such as the Adaptive Saccades Hypothesis above. Therefore, our first attempt focuses on modeling adaptive processes that simulate saccades and comparison of the resulting neural activation, in parlance of neural networks and learning theory (Latimer, 1996). In this paper, we consider a simplified version of the above-mentioned theory: a computational model is presented that detects the fundamental domain for translation symmetry of patterns on a flat surface parallel to the observer's plane of view (or the robot's camera screen). This model is robust with respect to noise, partial occlusion and some irregularity in the translation symmetry. Moreover, it lends itself to generalization to visual attention and theories of active vision. Finally, there is a generalization to hybrid adaptive implementations that incorporate support vector machines, a regression technique that has its biological justification in the finer anatomy of the superior colliculus. In a forthcoming paper, we present the generalization of the model below to incorporate support vector regression.

Perception of symmetry in the human brain has been the subject of various studies. The approach in neurobiology, the so-called bottom-up approach (see for example, Hubel 1995), proceeds to study the cascade of neuronal events from the onset of the stimulus as it progresses toward the emergence of the percept of symmetry (Julesz 1979; Troscianko 1987). In contrast, many experimental and cognitive psychologists take a "top-down" viewpoint (Rock 1963; Palmer 1978; Royer 1981; Van Gool 1990; Wagemans 1993). In the "top-down" construct one proposes a theory (based on the psychophysical measurement) that reconstructs the sequence of events in the brain leading to the observed behavior; see e.g. (Hogben 1976; Zimmer 1984; Herbert 1994). Alternatively one may propose a theory where both top-down and bottom-up reasoning, possibly with several building blocks of each kind. Such hybrid theories use each type of reasoning in a sequence that complements and improves partial results in individual steps. More generally, it is reasonable to propose a more integral (holistic) approach to study symmetry, for instance, by considering the higher-level cognitive functions, such as thinking, creativity, and problem solving (Van Gool 1990; Rentscher 1996; Varshney and Burrus 1997).

As a starting point, it is convenient, and not so unreasonable to begin with the assumption that any computational theory of perception of symmetry would most likely use the special cases of translation and reflection of planar periodic tilings as one of its building blocks; see e.g. (Royer 1981). As for reflection symmetries, there is a great deal of research that is partially outlined in (Tyler 1996) and elsewhere (e.g.(Gerbino 1991) (Wagemans 1992; Wagemans 1993; Wagemans 1996)). In fact, almost all articles in that volume are devoted to bilateral and reflection symmetry (e.g. (Marola 1989; Dakin 1996)) with a few exceptions dealing with circular and rotation symmetries (Palmer 1978). Translation symmetry seems, for the most part, not adequately covered in the literature (see (Corballis 1974; Bruce 1975; Baylis 1994)). As for connectionist models, back propagation model for detection of reflection symmetry are provided in (Baylis 1994; Latimer 1996). Dakin and Watt, on the other hand, have used spatial filters for detection of bilateral symmetry (Dakin 1996). Thus, for the remaining part of this paper, we concentrate on translation symmetry.

Review of Related Previous Results

In our previous work (Manske, et. al, 1999), detection of symmetry in periodic tiling was studied from the point of view of machine vision. The very systematic approach of the main algorithm to search and target selection in the image makes the model of (Manske, et. al, 1999) incompatible with the human visual search and natural eye movement. Nonetheless, a case was made for a dynamic model of symmetry detection where adaptive saccades in conjunction with visual attention play an important role. In this work, we present a computational model of saccadic target selection and simulate its action in the context of perception of global symmetry of tiling. This approach uses local (foveal) symmetry estimation via direct saccadic eye movements, in agreement with well-known biological models of human eye movement. However, the variant of spatial channels considered in this paper are motivated by wavelets and filtering, rather than the model based on spatial Fourier transform. From the point of view of computational efficiency and numerical convergence, the model is certainly superior to the Fourier alternative. While the computationally proposed approach has not been experimentally tested by biologists in the laboratory, the localized nature of wavelet supports appears to fit better with the Gabor-like filter models of receptive fields in the neural circuitry of low-level primate vision.

The motivation for a model of saccadic target selection finds its roots in the properties of the superior colliculus (SC). Studies of SC indicate that the nature of the information processing is highly distributed and integrated over a large spatial area. More specifically, the following is known: (1) In the SC, visual cells have large and overlapping receptive fields; (2) The SC has a layered architecture where retinotopic mapping is preserved; (3) The size of the visual fields increases systematically with depth of layers. It has been demonstrated that any particular point in the collicular motor map can be stimulated by visual input from a wide region of the visual space. Conversely, stimulation from a punctate peripheral source can generate activity over a wide area of the colliculus.

A Heuristic Discussion Of Symmetry Detection

We start with the consideration of the tiling of a rectangular subset of the plane s with the rectangular tile B. For a perfect (no partial tiles) and noiseless tiling, the regularity of the structure of this subset can be determined by solving the following problem: *Determine B with the smallest area such that* s *can be reconstructed from a number of horizontal copies and a number of vertical copies of* B. In this deterministic setting, the problem can be viewed as finding a basic element for s in the following way. The subset s is obtained form translates of this basic element B and the apparent large data volume of s can be compressed by finding the basic element B and constructing s from the translates of B. To address the same problem in the presence of noise and other imperfections, we must solve the search for this basic element in the statistical setting. In this setting we view s as a signal that we wish to represent in terms of B yet to be determined.

The traditional Fourier representation of a signal is based on the superposition of sinusoids. The sinusoids can be thought of as elements of a dictionary – a collection of parameterized waveforms. However, one may choose from a host of alternate dictionaries such as wavelets, chirplets, cosine packets and others. Each such dictionary D is a collection of elements ϕ_t, where t is a parameter, and is used to decompose a signal s in the form of:

$$s = \sum_t a_t \phi_t + C$$

Where C is a "error" term representing how well the signal s is approximated using the given dictionary. The goal of using alternative dictionaries is to obtain a sparse as well as robust representation of the signals. However, the sparseness requirement makes for a difficult optimization problem since minimizing the number of coefficients in the representation of **s** above requires a combinatorial check of all coefficient possibilities. Mallat and Zhang [15] proposed a general approach (called matching pursuit), based on the simple idea that one can start with an initial approximation and at each stage find the element in the dictionary such that its multiple minimized the residual C – a simple greedy algorithm. The algorithm is stopped after a few steps and the representation of the signal contains only a few elements of the dictionary. Of course, as it has been pointed out by Devore [6], this algorithm can fail badly for certain signals and non-orthogonal dictionaries. Basis pursuit as proposed by Chen [23] solves some of the problems by attempting to minimize the L^1 norm. However, most of the work mentioned so far and the reference therein considers the problem of choosing coefficients for a given dictionary.

The case of interest for us is to construct a (small) dictionary such that one of the items in the dictionary can be used to represent the information content of the entire signal. Loosely speaking, we wish to find an element for the dictionary such that any random selection of a segment of the entire signal effectively provides no additional information beyond what is available from the selected element from the dictionary. For this approach, we only borrow the spirit of the Mallat and Zhang algorithm in the sense that we start with an approximate solution and look for improving residual errors through iterations. However, the analogy stops at this level since our algorithm uses a multiscale search criteria and an information theoretic measure for its approach as will be described in the next section.

Principal component analysis (PCA) is a well-known statistical method for lowering the dimensionality of a data set by finding a set of basis, which contain "most" of the information in the original data set. To use the PCA approach to discover the tile B the single image object represented by **s** must be transformed into a collection of data. A straightforward approach would proceed as follows: *Choose a trial size for B and break* **s** *into a collection of trial Bs that will act as the collection of data for the PCA. Compare the results of PCA with those of another trial B and continue until a B is found that has the optimal size(this is reminiscent of Mallat's algorithm).* The computational difficulty with this approach stems from the large search space associated with the solution of the problem. There is also a conceptual difficulty which stems from our desire to use a dynamic process in which an approximate symmetry is detected and then continually refined - i.e. the use of coarse grain information to determine approximate symmetry and then refining by using fine-grained information.

We introduce a dynamic initial process based on entropy and operating on a "low-resolution" version of the signal, which can operate in parallel with "high-resolution" information to discover symmetries. Entropy is a measure of order and is related to PCA as we will see below. The initial process operates using the entropy measure for a continually refined coarse grained information. This process is used to determine the initial approximation for the size of B which is continually refined with the first process.

PCA and Entropy

Suppose that we have broken the rectangular domain D into a set of tiles. Each tile can be treated as a zero mean vector indexed by the variable t, X(t), define the covariance matrix R: $R = E\left[X(t)X^T(t) \right]$ (E denotes the expectation). Let $R = U\Lambda U^T$ be the eigendecomposition of R where $U = [u_1, \cdots, u_n]$ is an orthogonal matrix formed by eigenvectors $u_1, \cdots, u_n$, and $\Lambda = diag(\lambda_1, \cdots, \lambda_n)$ a diagonal matrix formed by eigenvalues of R. Then, in this new coordinate system, X(t) can be written as:

$$X(t) = a_1(t)u_1(t) + \cdots + a_n(t)u_n(t)$$

where $a_i(t) = X(t)u_i$, and the following properties are satisfied:

$$E[a_i(t)] = 0, \quad E[a_i^2(t)] = \lambda_i, \quad E[a_i(t)a_j(t)] = 0 \; for \, i \neq j$$

In this setting, given X(t), PCA provides a new coordinate system in which a_i are decorrelated. For the precise tile size any two adjacent blocks are maximally correlated. Therefore, decorrelation of the precise tile size means that a first eigen-value is significantly larger compared to the remaining eigen-values.

However, as we pointed out in the informal discussion, searching the large space of all tile sizes is not an efficient approach. To remedy the inefficiency we introduce a dynamic multi-scale process that uses entropy at a coarse grain level to obtain an initial approximation of the tile size. Given a set of discrete random variables (or symbols) denoted by x_i with probability of the event x_i denoted by $P(x_i)$, entropy is defined as:

$$H(x) = E[\log(P(x))] = -\sum P(x_i)\log(P(x_i))$$

The concept of mutual information between random variables x and y is related to the concept of joint entropy and is defined as:

$$I(x, y) = H(x) + H(y) - H(x, y) \quad where \quad H(x, y) = -E[E[\log(P(x, y))]]$$

where P(x , y) is the joint probability of events x and y. Notice that H(x , y) is zero when events x and y are statistically independent. It is important at this point to notice that when the distribution of the random variables follows a Guassian distribution, finding decorrelated components as done by PCA is the same as minimizing mutual information. Therefore, the two approaches of PCA and entropy are closely related. We also note that both methods are applicable to symmetric tiling without the presence of any projective transformations.

To use entropy in a dynamic process we require another ingredient. One must observe two important facts: 1) entropy as defined is invariant under a permutation of symbols. 2) entropy is subject to fluctuations when noise is present, 3) accurate computation of entropy requires the knowledge of distribution of the random variables. Although, the distribution is often taken to be synonymous with frequency, the use of frequency may introduce unacceptable inaccuracies. Furthermore, the invariance to permutations has the potential of giving false positives. We can overcome these problems in practice by introducing an filtering operator F. In the discrete case of images we are considering, the filtering operator replaces every pixel with an average of the pixels in a neighborhood:

$$F(i, j) = \frac{1}{(2s+1)(2t+1)} \sum_{l=-s,m=-t}^{l=s,m=t} F(i+l, j+m)$$

where s and t define the dimensions of the window over which averaging is performed. Since F captures information from a neighborhood it has the affect of reducing noise, reducing the affect of inaccuracies of estimating the distribution form the frequencies and reducing the affect of false positives resulting from permuted sets. Furthermore, from a biological view point, it is the statement that the coarsest level of information, that is what is available first, is used for estimates as soon as it is available.

Computational Formulation of Symmetry Detection

For our computational work we have used a series of images which contain both natural as well as synthetic symmetries. The images are stored as grayscale images with 256 levels of gray and are normalized to have intensity between values of zero to one. The algorithm is written in the Matlab computational environment and the core of the algorithm excluding the filtering portion runs very fast.

We describe how the overall process can be visualized in the following diagram where we have commented alongside the diagram.

Starting at the coarsest level of resolution we proceed to compute entropy for a window which is ¼ the size of the entire window.	*Entropy_decreasing = true;* *Initial_Window = 0.25*max(size(image));* *P_Entropy = large_number;* *while (entropy_decreasing)*
First filter the image to the initial window size compute entropy for a random window location and repeat for a second random window. if entropy continues to decrease, we want to continue with increasing resolution and measuring entropy,	*I = Filter with F in Initial_Window* *e1= entropy_of_random_window(I)* *e2= entropy_of_random_window(I)* *if e1 < p_entropy or e2 < p_entropy* * p_entropy = min(e1,e2);* * half the size of Initial_Window* *else*
If entropy begins to increase for both windows then we may be transitioning to a window of size smaller than the symmetric tile.	* if e1<e2 report e1* * else report e2* *endif*
First we optimize the PCA process for the horizontal line scans and then optimize PCA for the vertical line scans.	*w = width(reported window);* *w = optimize_PCA(w,I,'horizontal');* *h = height(reported_window);* *h = optimize_PCA(h,I,'vertical',w)*

The simulation results are illustrated in the figures below. From our simulation we have found that the entropy portion of the algorithm always overestimates the size of the symmetry window. This allows the PCA portion of the algorithm to start with the window as an upper bound. We also found that the entropy portion of the algorithm works just as well with non-rectangular tiles. We use this fact later to discuss how non-rectangular tiles detected. Finally we mention that we determine the size of the tile and not necessarily the tile itself. The choice of the tile has cognitive components that need to be encoded in a higher level algorithm.

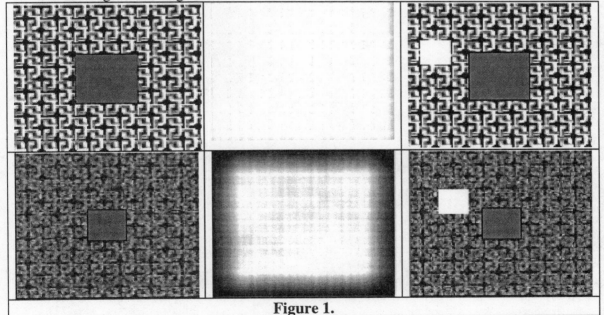

Figure 1.

Initial entropy minimization leads to an estimate of tile size. The tile is shown on the background of the image. Top image is the original. Bottom image is the noisy version of the top.	The entropy is measured after applying the averaging operator with decreasing window size. The two images show how images appear at the scale at which the tile size was determined.	Final estimate of the window size at the corresponding level of application for the averaging operator is shown in the smaller tile with the white background.

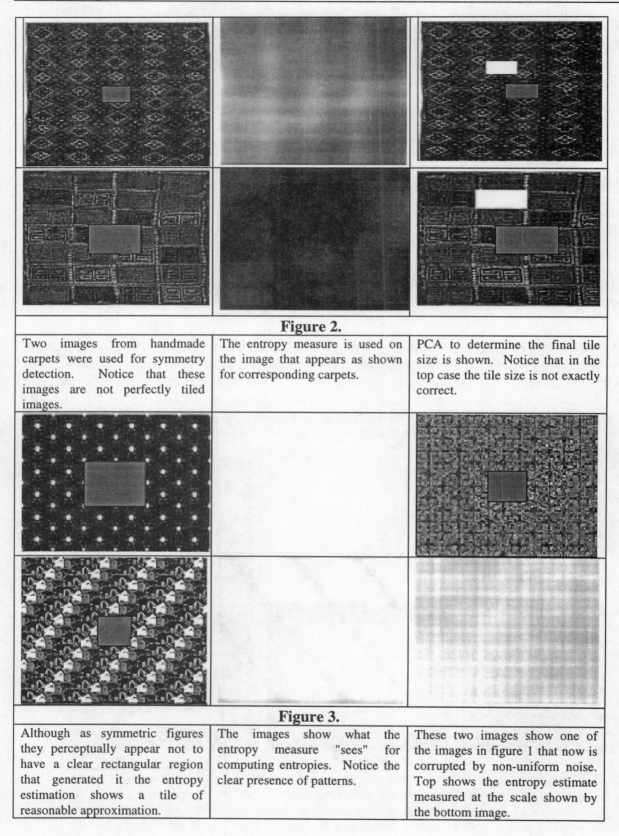

Figure 2.

| Two images from handmade carpets were used for symmetry detection. Notice that these images are not perfectly tiled images. | The entropy measure is used on the image that appears as shown for corresponding carpets. | PCA to determine the final tile size is shown. Notice that in the top case the tile size is not exactly correct. |

Figure 3.

| Although as symmetric figures they perceptually appear not to have a clear rectangular region that generated it the entropy estimation shows a tile of reasonable approximation. | The images show what the entropy measure "sees" for computing entropies. Notice the clear presence of patterns. | These two images show one of the images in figure 1 that now is corrupted by non-uniform noise. Top shows the entropy estimate measured at the scale shown by the bottom image. |

REFERENCES

[1] Assadi, A., Palmer, S., and Eghbalnia, H., Ed. (1998). Learning Gestalt Of Surfaces In Natural Scenes. Proceedings of IEEE Int. Conf. Neural Networks in Signal Processing.

[2] Bruce, V. G. a. M., M. J. (1975). "Violations of symmetry and repetition in visual patterns." Perception 4: 239-249.

[3] Corballis, M. C. a. R., C.E. (1974). "On the perception of symmetrical and repeated patterns." Percept. Psychophys. 16: 136-142.

[4] Dakin, S. C. a. W., R. J. (1996). Detection of bilateral symmetry using spatial filters. Human Symmetry Perception. C. W. Tyler. Utrecht, the Netherlands, VSP BV: 187-207.

[5] Davies, E. R. (1997). Machine vision : theory, algorithms, practicalities. San Diego, Academic Press.

[6] De Vore, R.A., Temlyakov, V.N. (1996). "Some remarks on greedy algorithms". Advances in computational mathematics (5).

[7] Haralick and Shapiro (1988). Computer and Robot Vision Vol. 1.

[8] Herbert, A. M., Humphrey, G.K. and Jolicoeur, p. (1994). "The detection of bilateral symmetry: Effects of surrounding frames." Can. J. Exp. Psychol. 48: 140-148.

[9] Hogben, J. H., Julesz, B. and Ross, J. (1976). "Short-term memory for symmetry." Vision Res. 16: 861-866.

[10] Hubel, D. (1995). Eye, Brain, and Vision, W.H. Freeman and Company.

[11] Julesz, B. a. C., J. J. (1979). "Symmetry perception and spatial-frequency channels." Perception 8: 711-718.

[12] Kanade, T. a. K., J. R. (1983). Mapping image properties into shape constraints: Skewed symmetry, affine-transformable patterns, and the shape-from-texture paradigm. Human and Machine Vision. J. Beck, Hope, B., and Rosenfeld, A. New York, Academic Press. 1: 237-257.

[13] Kurbat, M. A. (1996). A network model for generating differential symmetry axes of shapes via receptive fields. Human Symmetry Perception. C. V. Tyler. Utrecht, the Netherlands, VSP BV: 227-236.

[14] Latimer, C., Joung, W., and Stevens, C. (1996). Modeling symmetry detection with back-propagation networks. Human Symmetry Perception. C. W. Tyler. Utrecht, the Netherlands, VSP BV: 209-225.

[15] S. Mallat and Z. Zhang Matching pursuits with time-frequency dictionaries IEEE Trans. on Signal Process., 12(41), pp. 3397-3415, 1993.

[16] Manske, K., Assadi A., Eghbalnia, H. (1999). The Second Bridges Conference, Reza Sarhangi, Ed. (1999)

[17] Marola, G. (1989). "On the detection of the axes of symmetry of symmetric and almost symmetric planar images." IEEE Trans. Pattern Anal. Machine Intell. PAMI-11: 104-108.

[18] Palmer, S. E. a. H., K. (1978). "Orientation and symmetry: Effects of multiple, rotational, and near symmetries." J. Expo. Psychol: Human Percept. Perform. 16: 150-163.

[19] Poincare, H. (1943) "Science and Hypothesis". Dover Publications.

[20] Rentscher, I., Barh, E., Caelli, T., Zetzsche, C., and Juttner, M. (1996). On the generalization of symmetry relation in visual pattern classification. Human Symmetry Perception. C. W. Tyler. Utrecht, the Netherlands, VSP BV: 237-264.

[21] Rock, I. a. L., R. (1963). "An experimental analysis of visual symmetry." Acta Psychol. 21: 171-183.

[22] Royer, F. L. (1981). "Detection of symmetry." J. Exp. Psychol: Human percept. Perform. 7: 1186-1210.

[23] Chen, S.S. (1994). "Basis Pursuit - Ph.d Thesis" , http://www-stat.stanford.edu/~schen.

[24] Troscianko, T. (1987). "Perception of random-dot symmetry and apparent movement at and near isoluminance." Vision Res. 27: 547-554.

[25] Tyler, C. W., Ed. (1996). Human Symmetry Perception and its Computational Analysis. Utrecht, the Netherlands, VSP BV.

[26] Van Gool, L., Wagemans, J., Vandeneede, J., and Oosterlinck, A. (1990). Similarity extraction and modelling. Third Int. Conf. Computer Vision, Washington, D.C.

[27] Varshney, P. K. and C. S. Burrus (1997). Distributed detection and data fusion. New York, Springer.

[28] Wagemans (1996). Detection of visual symmetries. Human Symmetry Perception. C. W. Tyler. Utrecht, the Netherlands, VSP BV.

[29] Wagemans, J., Van Gool, L. and d'Ydewalle, G. (1992). "Orientational effects and component processes in symmetry detection." Q. J. Exp. Psychol. 44A: 475-508.

[30] Wagemans, J., Van Gool, L., Swinnen, V., and Van Horebeek, J. (1993). "Higher-order structure in regularity detection." Vision Res. 33: 1067-1088.

[31] Washburn, D. K., and Crowe, D. W. (1988). Symmetries of Culture. Seattle, WA, University of Washington Press.

[32] Weyl, H. (1952). Symmetry. Princeton, NJ, Princeton University Press.

[33] Zimmer, A. C. (1984). "Foundations for the measurement of phenomenal symmetry." Gestalt Theory 6: 118-157.

Structures: Categorical and Cognitive

Mara Alagić
Wichita State University
Wichita, KS 67260-0028
e-mail: mara@math.twsu.edu

Abstract

This paper is an inquiry into two paradigms of structures: cognitive and categorical. This investigation comes from the two special interests of the author - mathematics and learning - not just learning mathematics.
Insight into cognitive and categorical structures is what sometimes sets them apart from each other, but more often, it brings them closer together.

1 "Structure is what structure does." (Van Hiele, [11])

Jean Piaget [12], a Swiss biologist and psychologist, developed an influential model of child development and learning based on the idea that the developing child constructs increasingly sophisticated cognitive structures – moving from a few inborn reflexes such as crying and sucking to highly complex mental activities. Cognitive structure is a person's internal mental "map," a scheme or a network of concepts for understanding and responding to physical experiences within his or her environment. Schema is viewed as a connected collection of hierarchical relations, an organized structure of knowledge, into which new knowledge and experience might fit. Understanding of something is equated with assimilating it into an appropriate schema. The formation of schema is the brain organizing its own activity. Piaget's theories of the development of logico-mathematical structures are based on this reflective activity of the brain.

Gestalt theory emphasizes higher-order cognitive processes. The focus is the idea of 'grouping' – characteristics of stimuli cause us to structure or interpret a visual domain or problem in a certain way. The primary factors called the laws of organization, that determine grouping, are: proximity, similarity, closure, and simplicity. These factors can be explained in the context of perception and problem-solving. The essence of successful problem-solving behavior according to Wertheimer [14] is being able to see the overall structure of the problem. Directed by what is required by the structure of a situation for a crucial region, one is lead to a reasonable prediction, which like the other parts of the structure, calls for verification. Two directions are involved: getting a whole consistent picture, and seeing what the structure of the whole requires for the parts.

Piaget gave the following properties of structure: structure has totality, structure is achieved by transformations, and structure is auto-regulating. The van Hiele theory [13] puts forward a hierarchy of levels of thinking: visualization, analysis, informal deduction, deduction, and rigor. Van Hiele claims parallelism with Gestalt psychology by explaining that insight exists when a person acts adequately with intention in a new situation . He further describes that the most important property of structure is that structure can be extended because of its composition: "Structure is what structure does."

The above is a brief overview of a few structures and some of their properties, usually mentioned in discussions about learning and teaching.

2 Categorical Structures

A category $\mathcal{K}$ consists of a collection of *objects* and a collection of *arrows* (morphisms) satisfying certain conditions. Given arrows $g : X \to Y$ and $f : Y \to Z$ there is an arrow $f \circ g : X \to Z$ which we call the *composition* of f and g. For each object X there is an arrow $id_X : X \to X$, called the *identity on X*. The axioms for a category are:

$$\text{Composition is associative: } (f \circ g) \circ h = f \circ (g \circ h).$$
$$\text{Identity property: Given any } f : Y \to Z, \; f \circ id_X = f \text{ and } id_Y \circ f = f .$$

Category theory provides a consistent treatment of the notion of mathematical structure. Almost every known example of a mathematical structure with the appropriate structure preserving map yields a category. The classic example of category theory is *Set*, the category with sets as objects, functions as arrows, and the usual composition. What characterizes a category is its arrows and not its objects. Thus, the category of topological spaces with open maps is a different category than the category of topological spaces with continuous maps.

In all these cases the arrows are actually special sort of functions. That need not be the case in general: any entity satisfying the conditions given in the definition is a category. For example, an *ordered set* is a category with its elements as objects and one arrow for each $X \leq Y$, but none otherwise. A *deductive system* such that the entailment relation is reflexive and transitive is a category [10].

2.1 Universals

Category theory unifies mathematical structures in a second, and perhaps even more important, manner. Once a type of structure has been defined, it becomes essential to determine how new structures can be constructed out of the given one and how given structures can be decomposed into more elementary substructures. For instance, set theory allows us to construct Cartesian product. For an example of the second sort, given a finite abelian group, it can be decomposed into a product of some of its subgroups. In both cases, it is necessary to know how structures of a certain kind combine. The nature of these combinations might appear to be considerably different when looked at from too close. Category theory reveals that many of these constructions are in fact special cases of objects in a category with what is called a "universal property". From a categorical point of view, a Cartesian product, a direct product of groups, a product of topological spaces, and a conjunction of propositions in a deductive system are all instances of a categorical concept: the categorical product.

The universal property of product: Any arrow $h : W \to X \times Y$ from an object W is uniquely determined by its composites $p \circ h$ and $q \circ h$, where $p : X \times Y \to X$ and $q : X \times Y \to Y$ are 'projections'. Conversely, given W and two arrows $f : W \to X$ and $g : W \to Y$ there is a unique arrow h which makes the corresponding diagram commute, namely $h = (f, g)$.

Thus, given X and Y, (p, q) is "universal" because any other such pair (f, g) factors uniquely (via h) through the pair (p, q). This property describes the product $(X \times Y, p, q)$ uniquely (up to a bijection).

Many properties of mathematical constructions may be represented by universal properties of diagrams [10].

2.2 Functors: Trans-structuring

Another crucial aspect of category theory is that it allows to see how different kind of structures are related to one another. For instance, in algebraic topology, topological spaces are related to groups by various means (homology, cohomology, homotopy, K-theory). It was precisely in order to clarify how these connections are made that Eilenberg and MacLane invented category theory [10]. Indeed, topological spaces with continuous maps constitute a category and similarly groups with group homomorphisms. In the very spirit of category theory, what should matter here are the arrows between categories. These are given by functors and are informally structure preserving maps between categories.

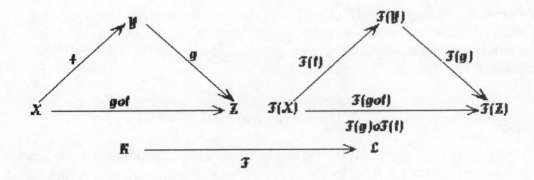

Figure 1: Functor $\mathcal{F}: \mathcal{K} \to \mathcal{L}$

A *functor* from $\mathcal{K}$ to $\mathcal{L}$ is a function from class of arrows of $\mathcal{K}$ to the class of arrows in $\mathcal{L}$ preserving identities and composition (see Figure 1):

If *id* is a $\mathcal{K}$-identity, then $F(id)$ is an $\mathcal{L}$-identity.
$$F(f \circ g) = F(f) \circ F(g), \text{ where } g \in Hom(X, Y) \text{ and } f \in Hom(Y, Z)$$

It follows immediately that a functor preserves commutativity of diagrams between categories. Homology, cohomology, homotopy, K-theory are all example of functors.

2.3 One More Bridge: Natural Transformation

There are in general many functors between two given categories and it becomes natural to ask how they are connected. For instance, given a category $\mathcal{K}$, there is always the identity functor from $\mathcal{K}$ to $\mathcal{K}$ which sends every object of $\mathcal{K}$ to itself and every arrow of $\mathcal{K}$ to itself. In particular, there is the identity functor over the category of sets.

Suppose we have two functors F and G from the category $\mathcal{K}$ to the category $\mathcal{L}$ and an arrow $f : X \rightarrow Y$ in $\mathcal{K}$. A *natural transformation* n from F to G consists of a family of arrows, an arrow $n(X) : F(X) \rightarrow G(X)$ for each object X in $\mathcal{K}$, such that $n(Y) * F(f) = G(f) * n(X) : F(X) \rightarrow G(Y)$ (the corresponding diagram commutes for every such f).

An example would be "abelianization", which maps a group H to the abelian group $H/[H, H]$. If F were the fundamental group and G were the first homology group, we could say that abelianization is a natural transformation from F to G [10].

2.4 Is That All?

The above notions constitute the elementary concepts of category theory. However it should be noted that they are not fundamental notions of category theory. These are arguably the notions of limits/colimits which are, in turn, special cases of what is certainly the cornerstone of the theory, the concept of *adjoint functors*. We will not present the definition here. Adjoint functors permeate mathematics and this quality of spreading through is certainly one of the mystifying facts that category theory reveals about mathematics and probably thinking in general. Universality may also be described in terms of adjoint functors expressing the objective dialectical equilibrium.

In this manner, category theory provides means to circumscribe and study what is universal in mathematics and other scientific disciplines. Also, by identifying logic with the study of what is universal, category theory supplies the means to describe such logic, the objective logic of the discipline in question [10].

3 Cognitive Structures

Cognitive science is the interdisciplinary study of mind and intelligence, embracing philosophy, psychology, artificial intelligence, neuroscience, linguistics, and anthropology. Its intellectual origins are in the mid-1950s when researchers in several fields began to develop theories of mind based on complex representations and computational procedures. Its organizational origins are in the mid-1970s when the Cognitive Science Society was formed and the journal Cognitive Science began. Cognitive science is revolutionizing our understanding of ourselves by providing new accounts of human rationality and consciousness, perceptions, emotions, and desires. It explicates how structures of different kinds are related to one another as well as the universal components of a family of structures of a given kind. Philosophically, it can be thought of as constituting a theory of concepts. It also gives new prospective on some traditional philosophical questions, for instance on the nature of reference and truth.

Modeling the mind as an information-processing machine is a result of a large body of research about computer knowledge. It has led to much psychological and pedagogical insight, but has sharply limited ability to predict human behavior and learning nonetheless. Cognitive theory describes memory storage and recall structures resembling Piaget's schema and describes regular inclusions and revisions to these structures similar to Piagetian assimilation and accommodation [11].

Cognitive science raises many interesting methodological questions: What is the nature of representation? What role do computational models play in the development of cognitive theories? What is the relation among apparently competing accounts of mind involving symbolic processing, neural networks, and dynamical systems?

3.1 Mental Representations

There is much disagreement about the nature of the representations and computations that constitute thinking. Thinking can best be understood in terms of representational structures in the mind and computational procedures that operate on those structures. This central hypothesis of cognitive science is general enough to encompass the current range of thinking in cognitive science, including connectionist theories which model thinking using artificial neural networks [11]. Most work in cognitive science assumes that the mind has mental representations analogous to computer data structures, and computational procedures similar to computational algorithms. The mind contains such mental representations as logical propositions, rules, concepts, images, and analogies, and uses mental procedures such as deduction, search, matching, rotating, and retrieval. "Connectionists" have proposed novel ideas about representation and computation that use neurons and their connections as inspirations for data structures, and neuron firing and spreading activation as inspirations for algorithms. Cognitive science then works with a complex 3-way analogy of mind, brain, and computers. Mind, brain, or computation can each be used to suggest new ideas about the other two. Different kinds of computers and programming approaches suggest different ways in which the mind might work. Therefore, there is no single computational model of mind. The first computers are serial processors, performing one instruction at a time, but the brain and some newly developed computers are parallel processors, capable of doing many operations at once.

Philosophy, in particular philosophy of the mind, is part of cognitive science. From a naturalistic perspective, philosophy of the mind is closely tied in with theoretical and experimental work in cognitive science. Metaphysical conclusions about the nature of mind are to be reached, not by a priori speculation, but by informed reflection on scientific developments in fields such as computer science and neuroscience. Similarly, epistemology is not a standalone conceptual exercise, but depends on and benefits from scientific findings concerning mental structures and learning procedures. It is an empirical conjecture that human minds work by representation and computation. Although this computational-representational approach to cognitive science has been successful in explaining many aspects of human learning, critics of cognitive science have offered such challenges as: the emotion challenge, the consciousness challenge, the world challenge, the social challenge, the dynamical systems challenge, and the mathematics challenge [9].

3.2 Architecture

A very restricted family of structures that provide the frame within which the cognitive processing in the mind take place is called architecture [11]. If the consideration is restricted only to symbolic structures, architecture is closer to structures analyzed in computer science. Viewing the world as constituted of systems whose behavior is observed is part of the common conceptual apparatus of science - a system of given structure producing behavior that performs a given function in the encompassing system. The notion of architecture in this sense supplies the concept of the system that is required to attain flexible intelligent behavior by invoking most of the psychological functions - perception, encoding, retrieval, memory, composition and selection of symbolic responses, decision making, motor commands and actual motor responses.

The role of architecture in cognitive science is to be the central element in a theory of human cognition. A theory of the architecture is a proposal for the total cognitive mechanism. It is reasonable to say that the cognitive architecture is realized in neural technology and that it was created by evolution.

3.3 The nature of cognitive architecture: Functions

Since the architecture is defined in terms of what it does for cognition, the nature of the cognitive architecture is given in terms of functions, rather than structures and mechanisms. What are some functions defining this nature of cognitive science?

Memory - composed of structures that contain symbol tokens.

Symbols - patterns that provide access to distal symbol structures

Operations on symbols - processes that take symbol structure as an input and produce (compose) new symbol structures as output; a sequence of symbol operations occurs on specific symbol structures.

Operations - can construct symbol structures that can be interpreted to specify further operations to construct yet further symbol structures.

Interpretation - processes that take symbol structure as an input and produce behavior by executing operations

Interaction with the external world - perceptual and motor interfaces; real time demands for action

One additional consideration that is specific to the nature of human cognition is that *the human mind* can carry out a large number of constructions that seem very natural and so universal that they must be severely constrained [9].

4 Conceptual Tool: Categorical Logic

Natural logic involves the simultaneous addressing of constancy and change, because change is realizable only relative to constancy. In the logic of types and kinds, predicates and modal connectives express change. Categorical logic with inherent functoriality, is appropriate to deal with such constancy and change. In the same manner, categorical logic is the convenient mathematical environment for handling that functor between linguistic structures and the structure of the non-linguistic well as interpretation.

The categorical logic, the study of logic with the help of categorical means, has produced many important results. Suffice it to mention the generalization of Kripke-Beth semantics for intuitionistic logic to sheaf semantics by Joyal. Ellerman 1987 has tried to show that category theory constitutes a theory of universals which has properties radically different from set theory considered as a theory of universals. If we move from universals to concepts in general, we can see how category theory could be useful even in cognitive science. Indeed, Macnamara and Reyes have already tried to use categorical logic to provide a different logic of reference [9].

The logic of types and kinds takes on a mathematical life of its own, it retains a structural harmony of its own. It is discovered by studying the category of kinds. The logic of kinds is expressed in terms of categorical logic. Categorical logic is intuitionist, though it can also be classical when the occasion arises. Natural logic is many-sorted, and categorical logic, because it treats arrows as basic, seems specially adapted to deal with many-sorted systems.

4.1 Topos

The 'element-free' formulation of mathematics provided by category theory is most strikingly realized by applying it to set theory itself. Formally, an *elementary topos* $\mathcal{E}$ is a finitely complete category with exponentials and a subobject classifier Ω. A *subobject classifier* or truth-value object in a category with a terminal object is an object Ω together with an arrow $\top : 1 \to \Omega$ called the truth arrow, from the terminal object 1, such that the diagram has a universal (pullback) property: For each arrow $m : B \to A$ there is a unique arrow, called the characteristic arrow of m, $\chi(m) : A \to \Omega$ such that $\top \circ b = \chi(m) \circ m$.

More precisely, (1) $\mathcal{E}$ has pullbacks and a terminal object (and therefore, all finite limits), (2) $\mathcal{E}$ is cartesian closed, and (3) $\mathcal{E}$ has a subobject classifier.

In any topos $\mathcal{E}$ one can give natural definitions of arrows ('logical operations' in $\mathcal{E}$), $\sim: \Omega \to \Omega$; $\vee$, $\wedge$, $\Rightarrow: \Omega \times \Omega \to \Omega$ in such a way that, if we regard these arrows as algebraic operations on Ω, the resulting (Heyting) algebra satisfies the laws of intuitionistic propositional logic. In this sense intuitionistic logic is 'internalised' in a topos. With some justice, then, we may regard a topos as an instrument for reducing logic to mathematics, the remarkable thing being that the logic obtained is not (in general) classical, but intuitionistic. Thus category theory, far from being in opposition to set theory, ultimately enables the set concept to achieve a new universality [6].

4.2 Cognition $\leftrightarrows$ Categorical Logic

Category theory can also be viewed as a foundational discipline capable of clarifying and sometimes expanding our understanding of mathematical knowledge and its applications. This change is mainly to the discovery of F. W. Lawvere that some categories may be viewed as universes of variable/cohesive sets, capable of modelling theories that lack models in the more rigid universe of constant sets. Already, category theory has been applied to a variety of subjects ranging from physics to linguistics.

Category theory could be a conceptual tool in the study of cognition. The thesis is that the explicit adequate development of the science of knowing will require the use of the mathematical theory of categories [8].

Category theory has developed a variety of notions in order to provide a guide to the complex constructions of the concepts and their interactions which grow out of the study of space and quantity. Galileo's insight is that physics and mathematics mutually constrain each other; Chomsky's insight is that psycholinguistics and linguistics constrain each other. J. Macnamara suggests that cognition and logic constrain each other in the same manner and he is contributing that insight to the father of logic, Aristotle. This relation can be expressed as Cognition $\rightleftarrows$ Logic. And, in the context of this paper, without collapsing to a single subject, **Cognition $\leftrightarrows$ Categorical Logic.**

Macnamara [9] shares his vision that in cognition we are in the year 1690. Our calculus (categorical logic) has been invented. A deep and satisfying theory of the human mind will be developing and replacing tendencies in 'cognitive studies' that are unworthy of their subject.

References

[1] M. Alagic, *Abstraction, Reflection, and Learning* (abstract), Bridges - Mathematical Connections in Art, Music, and Science, Conference Proceedings, pp. 301. 1999.

[2] M. Alagic, *A Visual Presentation of Rank-Ordered Sets*, Bridges - Mathematical Connections in Art, Music, and Science, Conference Proceedings, pp. 237-244. 1998.

[3] S. Alagic & M. Alagic, *Order-sorted Model Theory for Temporal Executable Specifications,* Theoretical Computer Science, Vol. 179, no. 1-2, pp. 273-299. 1997.

[4] M. Alagic, & S. Neal, *From Concrete to Abstract: Reaching for Higher-Order Thinking Skills Through Geometry.* Paper presented at the NCTM Central Regional Conference, Topeka,1999.

[5] J. L. Bell, *Category theory and the foundations of mathematics*, British Journal of Philosophy of Science, Vol. 32, pp. 349-358. 1981.

[6] J. L. Bell, *Toposes and Local Set Theories, Clarendon Press,* Oxford, 1988.

[7] L. D. English, *Mathematical Reasoning Analogies, Metaphors, and Images*, Lawrence Erlbaum Inc. 1997.

[8] F. W. Lawvere, *Tools for the Advancement of Objective Logic: Closed Categories and Toposes,* In J. Macnamara & G.E.Reyes

[9] J. Macnamara & G. E. Reyes, *The Logical Foundations of Cognition*, Oxford University Press, New York, 1994.

[10] S. Mac Lane, *Categories for the Working Mathematician*, Springer Verlag, New York, 1971.

[11] A. Newell, P. S. Rosenbloom, & J. E. Laird, *Symbolic Architecture for Cognition*, In Foundations of Cognitive Science. M. I. Posner, (Ed.). MIT Press, Cambridge. 1989.

[12] J. Piaget & B. Inhelder, *The Child's Conception of Space*, Norton, New York, 1967.

[13] P. Van Hiele, *Structure and Insight A Theory of Mathematics Education.* Academic Press, Inc.1986

[14] M.Wertheimer, *Productive Thinking* (Enlarged Ed.). Harper & Row, New York 1959.

Bridges between Antiquity and the New Turkish Architecture in the 19th Century

Zafer Sagdic
Faculty of Architecture, History of Arcitecture Dpt.
Yildiz Technical University
Yildiz, 80750, Istanbul, TURKIYE
sagdic@ yildiz.edu.tr

Abstract

From the beginning of the 19th century, the Ottoman Empire went through a phase of intensive economic and socio-political transformation aimed at modernising the old system. A series of social and institutional reforms based on the Western models, was attempted in order to re-structure the Ottoman Empire. One of the major areas where this transformation took place was the New Turkish Architecture.

The term "New Turkish Archtecture is used to describe the products of a movement which claimed to be a big step forward and which was predominant in the Tukish Architecture in the 19th century, especially during the reigns of Sultan Abdulmecid (1839-1861), Sultan Abdulaziz (1861-1876), and Sultan Abdulhamid the 2nd (1876-1909).

Throughout the 19th century, it can be seen that architecture was dominated by Neo-classical features. In this period, archiects both of muslim and non-muslim origin, attempted to trigger a renewal movement through the use of mainly traditional Ottoman motifs. Thus, local elements were exploited within the frame-work of the trends and schemes current in Western architecture.

These trends included certain principles previously tried out by the French "Beaux-Arts" school, and the Gothic style, which was adopted in many Western countries with a concern for national expression, and the Orientalist movement, which encourged the adaption of the traditional Ottoman motifs.

Besides the fact that the new designs were in line with modern Western concepts, they expressed to be accepted in the eyes of the West (Usul-i Mimari-i Osmani contained theoretical ideas and incorporated Ottoman Architecture in the context of the architecture of the civilised world, tracing its history back to antiquity. Traditional styles were approached with a reference to the Antiquity and they were subjected to new interpretations).

Thus, some bridges between the Antiquity and New Turkish Architecture in the 19th century were established: These bridges included the integrations of the Antique Western architectural elements were seen/ identified in the Traditional Turkish Architecture; but nature and historical elements were maintain in local practices.

It can be said that with the use of elements from the antiquity the New Turkish Architecture was re-structured in the name of modernism and transformed as a whole in the eyes of a European-centered history of theory and style.

1. The First Bridge between Usul-i Mimari-i Osmani and Antiquity

The first bridge between Usul-i Mimari-i Osmani and the Antiquity: Usul-i Mimari-i Osmani was the name given to the architectural order which was re-arranged for the Vienna World Exibition in 1873. Within the arrangements made for the Vienna World Exibition, the most important architects of the Ottoman Empire prepared a book, which did not only have texts about the history and basic principles of Ottoman Architecture, but which also contained many drawings of important buildings. Thus, the book was also named "Usul-i Mimari-i Osmani", which was written in three languages of the Empire: Ottoman Turkish, German and French.It can be said that <u>Usul-i Mimari-i Osmani</u> established bridges between "The Ten Books" of Vitruvius, "The Four Books" of Alberti, and "The Ten Books" of Palladio as the conceptual approach to architectural design. In all of these books, the system of architectural orders relating to their respective periods were described and the principles of creation were examplified with different architectural compositions.

As is well known, Vitruvius wrote about the orders of antiquity in the Chapter on the Principle of "Décorum". In this chapter the connection between the specific orders and the specific divine characters (the contents) was explained. Unornamented Doric order was created for some gods, such as Athena, Ares, and Heracles to show the dominant masculine aspect of their characters. Florally ornamented and feminine Corinthean order was created for some goddesses such as Afrodite, Flora, Proserpina and for the nymphes. Finally, graceful-temples of semi-ornamented Ionic order were created for Hera, Artemis, and Dionysos because of their powerful and graceful-looking characters.

Alberti explained the nature of there orders: Doric order represented power and strength; the Corinthean order represented beauty and coquettish desires, and the Ionic order represented the graceful-style, which took its place in the middle of the other two orders. Palladio re-examplified the principle of propriety in the principle of "Décorum" as mentioned by Vitruvius.

In Usul-i Mimari-i Osmani under the third sub-heading In the first chaper of (Technical Documents) are listed three orders in the Ottoman architecture: the Mahruti order (échanfrine-schragkantig); the Müstevi order (bréchiforme-breccienförmig); and the Mücevheri order (crystallisé-kristallförmig). In this book it is claimed that there are specific bridges between Ottoman Architecture and the antiquity. This relationship was not limited to the descriptions of the architectural orders; the selection of the orders or the milieu were created according to the principles of "Décorum" in the Antiquity.

The Mahruti order referred to as the Doric order, was used in tekkes (the dervish lodges), in arastas (commercial centers), in stores and in buildings where simplicity should be emphasized. The Mücevheri order, which took its reference from the Coritnhean order, was generally used in the buildings where magnificence and an imposing appearance were desired. The Müstevi order, which was in the middle of the two other orders like the Ionic order, was used in the arcades of the buildings and in tombs.

2. The Second Bridge Between Usul-i Mimari-i Osmani And Antiquity:

The principles set by the Usul-i Mimari-i Osmani for using different orders on the same façades and within the same buldings are created conforming to the contepts in antiquity. Vitruvius wrote that the stoas which covered the squares and the coloumns on the ground-floor should be higher and thicker than the columns on the upper floors. According to Palladio, the strongest order is the one in which the ground-floor was created to carry all the load of the composition as a whole. In the antiquity, the Doric order, the Ionic order, the Corinthean order, and the Composit order were all used in this way. Thus, in Usul-i Mimari-i Osmani, the Mahruti order, the Müstevi order, and the Mücevheri order were used according principles applied in antiquity. (However, different variations on the arrangement of these orders could be seen in some examples of Usul-i Mimari-i Osmani).

In Usul-i Mimari-i Osmani, the Ottoman Architectural orders were presented as the original elements of the national architectural dialec, which was brought to the peak by Architect Sinan. Thus, it can be said that the main aim of the Usul-i Mimari-i Osmani was to create an original national style, as did the other nationalities and that the Usul-i Mimari-i Osmani was expected to be a part of the history of Western architectural theory. It was also emphasized that the national dialec of architecture had the same principles with the international architectural dialectics. Thus, Alberti adopted the same approach when writing his book; he claimed that the Composit order was Italian, in order to prove that the Italians did not owe anthing to foreigners.

3. The Third Bridge between Usul-i Mimari-i Osmani and Antiquity:

It was also pointed out in <u>Usul-i Mimari-i Osmani</u> that the Ottoman Architecture had some connections with mathematics in the same way as the architecture of Antiquity had some relationships with mathematics.

In general, it was can be said that both <u>Usul-i Mimari-i Osmani</u> and the Architecture in Antiquity, used some common systems of proportion.

Now that we are celebrating the 700th anniversary of the establishment of the Ottoman Empire, it would be timely and useful to study the unique and magnificent architecture of this Empire as well as its military and political achievements on three continents.   Within Ottoman architecture, the examples of the 19th century architecture occupy a particular place, especially regarding its aesthetic qualities and highly advanced structural characteristics.   The relationship between design and mathematics is also significant in these examples and deserved to be studied in depth.   In order to show why, a general review of the relationship between architectural design and mathematics is helpful.

Mathematics can be roughly divided into two main parts: one is "practical mathematics", such as the operations we use in daily life; the other is "pure mathematics", which is used in establishing the complex mathematical relationships in the positive sciences. The study of the relationship between architectural design and mathematics depends on "pure mathematics".

We must first accept that all the objects in nature and the relationships between these objects are governed by rules of geometry. As Galileo stated, nature has a certain mathematical design: "the book of nature can only be read by those who know its language, which is mathematics". Thus, if we accept "pure mathematics" as a kind of game, it can be said that nature is the medium in which this game is played, and the constituent parts of nature are the symbols used in mathematics; in other words, they are the pawns used in the game. Throughout the game, by further developing the metaphor, we can assume that the single symbols will come together to form groups of symbols. As a result of these formations, we can see the geometrical rules governing the symbols, that is, the groups of objects existing in nature. To put it more clearly, the rule(s) governing the relationship between architectural design and mathematics can be found among the rules of governing the relationships in nature.

Architects, while analyzing the designs of various buildings, almost always turn to nature, observing it carefully and transforming their observations into design elements. Consequently architects create their designs by studying the geometrical rules that establish the various natural correlations with a concern for architectural style. In that sense, architects can be defined as the organizers of the relationships between the forms and functions of buildings. As Monroe Beardsley said, "The form of an aesthetic object is the total web of relations among its parts". In this way, the geometrical rules in nature can be taken as the rules of "Beauty". It is known that, throughout the history of design, architects have always created their designs to create "artifical environment" within the framework of the geometrical rules.

In the light of this preface, the question of how the mathematical rules (symmetry, proportion, geometry, etc.) affected the 19th century Ottoman Architecture can be answered more clearly. However, consider the question in a certain framework, a few more basic questions should be answered. What are the structural and aesthetic rules that were used in the 19th century Ottoman architecture? Why is it still important today to take these rules into consideration?

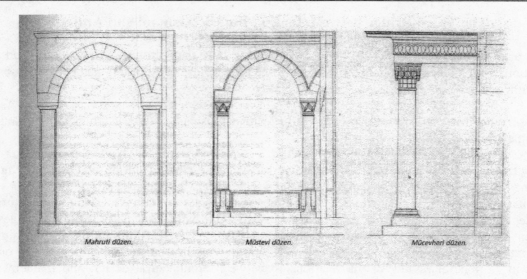

Fig.1 The mahruti order Fig.2 The müstevi order Fig.3 The micevheri order

Generally, it can be said that in the examples given in <u>Usul-i Mimari-i Osmani</u>, the width of the capital on a column inside the mosque is taken as the module. The height of the columns used in monumental buildings is usually 10 to 18 times the radius of the capital. In all the orders used for columns, the radius at the bottom is the size of six modules and the radius at the top is five-and-a-half modules. In this connection, it might be useful to remember the orders used in columns: the conical order (*tarz-i mimari-i mahruti*); the multiple-plane order (*tarz-i mimari-i müstevi*); the stepped/crystal order (*tarz-i mimari-i mücevher*i). In order to define these orders briefly, it can be noted that the conical order has columns whose maximum height is six modules. In the multiple-plane order, the height of the whole column, including the base and the capital is ten modules. The stepped/crystal order, which is both spectacular and sophisticated, the maximum height of the column together with the base and the capital is eighteen modules.

It is also known that there is a relationship between mathematics and the dimensions of the colums used in antiquity. From the antiquity onwards, columns have been reduced in diameter towards the capital with a taper unevenly distributed over the height of them. Altough, there were some varieties the upper diameter of columns were usually 0.85 of the lower diameter in all of the there orders of antiquity, the Doric, the Composit, the Corientihean.

Fig.4. Yildiz Valide Sultan Mosque

Fig.5. The facade of Yildiz Valide Sultan Mosque

Fig.6. The Sadabad Mosque

Fig. 7. The details from the Dolmabahçe Palace

Fig. 8 The details from the Dolmabahçe Palace

Fig.9 The Gate of the Dolmabahçe Palace

From examples are given above, it is cleary identified that the 19th century Ottoman Architecture, the Usul-i Mimari Osmani was the interpretation of the Antiquty and the Classical Ottoman Architecture, which was designed by geometrical rules. When we look into the relationship between mathematics and the architectural design found in the monumental architecture buildings of the Classical Ottoman Period, it will be observed that the certain proportional relationships were used, the existence of which cannot be denied. According to Arpat[5], these proportional relationships can be divided into two groups: those that use modules or some religious-symbolic figures as principles arrangements; those that use proportions. It can be seen that generally in the mosques of the 19th century, there were used a modular network obtained by placing an octagon within a circle with a radius of 68cm (3 *arshin*) or multiples of this unit-measure. It has also been established that the main module commonly used in all the mosques mentioned above are multipled by 3 (that is 3x3=9 *arshin*). This measure, 204cm (9 *arshin*), was also employed in establishing the design principles of such elements as the levels in buildings, the overhangs of the eaves of domes, etc. (Figures 1, 2 and 3). Another fact that it is important to emphasize here is that, in the said period, the design principles governing monumental architecture both European and Ottoman depended on some symbolic values along with functionality. In Christian architecture, the number 3 and its multiples have been used symbolically in organizing space because of the Holy Trinity. In Islam, it is believed that there are three spirits: the good spirit, the evil spirit, and a third spirit that tempts people to evil-deeds. Moreover, in the monumental religious buildings of both Christianity and Islam, a centralized plan, believed to represent the monotheistic belief in the organization of space, is commonly used.

It can be easily understood from examples are given above that 19th century Ottoman Architecture defines the sophisticated relationship between the architectural design and mathematics. A specific space organization always was created in light of the structural rules and mathematical rules. By reflecting the idea of functional design onto the creation of monumental buildings in the pre-modern world, he became one of the most important and interesting master-builders in history. It may be said that, even today in our modern age, Ottoman Architecture's compostions should be observed because of their rules of functionality reflect the relationships between architectural design and mathematics.

References

Arpat, A., "Osmanli Dini Mimarisi'nde Modul ve Duzenleyici Geometri", *MTRE Bulteni*, no. 13-14 (1981), pp .29-35.

Camlibel, N. *Sinan Mimarliginda Yapi Strukturunun Analitik Incelenmesi* (Istanbul: Yildiz Technical University1998).

Edhem, P. *Die Ottomanische Baukunst*, (Constantinople, 1873).

Egli, E. *Sinan der Baumeister Ossmanischer Glanzzeit* (Stuttgart: Verlag für Architektur Erlenbach), 1954.

Freely, J. And A. Burelli. *Sinan* (New York and London: Thames & Hudson, 1992).

Goodwin, G. *Ottoman Architecture* (London: Thames & Hudson, 1997).

Haidar, S and H. Yazar. "Implict Intentions & Explicit Order in Sinan's Work, II", *Uluslararasi Turk ve Islam Tarihi Kongresi,* Cilt no. 2,   (Istanbul:Istanbul Technical University1986), pp.29-42.

Kuban, D. *Sinan Dünya Mimarisindeki Yeri, Mimarbasi Koca Sinan* (Istanbul: Vgmy, 1988), pp.581-624.

Sagdic, Z., "Mathematics & Design Relationship in Architecture", *Mathematics & Design 98*, Javier Barrallo, ed. (Proceedings of the Second International Conference1-4 June 1998) (San Sebastian, Spain: The University of Basque Country, 1998), pp. 383-391.

Saner, T., "19. Yüzyil Yeni Türk Mimarliginda Antik Anlayisi", *Yapi 211*, pp.70-77.

Schemmel, A. *Sayilarin Gizemi* (Istanbul: Kabalci 1998).

Soylemezoglu, K., *Istanbul Rustem Pasa Camii*, II. Uluslararasi Turk ve Islam Tarihi Kongresi, Cilt no.2 (Istanbul: Istanbul Technical University, 1986) pp.105-114.

Tuncer, N. *Klasik Dönem Osmanli Mimarisi'nde Iç mekan ve Cephelerde Oran.* Ph.D. thesis, Yildiz Technical University, Istanbul, 1997.

Humor and Music in the Mathematics Classroom

James G. Eberhart
Department of Chemistry
University of Colorado
Colorado Springs, CO 80933
E-mail: jeberhar@mail.uccs.edu

In the first meeting of my class in Quantitative and Qualitative Reasoning, many of the Liberal Arts students in this required course willingly admit to having math anxiety. With a large lecture hall full of students it isn't possible to deal individually with this problem, but it is necessary to respond in some way to these feelings. First, I assure my students that help with the course is available from me and also from two Learning Centers we are fortunate to have on our campus. Second, I tell them about a student in last semester's class who was so anxious that he wasn't able to sleep through any of my lectures. We all laugh a little and some of the tension is dissipated.

I also assure my students that you don't have to be a genius to succeed at mathematics. I tell them (tongue-in-cheek) that I didn't pass most of the math courses I took as a student. But I reassure them that I was always at the top of the group who failed.

There are two ways I attempt to help my students with their anxiety. The first is to try to guide the students into successful experiences in the course to build their self-confidence and convince them that they are capable of doing mathematics. The second approach, which is the focus of this presentation, is to inject appropriate humor into the classroom in an effort to relieve stress, anxiety, and antipathy, and to bring a little fun and interest to the experience.

Classroom humor comes in many forms and can include stories, jokes, cartoons, limericks, and yes, even songs. I'll provide you with some of my favorite resources for classroom humor and some samples of each. We will also consider how to fit a teachers favorite humor into the appropriate place in the course, and how to avoid some of the pitfalls of classroom humor.

The use of music as a vehicle for humor will be given particular emphasis. Yes, there are a few humorous songs with a mathematical theme that can be brought to class via CD and boombox. And if you play a musical instrument which is good for accompanying singing, you can summon your courage, bring it to class, and sing a song for (or with) your students. Generally speaking, students are a wonderful audience to amateur musicianship. They are delighted you are not lecturing on a new topic at the moment, and most of them are too polite to get up and leave.

I have found in the last few years that writing a song to meet your own classroom needs is not as difficult as it might sound. A simple procedure for this activity will be shared with attendees.

I wanted to come up with a song for my math students, which acknowledged how hard most of them were working, despite the fact that some would no doubt have preferred a root-canal session to taking a math course. A blues tune seemed the perfect choice. I selected the melody of a traditional song which is a favorite with many bluegrass bands and old-time string bands called Brown's Ferry Blues (Brown's Ferry is a small town in northern Alabama). The tune is called "The Mathematics Blues." I'll share it with you (with guitar accompaniment) and encourage attendees to join in. The chorus line, which is sung twice in each verse, goes "Lord, Lord, I've got the Mathematics Blues." All of my students (whom I encourage to sing along) seem to relate with delight to that sentiment. (I can't imagine why.)

If appropriately used, a little humor in the classroom can go a long way toward relieving the anxiety and stress of students and teachers alike. It can also put a more human face on the entire mathematical endeavor. As Mary Poppins rightly told us many years ago, a spoonful of sugar really does help the medicine go down.

The Development of Integrated Curricula:
Connections between
Mathematics and the Arts

Virginia Usnick
College of Education
University of Nevada, Las Vegas
Las Vegas, NV 89154-3005
Email: vusnick@nevada.edu

A current trend in elementary education is the development of integrated curricula; that is, using the content or processes of one area of content to develop concepts in another area of knowledge. For example, many elementary teachers use children's literature as springboards into mathematics lessons. In this workshop teachers will investigate connections between mathematics and the arts by participating in activities designed to be used with elementary-aged students. Specifically, we will explore probability and the generation of abstract art.

The Golden Ratio and How it Pertains to Art

Michael J. Nasvadi and Mahbobeh Vezvaei
Mathematics and Computer Science Department
Kent State University
Kent, Ohio 44242, USA
E-mail: vezvaei@mcs.kent.edu

The Golden Ratio is one of the great mysteries of nature. The human form is part of nature, thus the Golden Ratio is in fact part of the most human bodies. One person might seem more proportional than another depending on how many times the Golden Ratio appear in his/her body measurements.

Our purpose for this presentation is to examine the Golden Ratio and how it pertains to art. We will first define what we call the Golden Ratio (Φ). Next we will inspect a manifestation of Φ in the Fibonacci Number Sequence. Then we will investigate the works of Leonardo DaVinci and show some examples of how Φ relates to artistic flow and beauty. Finally, we have conducted our own study on an independent model. We will see if our model fits in the mold of the Golden Ratio.

This is just the first study conducted by what we call the (Science, Math, and Art) SMART Association. In the future we plan to investigate other mathematical factors in art such as the tessellation, cubism, and fractals. We also have plans to open the SMART Association home-page on the World Wide Web, where users can research art and mathematics through the use of interactive scripts.

The Art and Mathematics of Tessellation

Travis Ethridge
Department of Mathematics and Computer Science
Southwestern College
Winfield, KS 67156

This workshop will provide participants with knowledge that the combination of mathematics and art can be useful in developing interest and understanding for a variety of other subjects. The connections developed from the creative use of geometry to develop artistic patterns, called tessellations, can be connected to the patterns existing in science, English, history, and, of course, math and art. This workshop was developed to provide educators with information about tessellations and offer a variety of methods, materials, and opportunities to integrate this amalgam of mathematics and art to any class curriculum.

Biological Applications of Symmetry for the Classroom

Patrick Ross
Department of Biology
Natural Science Division
Southwestern College
Winfield, KS 67156

Many organisms exhibit a superficial level of bilateral symmetry. However, upon closer examination, many individual characteristics exhibit some levels of asymmetry. Some well studied examples include tail feathers in bird, wing veination patterns in insects, and human facial characteristics. In many cases, the degree of asymmetry is thought to be a measure of developmental precision, thereby allowing an assessment of the degree to which the genome can resist the variability present in the environment. Higher levels of asymmetry have been found in populations subjected to poor environmental quality and other stressful situations. In addition, experiments on mating behavior have shown a preference for mates with more symmetric characteristics. Participants will be introduced to the concept of fluctuating asymmetry. Participants will measure and quantify the degree of asymmetry on a variety of characteristics. Applications in the areas of assessment of environmental stress and mating behavior will be presented that can be used in classroom situations.

Exploring Technology in the Classroom

Terry Quiett
Center for Academic Technology
Memorial Library
Southwestern College
Winfield, KS 67156

In this workshop we utilize readily available software to visualize how science and art intersect to create new paths to learning. Participants in this workshop will explore how software can enhance teaching. They learn about presentation software (PowerPoint), web page creation software (FrontPage 2000), basic design elements, and have the opportunity to enhance their learning through group discussions. Handouts will be supplied.

On Visual Mathematics in Art

Clifford Singer
510 Broome Street
New York, New York 10013
212 431 4408
Email: CliffordhS@aol.com

In the progress of this paper I have interspersed some reflections that we can collect into a general point of view on my art works. While the geometric style in the broadest sense can be classified in the art world as 'geometric abstraction', what actually takes place on the picture plane between form, space and color goes far beyond the limitations of such arbitrary labels. A complex underlying structure composed of straight lines, arcs, circles, elliptical sections, parabolas, hyperbolas, epicycloids, and spirals informs the viewer with a sense of controlled order and disciplined rational thinking reaching for a formalized mathematical perfection. In the final analysis, this geometrical structuring serves as no more than a foil for sensations for a more intangible and metaphysical (absolute) scope of vision. It is not the intention to present the viewer with a tasteful static equilibrium but to spark the imagination with suggestions of the powerful forces and processes operative on a universal scale. The work evokes ceaseless mobility, speed, time, space, growth, compression, tension, attraction, and the infinite nature of pure structural elements. The geometry provides the fundamental methodology for this dynamic expressive language. It is the personal vision that transforms this familiar vocabulary in mathematics into a unique transcendental syntax. If the circles were imagined on a larger scale they could represent the great circle that would bisect the celestial sphere or earth.

In understanding geometrical form there is classification of elements, functions, nonlinear approaches, and discussion of the accuracy of measurements, and of symmetry. There exists an internal configuration, and it is therefore possible to gain visual information from the order of the picture, its aesthetic qualities as well as the actual graphing and the pictures' corporeity. The latitude of geometrical method has many different examples of variations in the intensity of the forms and qualities: for example the dynamics of motion, gravity, flotation, and corpuscular unity. An intuitive design sense in conjunction with a practical mensuration is a major point. In this notion we see the continuance of a primordial condition that has carried from a first principle or element. Expressing a constant as the horizontal axis (longitude) in the graphing of the picture, and the variable (the varying intensity of quality throughout the picture) is the height on the vertical axis (latitude). The perimeter closes all of the formed figuration that represented the configuration or distribution of the varying qualities.

That a complex formalism is the foundation behind my paintings is a premise essential to the work in its constructible methods. Within the doctrines of the philosophy of mathematics, metamathematics includes questions of both semantics and syntax such as whether axiomatic systems are independent, consistent or complete. A transcendental extension field exists with applications to my art works. There is a duality of spatiotemporal reality between the physical and geometrical configuration that interpenetrate in superposition. Physical laws are simply applied to the geometrical configurations. The hyperbola can increase to infinity and therefore converge towards zero. The branch of the hyperbola bends more and more towards the x axis as the angle of the asymptotes become more and more obtuse. To look at far horizons we can imagine the far reaching parallel to our concept of nature and the physical world. In the totality of natural phenomena by enhanced approximations in the paintings we approach a system of reference x, y, z, t space and time by which these visual phenomena present themselves in agreement with geometrical rules and laws. H. Minkowski had introduced this fundamental axiom: '*The substance at any world point may always, with the appropriate determination of space and time, be looked upon as at rest.*' In painting it is my goal to capture a fragment of space and time although there is a duality where it is also rooted in singularity. In non-Euclidean space the geometry I have conceived consisting of the picture plane demonstrates an independent singular universe. Geometrical space and structure is adequately explained in that we consider points, curves and geometrical surfaces as minute particles, fine paths or thread, and thinly veiled surfaces.

Collinearity, (of a set of points lying in the same straight line) a recurrent theme is never coincidental but is an inevitable outcome, forced in the same way that it occurs in the classical Theorems of projective geometry, such as those of Pascal, Brianchon, Desargues, and others. These mathematicians recognized that spatial situations that produce collinearity were invariably the result of deep underlying geometric truths. The incidence of a point on a line is invariant under the projective rules. If three or more points are collinear along a line, then incident with a straight line, the images will thus be collinear. Therefore, the characteristics of incidence, collinearity, and concurrence are principle requisites of my work. So, the geometry focuses as to the overall collinear connective intersections and edificial detail of the inner organization.

Everything around us has a history or combination of historical influences is noted or emphasized. As an artist there are issues such as historic perspectives, ranges of influences, and importantly the sources of reference that are available to choose from which includes aesthetic and/or rational decisions. For my work in art, I have gone from art to mathematics or shall we say an integration of the two disciplines in tandem. I had never imagined that geometry would be a commodity or tangible product with economy. I have always considered geometrical paintings as windows or devices for contemplation and thought.

Clifford Singer, *Continuity,* 1999, Acrylic on Plexiglas, 36 x 36 inches

A Bridge for the Bridges

Jason Barnett
1072 Dayton Ave,
St. Paul, MN 55104, USA
E-mail: jasonbarnett@earthlink.net

I was inspired by M.C. Escher in some of his work where he represented realistic environments at the same time adding confusing and structurally impossible elements, which in my mind, adds interest and intrigue. As a sculptor I believed it would be interesting to try to create some of these types of structures in three-dimensional space. I attempted in the two following models to bring similar ideas of impossible structures and gravitational shifts into real space. I believe that these structures have interest on both aesthetic and mathematical levels.

In relation to the *Bridges* conference where the fields of mathematics and aesthetics blend, I believe that it would be fitting to attempt to build a structure, or more appropriately a bridge, similar to my models, or Escher's prints on the Southwestern College's campus. This would make a timeless expression of the great thoughts and theories that come to life here at *Bridges*.

Bridges 1999